Analysis and Prediction of SOIL BEHAVIOUR

Analysis and Prediction of SOIL BEHAVIOUR

T S Nagaraj (PhD, DSc.)
B R Srinivasa Murthy (PhD)
A Vatsala (PhD)
Professor
Department of Civil Engineering
Indian Institute of Science
Bangalore, Karnataka

Publishing Globally

NEW AGE INTERNATIONAL (P) LIMITED, PUBLISHERS
New Delhi • Bangalore • Chennai • Cochin • Guwahati
Hyderabad • Kolkata • Lucknow • Mumbai
Visit us at **www.newagepublishers.com**

Published by New Age International (P) Ltd., Publishers
First Edition: 1994
Reprint: 2015

BRANCHES

- **Bangalore** 37/10, 8th Cross (Near Hanuman Temple), Azad Nagar, Chamarajpet Bangalore-560 018 Tel.: (080) 26756823, Telefax: 26756820 **E-mail: bangalore@newagepublishers.com**
- **Chennai** 26, Damodaran Street, T. Nagar, Chennai-600 017 Tel.: (044) 24353401, Telefax: 24351463 **E-mail: chennai@newagepublishers.com**
- **Cochin** CC-39/1016, Carrier Station Road, Ernakulam South, Cochin-682 016 Tel.: (0484) 2377303, Telefax: 4051303 **E-mail: cochin@newagepublishers.com**
- **Guwahati** Hemsen Complex, Mohd. Shah Road, Paltan Bazar, Near Starline Hotel, Guwahati-781 008 Tel.: (0361) 2513881, Telefax: 2543669 **E-mail: guwahati@newagepublishers.com**
- **Hyderabad** 105, 1st Floor, Madhiray Kaveri Tower, 3-2-19, Azam Jahi Road Near Kumar Theater Nimboliadda, Kachiguda, Hyderabad-500 027 Tel.: (040) 24652456, Telefax: 24652457 **E-mail: hyderabad@newagepublishers.com**
- **Kolkata** RDB Chambers (Formerly Lotus Cinema) 106A, 1st Floor, S.N. Banerjee Road, Kolkata-700 014 Tel.: (033) 22273773, Telefax: 22275247 **E-mail: kolkata@newagepublishers.com**
- **Lucknow** 16-A, Jopling Road, Lucknow-226 001 Tel.: (0522) 2209578 4045297, Telefax: 2204098, **E-mail:lucknow@newagepublishers.com**
- **Mumbai** 142C, Victor House, Ground Floor, N.M. Joshi Marg, Lower Parel Mumbai-400 013 • Tel.: (022) 24927869, Telefax: 24915415 **E-mail: mumbai@newagepublishers.com**
- **New Delhi** 22, Golden House, Daryaganj, New Delhi-110 002 Tel.: (011) 23262368, 23262370, Telefax: 43551305 **E-mail: sales@newagepublishers.com**

ISBN: 978-81-224-0633-7

₹ 495.00

C-14-10-8009

Printed in India at Glorious Printers, Delhi.

PUBLISHING GLOBALLY
NEW AGE INTERNATIONAL (P) LIMITED, PUBLISHERS
7/30 A, Daryaganj, New Delhi-110002
Visit us at **www.newagepublishers.com**

PREFACE

Of the civil engineering construction materials, soil is unique in being a natural material, most often engineered, as it exists, unlike other processed or manufactured materials like concrete or steel, which can be processed to suit specific requirements. Hence, property characterization of the soil is an important aspect of geotechnical engineering practice. The determination of properties by testing often needs to be supported by predictions mainly because of two reasons. One, the volume of soil that can be tested in any project within the time constraints and economical feasibilities is only a very small fraction of the total volume of soil involved. Secondly, the present day construction practice is such that the cost of savings for elaborate testing may be outweighed by savings in time by prediction with accuracy acceptable at engineering level.

This book is intended to bring out the possibilities of generalizations of behaviour of soils and hence of predicting the required engineering properties without elaborate testing. We recognize that a single approach cannot be evolved for all soil types and hence the necessity for classifying soils into different categories and to use appropriate model for each. First of all, based on mechanism of stress transfer and interaction between the phases, two obvious classes, the fine grained and coarse grained soils, have been differentiated. The discussions bring out that because of identical mode of stress transfer, the mechanical behaviour i.e., compressibility, shear strength, permeability etc. can be generalized for all fine grained soils, enabling the prediction of behaviour of such soils with just the knowledge of certain state and index properties. The subsequent chapters discuss the characterization of specific soil states starting from the saturated uncemented soils, both normally and overconsolidated, through cemented saturated soils to partly saturated soils. The behaviour of coarse grained soils, the importance of microfabric effects, and the difficulties in possible generalizations are discussed. Perhaps this is the unique feature of this book that the division of the chapters is based on different soil states, all the mechanical behaviour being discussed under each soil state, whereas in most of the text books, the different chapters concern with different mechanical behaviour like consolidation, shear strength etc, without much detailing about the soil state. The subject covered in this book is mostly a synthesis of the research work carried out at the Indian Institute of Science over the last decade which has been published in refereed journals and also the related work of other researchers world over.

The book will be of interest to both academicians and practising engineers, researchers and post graduate students. It may also serve as a text book for under-graduate, students having an adequate background of basic principles of soil machanics.

Bangalore
21-4-1994

T.S. NAGARAJ
B.R. SRINIVASA MURTHY
A. VATSALA

ACKNOWLEDGEMENTS

To acknowledge all who have contributed to this book individually is a formidable task. In the development of this book, we were greatly benefited by our association with many graduate students who have contributed to the subject matter significantly. Our research and teaching efforts in this broad area have further helped in an in-depth treatment of this subject and the coverage provided in this book.

At all stages of writing of this book, we have received a great deal of assistance from numerous persons. Thanks are due to our esteemed colleagues of the Department of Civil Engineering for their help at appropriate stages. We place on record our sincere thanks to Mr. A.V. Narayan, Draughtsman, Department of Metallurgy of the Institute who meticulously prepared all the illustrations needed for the book and Mr. P. Raghuveer Rao, Laboratory Assistant of the Department of Civil Engineering for formatting the entire text, tables and references on a personal computer.

We place on record our sincere thanks to the Curriculum Development Cell established at the Indian Institute of Science, Bangalore, India by the Ministry of Education and Culture, Government of India, for providing financial support for the preparation of this book.

Finally, we alone remain responsible for any lapses and errors that might have crept into these pages, despite our best efforts to avoid them. We look forward to the comments and suggestions from the readers.

Bangalore, India

T.S. NAGARAJ,
B.R. SRINIVASA MURTHY
A. VATSALA

CONTENTS

PREAMBLE

***"Synthesis unrelated to science is empty
Prediction unexplained by basics is blind"***.

Engineering is an act of decision making under uncertain premises. It is mainly an expression of judgment. To exercise the right judgement during various phases of professional activities such as

— cost
— construction time
— acceptability, and
— performance

the engineer is constantly required to predict. In fact prediction is an integral component of practice.

Prediction has a long and renowned history. Throughout the historical time the demand for prediction has not diminished. We apparently feel that if the happening of events can be predicted we can resort to corrective measures appropriately. Hence prediction has always played a significant role in nearly every aspect of human endeavour. Prediction being a key step in civil engineering practice, its role and significance in geotechnical engineering needs hardly be questioned.

As a civil engineering material, soil is just as important as concrete and steel. All man made structures except those which float or fly are supported by natural soil or rock deposits. This might be realized either by load transfer through many components or directly as in the case of highway and airfield pavements. In many cases, the transfer of load might be between earth and earth (embankments and dams) or between masonary wall and earth (retaining walls). The earth materials and water support, exert pressure on, or are utilized in construction, thus affecting the safety of all structures. But, generally soils are more or less taken for granted by the common person. It implies that it makes up the ground on which we live and makes us dirty. Beyond these general observations most people are not overly concerned with soil. There are however certain people who are invariably deeply concerned. These include geologists, hydrologists, mining and geotechnical engineers, agronomists and others alike.

It is quite often necessary for the engineer to characterize and determine the engineering properties of materials contemplated for practical usage. Manufactured and synthetic materials, within limits, fulfill any stipulated requirements. Tests on few representative samples may be adequate to use large quantities of similarly composed materials. It would also be possible to determine accurate and complete information by appropriate tests under controlled and simulated conditions. Most often the engineer can and does predict with a high degree of confidence how these materials will perform in a given situation without the risk of facing unexpected behaviour which might lead to unsatisfactory performance and even failure. The story with soils is quite different. The inherent nature and the diversity of geological processes in soil formation are responsible for the wide variability in the state of the soil. Only a minute fraction of soil can be sampled and tested because of practical and economical constraints. For example even if the spacing of bore holes is 10m and 50mm sample is tested every second meter, only one millionth of the total volume of soil would have been explored. In contrast to many other branches of engineering where one would normally specify the requirements of materials to be used, geotechnical engineer usually has to adjust his design to accommodate the prevailing properties of in-situ soils. Hence innovative approaches which can effect significant economies, can be practiced, if the soil variability in terms of soil parameters due to erratic conditions can be realistically arrived at.

Practical Requirements

Study of field situations, realistic assessment of soil properties, analysis and design, and construction are the different phases in the practice of geotechnical engineering. In the realistic assessment of engineering properties of soils the frequent questions for which answers are sought are:

(1) What is the state of the in-situ soil? i.e., is it saturated and normally or overconsolidated or sensitive or stiff cemented?
(2) What are the values of compression and recompression indices and magnitude of compression in all the above soil states at different stress levels?
(3) What is the magnitude of preconsolidation pressure in uncemented soils and the cementation bond strength in a soft and sensitive clay?
(4) What is the shear strength of the soil and how does it change due to consolidation under an applied load?
(5) What is the stress-state permeability relationship of soils and its significance in environmental geotechnique?
(6) How the engineering behaviour of residual soils can be characterized and how are they different from sedimented soils?
(7) In partly saturated soils what are the magnitudes of heave or collapse upon inundation?
(8) What is the shear strength and/or compressibility of a partly saturated soil and its change upon loading and inundation?
(9) How compaction characteristics of soils can be assessed and their implications in field compaction control?
(10) How the engineering properties of layered soils can be characterized and assessed?

And several such other questions.

Need for Predictive Models

It has been the fervent desire of geotechnical engineers to find simpler and quicker methods of testing, using the data of which the above questions can be satisfactorily answered. Any attempt to interpret, generalize and predict soil behaviour will go a long way to:

(i) form an independent check on the laboratory investigations,
(ii) increase the level of confidence to handle extensive test data,
(iii) optimize time and cost to obtain acceptable engineering parameters,
(iv) monitor progressive changes in ground engineering and
(v) resort to observational approach in geotechnical engineering more effectively.

To achieve these objectives the two approaches pursued by researchers and engineers are:

(a) An engineering approach in which empirical methods of correlation, essentially based on experimental observations are developed.
(b) Micro-mechanisitic approaches which have a strong bias towards the basic sciences and consider the nature and equilibrium of forces at micro level and associated interactions between different phases of the multiphase systems.

A large number of scientific publications of the recent past testify the involvement in the above approaches. Although both the above paths have made great strides independently, they have not been always complimentary to each other. In the first approach of property correlation through empirical means, a scientific basis cannot always be found. In the latter approach, macrobehaviour characterized in terms of microparameters are not always amenable for measurement. Perhaps the most desirable situation is:

(i) to identify the basic mechanisms responsible for the observed behaviour
(ii) to link the empirical and basic approaches, both behavioral and parametric and
(iii) to formulate appropriate physical models for prediction of engineering properties within the limits of accuracy acceptable at engineering level.

Such an approach may be desirable than to attempt exact micro- analysis with the parameters which are difficult to measure and link with macro- behaviour. The cardinal aim could be to establish new meaningful connections between scattered facts and hence uncover new patterns so as to result in a holistic approach of analysis and prediction.

The geotechnical engineering practice is presently under phenomenal change. Soil reinforcement technique using geosynthetics might provide satisfactory means to solve many practical problems. Hence emphasis might be on covering large areas under investigation as quickly as possible. Savings in time might outweigh economy that might otherwise be achieved by extensive testing to determine the soil parameters. What one looks for may be immediate solutions by use of phenomenological approaches to predict soil behaviour and exercise judgment than elaborate testing.

This treatise is an attempt to achieve these objectives in tune with this new trend. It is stressed that our intention is not to aevelop an universal approach to analyze and predict soil behaviour. It is more realistic to identify the state of soil complexity and to use appropriate phenomenological model to predict the behaviour. It is this aspect which makes this treatise different from the rest. The later chapters have been based on the physical state of soil complexity starting from the simple state of normally consolidated to the most complex system of layered/ fissured/ varved clays. The input parameters for prediction in most of the cases are only the commonly measured inferential and state parameters.

It has been the case in most of the disciplines that each decade sets forth new basic developments and ideas, in an attempt to attain required scientific stature. In tune with this trend, this treatise is an attempt to synthesize the observed response of soils with the scientific and technological developments according to the needs of engineering. Most of the present day graduate level of soil mechanics will not be presented in this book. The brief coverage provided, as background information, pertains to soils as engineering materials and soil properties and their determination.

1

SOILS AS ENGINEERING MATERIALS

1.1 Introduction

The geotechnical engineer, in pursuing his field of endeavour has many diverse encounters with rocks and soils, being the materials of geological origin. In nature, rock formations and soil deposits of widely varying characteristics can very well be expected. This is due to the inherent nature and diversity of geological processes involved in such formations. As such, a geotechnical engineer is confronted with greater degree of uncertainty in assigning the most probable engineering properties to these geological materials, than of other processed and manufactured construction materials.

1.2 Soil Formation

Rock formations form the basic materials from which soils are derived. Geologists have classified rocks on the basis of origin and method of formation into three major groups, viz, igneous, sedimentary and metamorphic. Igneous rocks are formed by the solidification of the molten magma injected upwards towards the earth's surface. Sedimentary rocks are derived from weathering and denudation of older rocks, prominently by weathering agents such as water, ice and wind. Subsequently with increased overburden from continued deposition, sediments get densified and cemented due to stress, time and environment resulting in sedimentary rocks. High temperature and pressure induce metamorphic state to igneous and sedimentary rocks. Such rock formations are reckoned as metamorphic rocks.

In a broad sense soil formation is due to weathering occurring continuously and relentlessly throughout the geological time scale. Weathering generally denotes processes of mechanical degradation or chemical decomposition of rocks or a combination of both the modes caused mainly by temperature changes, frost action, chemical action of percolating water, effects of plant roots to name a few. In a general sense the atmosphere, biosphere and hydrosphere are the reactors in a vast chemical process with the surface of lithosphere to produce the earth's crest. Each reaction consists of a complex chemical process that is intimately associated with the physical process. The reaction is cited as (Carroll, 1970):

(Atmosphere + Biosphere + Hydrosphere)
+ lithosphere $\longrightarrow$ weathered lithosphere
+ residual materials + dissolved chemical elements

Weathering is at its highest intensity where interfaces between the hydrosphere, biosphere and lithosphere overlap. Weathering is a spontaneous reaction involving geologic material and is a change towards the decrease in the free energy of the system.

1.2.1 Physical Weathering

Physical weathering takes place when solid rocks are broken into fragments with little or no chemical change in the rock itself. Mechanical disintegration yields mineral components of soils. Temperature change, by itself, independently can disintegrate rocks into fragments by exfoliation. Rocks may also be broken and degraded by the wedging action of ice and spreading roots. Water, wind and moving ice (glaciers) although play their primary role in the transport of soil material from one location to another, are powerful agents in scouring and eroding rock formations. The size and shape of fragmented products are influenced by the structure as well as texture of parent rock material and cementing agents (Carroll, 1970). In igneous and metamorphic rocks the grain size and packing influence the disintegration process. Due to the thermal expansion of the constituent minerals, in coarse grained granular rock, disintegration is more rapid than in one of finer grains.

1.2.2 Chemical Weathering

Apart from physical degradation, surface chemical reactions between rocks and environment might take place. In chemical weathering, original rock minerals are transformed into new minerals by chemical reactions. There are two types of chemical weathering. If chemical reactions are inorganic it is regarded as geochemical. Another type is pedochemical weathering, reflecting the biological activity as a result of both organisms and organic solutions reacting with rocks (Carroll, 1970).

The magnitude of chemical weathering is dependent on the amount of water with or without electrolytes that is available to leach rocks and on the drainage characteristics of rock material. The reactions between silica, alumina and water determine the characteristics of the resultant soil minerals. Silica is slightly soluble at all pH values, whereas alumina is soluble only below pH 4 and above 8.5. Water with its content of cations and anions is the reagent that causes hydrolysis, a reaction between a salt and water to yield an acid and a base. The amount of salt dissociated in water, indicated by the pH value, reflects the potential of chemical environment. Clays, laterites and bauxites are formed by hydrolysis. Other chemical processes of decomposition are carbonation in which minerals are attacked by carbonic acid and oxidation in which oxygen combines with minerals of rock.

The climatic conditions play an important role in initiation and intensity of weathering. In arid regions, for example, physical weathering as effected by temperature changes predominates and fine weathered material is removed mainly by wind. In temperate regions, freeze - thaw is the main cause for degradation. In tropical regions, a warm and humid climate supports deep-reaching chemical weathering which results in chemical decomposition of rock components. According to the character of chemical weathering, soils are distinguished. But in reality the physical and chemical weathering processes generally operate simultaneously. Depending on the climate one of them may predominate in a particular location.

1.3 Soil grouping based on Origin

Soils are divided into two broad classes on the basis of origin (i) residual - those derived insitu from bed rock and (ii) transported - those deposited by wind, water and ice. Both classes exhibit a wide variety and range of physical and chemical characteristics in as much as the varying characteristics of parent rock materials.

Where topography of rock is approximately horizontal, weathering can take place without the soil being moved from its original position. Soils of this type are residual soils. The terminology is self explanatory in the sense that soils are actual residues of original rock with all soluble materials being progressively leached out by the long continued seepage of ground water. The alteration of the original rock becomes less with depth until finally the natural unaltered rock is reached.

Although the gradual downward transition from the soil to rock is a distinctive feature of residual soils, some residual clays show no evidence at all of their origin. Residual soils such as laterites possess peculiar characteristics due to the absence of organic materials in the top soil and the consequent lack of humic acid to act as a leaching agent. Such soils contain high alumina and iron oxide. They are soft when moist but harden on exposure to atmosphere. Laterites are found in tropical areas such as India and parts of South America.

In the second phase of the geological cycle most of the products of erosion and degradation are transported and deposited by wind, water, ice and gravity. Dominant textures of these soils vary with the method of deposition and with the distance transported. The shifting movement of dune sand in desert and semi- arid regions is an example of wind - borne soil movement. Such soil deposits are identified as aeolian deposits. A special type of these deposits, widely distributed throughout the world is known as loess. They are generally brown in color with particle size range between narrow limits of 0.01 and 0.005 mm. Sometimes it can also have some traces of clay. Loess owes its engineering characteristics largely to the way in which it has been deposited with metastable structure due to loose fabric of the soil deposit. Carbonate cementation is often responsible for retention of this loose fabric.

The size of the particles which can be moved by water is a function of velocity of water. Coarser particles are deposited first in the upper regions of river due to steep gradients, then sand and the silt sized particles become an intolerable load. Sorted deposits of these soil fractions are formed. Quiescent conditions are required for deposition of the clay particles. Such conditions prevail in lakes, ponds and delta areas. Examples of such deposits in United States are filled valleys and basins in the west of Rocky mountains, the great outwashes covering the great plains, the extensive Atlantic and Gulf coastal plain (Leonards, 1962) and the Indo-Gangetic alluvium plain of India (Wadia, 1957).

Glacial drift contributes to form transported soil deposits. Such deposits are found in large portions of North America as well as Northern Europe and Asia. The counter part in India is found in Himalayan region. Till is an example of unstratified glacial deposits of clay, silt, sand, gravel and boulders. Sorting of rock debris has also been possible by flowing of water due to melting of enormous quantities of ice. Table 1.1 details the grouping of soils based upon mode of transportation and deposition.

Table 1.1 Transporting agents and Soil deposits

Soil deposit	*Mode of Transport*	*Depositional Environment*
Alluvium Marine Lacustrine	Water	Flowing water Quiet brackish water Fresh water
Loess Dune sand	Wind	Arid and coastal lands
Moraine or Till Drumlins Eskers	Ice	Glacial
Colluvium Talus	Gravity	Below slide area Base of cliff
Muck or peat	In-situ	Swamp and Marsh deposits – organic in nature

The following clear differences between residual and transported soils have been highlighted (Vaughan et al., 1988). In residual soils the particles and their arrangement evolve progressively as the consequence of chemical weathering. Quite often residual cementation would be left behind due to various degrees of weathering. Little (1969) presented a typical profile indicating descriptive terminology to reflect the degree of residual cementation (Fig. 1.1). In transported soils the preformed particles are transported and deposited in different land forms. Generally the ideas of classical soil mechanics are based on the specific soil types transported and deposited by water. After deposition, the particulate material is acted upon by effective stresses involving typically an increase due to increasing overburden (normal consolidation) and a subsequent decrease due to removal of overburden (overconsolidation).

1.4 Clay minerals as weathered products

The fraction of soil which determines the physical behaviour most decisively is the colloidal clay. It contributes maximum to the total specific surface of the soil even

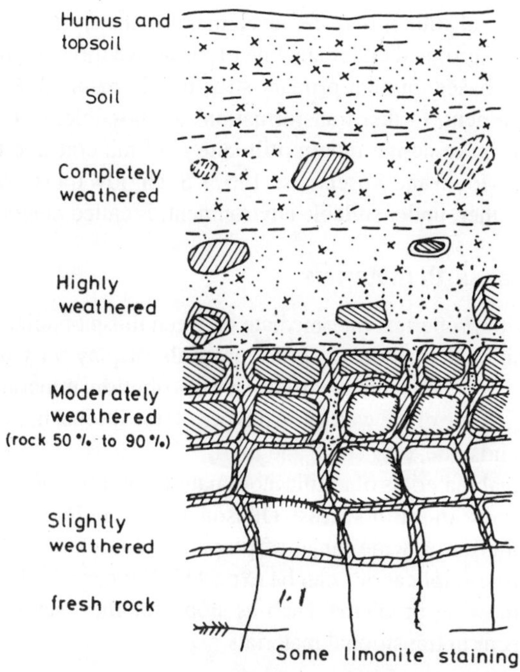

Fig. 1.1 Typical tropical residual soil profile.

if the percentage of clay in soil is far less than sand and silt fractions. Greater the specific surface of the soil the physico-chemical potential of the soil is generally higher.

In the absence of sophisticated tools like X-ray diffraction, infrared spectrometry adopted for clay mineral identification, earlier it was thought that kaolinite clay mineral was the essential constituent of all colloidal fraction of soils. Those not fitting to this composition were regarded as impurities (Grim, 1969). Now, most clay minerals can be grouped into any one of the three main groups, viz., kaolinites, montmorillonites and illites. Clay minerals are basically the hydrated silicates of aluminium, magnesium and iron. In an acid environment, sodium, potassium, calcium, magnesium, and iron are mainly leached leaving behind the hydrogen ions in clay forming process. If the ratio of silica to alumina is low in the bed rock, clay minerals of kaolinite group are formed. For the higher silica to alumina ratio in the parent material with the chemical reactions in the alkaline environment, the clay minerals formed are of montmorillonite group. When potassium ions fit into crystal lattice structure of montmorillonite under suitable conditions the resulting clay mineral is illite. In practice these above cited conditions are related to the nature of electrolyte that leaches out of the weathered products and the drainage conditions. Kaolinites tend to form wherever plentiful supply of acid water and good drainage conditions are encountered. On the other hand alkali water and poor drainage conditions contribute to the formation of montmorillonite and illite clays. Although a wide variety of clay minerals can be identified due to the substitutes of different elements in the crystal lattice structure, most of these can be grouped under any one of the above three main groups. This grouping is mainly based on their primary structure. In nature due to weathering of natural clay sediment, further transformations are possible. Illite clay mineral can decompose to more active montmorillonite clay mineral due to low pH of marine sediments (Moum and Rosenquist, 1967). Such slow transformation of one clay mineral to another, under suitable environment, is called diagenesis.

1.5 Transitional earth materials

Earth materials with properties intermediate between those of soils and rocks are regarded as transitional materials. These materials display varying degrees of sensitivity to water dependent upon particle size distribution, mineralogy, type and degree of bonding. Some of the well known types of transitional materials are, clay shale, silt stone, mudstone, clay stone and marl.

Since transitional materials of argillaceous nature comprise 50 to 75 percent of the sedimentary rock in earth's crust (Judson et al., 1976) with widespread geophysical distribution, a wide range of properties and characteristics and the resulting engineering implications can be expected. Further a high percentage of geotechnical engineering problems, such as slope stability, settlement, bearing capacity failure occur in transitional materials.

1.6 Soil Deposits of India

Under different conditions of climate, topography and action of various weathering agents, several soil types have been encountered in India. Figure 1.2 is a map indicating major soil deposits.

The marine deposits of India are mainly in tidal flats along the coast. Cochin area on the west coast and the Vishakapatnam harbour area on the east coast are important regions. These clays are slightly sensitive and are essentially inorganic in nature. One of the major regional deposits of soils, commonly designated as Black Cotton soils cover an extensive area of 300,000 sq. km. which is one fifth of total land area. The underlying bed rock of black cotton soils is to a large extent basaltic. Due to high active clay fraction present in such soils and the prevailing tropical environment, unacceptable volume changes due to heave or shrinkage can take place. Such expansive soils, inevitable problematic geotechnical materials, are encountered in many parts of the world. In India Laterites form another important regional deposit covering an area of about 100,000 sq. km. They are formed by the decomposition of rock, removal of silica and bases in solution and accumulation of aluminum and iron sesquioxides, titanium, magnesium, calcium and other amorphous products. In large parts of Indo-Gangetic river plains, alluvium deposits are encountered. The thickness of such deposits in many places is more than 100 meters with alternating layers of sand, silt and clay deposits. Most of the soils found in large parts of Rajasthan are wind blown deposits formed under arid conditions with hardly any rain-fall.

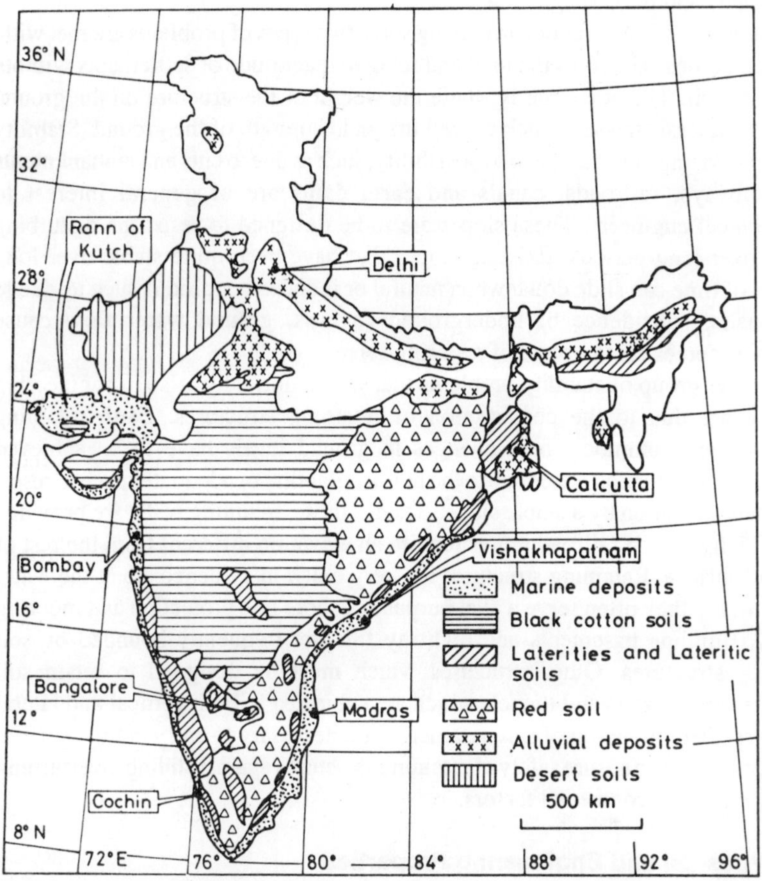

Fig. 1.2 India map indicating major soil deposits.

1.7 Practical problems with soils

The first step in any construction activity is to ensure that an adequate support to the structure has been provided to transfer to the substratum the total weight of the structure, live load and all other forces which may act on the structure. The substructure thus must perform its function in such a manner that it ensures a margin of safety against collapse of the structure and that its settlements or displacements are within the limits determined by structural, functional, aesthetic and psychological considerations. The structural aspect refers to prevention of the total collapse. A situation of functional impairment arises, for example, by excessive uniform settlement of the structures. Settlements in high water content soft soils can be of the order of one to two meters. Utilities and drainage lines might have to be relaid to restore the functional needs. A structure, for example, might not suffer complete failure but might undergo excessive settlement or differential settlement to develop different degrees of cracks. These might impair the utility of the structure on aesthetic and psychological factors.

In any type of construction involving soils, two types of problems are met with. The first of these concerns the total and relative magnitude of settlements without failure of soil. The other one is where the weight of the structure on the ground generates lateral stresses which exceed the yield strength of the ground. Stability problems arising out of the second possibility, such as due to cuts and embankments for highways, railroads, canals and earth dams are of general interest to geotechnical engineers. These slopes are to be designed to resist the disturbing forces over long periods. Existing slopes that have been quite stable over long periods of time can slide down when natural or induced conditions such as due to earthquake, subsidence of underground caverns, ground water flow cause disturbing forces far in excess of restoring forces.

Another group of stability problems that would normally arise in engineering practice are due to the construction of retaining structures. Earth retaining structures are common where changes in ground levels have to be achieved abruptly. This situation arises due to the non-availability of adequate space to resort to gradual transition by a stable slope. Besides these, retaining walls are provided where highway and railroad routes require an elevation different from the rest of the land surface. Retaining structures are very much in evidence along the water fronts, where they often serve a dual purpose of shore line protection and mooring facility. Building basements and highway fills are frequently bounded by soil retaining structures. Other structures which must be designed to retain soil pressures are culverts and tunnels which are subjected to both vertical and lateral pressures. Determination of soil parameters are necessary to carry out the general stability analysis to ensure safety of retaining structures against sliding, overturning and for other environmental factors.

1.8 Physical and Engineering Properties

Design for stability, in geotechnical engineering when reduced to its simplest form, can be visualized as a comparison between loading and the ability of soils to withstand the same at permissible levels of deformation. Consequently rational approach to practical problems involving stability, settlement and flow, compressibility, strength and permeability characteristics are to be determined and expressed in terms of appropriate parameters. Soils primarily being particulate in nature can be regarded as a multicomponent multiphase system with compositional and structural discontinuities. While, they are considered as a system, the relative proportions of solids, liquid and gaseous phases in unit volume play a dominant role in the soil behaviour determined at engineering level. Hence, water content, w voids ratio, e and degree of saturation, S_r become the basic parameters of the physical state of soils. From the phase diagram shown in Fig. 1.3 the following weight-volume relationships can be obtained.

$$W = W_s + W_w \tag{1.1}$$

and also

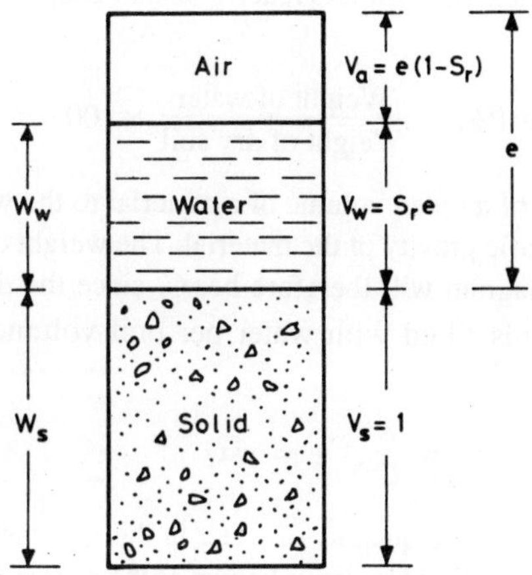

Fig. 1.3 Phase diagram of soil.

$$V = V_s + V_w + V_a = V_s + V_v \tag{1.2}$$

$$V = V_s + V_w \text{ for saturated soils} \tag{1.3}$$

Where W_s = Weight of soil solids
W_w = Weight of water
V_s = Volume of the soil solids
V_w = Volume of water
V_a = Volume of air

Void volume: V_v is the volume not occupied by solids in unit volume is known as void volume. It may be occupied by either water or air, or a combination of both.

Void ratio, e is defined as the ratio of the volume of voids, V_v to the volume of solids:

$$e = \frac{V_v}{V_s} \tag{1.4}$$

Another way of expressing the quantity of voids is to relate the void volume to the total volume.

Porosity, n is defined as the ratio of voids to the total volume.

$$n = \frac{V_v}{V} \tag{1.5}$$

The relationship between void ratio and porosity is expressed as,

$$e = \frac{n}{1-n} \tag{1.6}$$

Water content: In soils, the void volume may be filled by water and/or air. The water content of a soil is measured by weighing a sample before and after it is heated to 105° for sufficient time for it to reach constant weight. The water content is expressed as

$$w\,(\%) = \frac{\text{Weight of water}}{\text{Weight of dry soil}} \times 100 \qquad 1.7$$

The ratio of weight of a given volume of a material to the weight of the same volume of water is specific gravity of the material. The weight of one unit of solid volume in the phase diagram will therefore be, G, since the density of water is unity. In saturated voids filled with water per unit volume of solids is void ratio. Hence.

$$w = \frac{e}{G}, \quad e = wG \qquad 1.8$$

In a partly saturated soil the quantity of water in the soil may be expressed as a fraction of the void volume. This fraction is known as the degree of saturation, S_r.

$$S_r = \frac{V_w}{V_v}, \quad V_w = S_r e \qquad 1.9$$

Percentage saturation $= 100\, S_r$

For perfectly dry soil, $S_r = 0$

For saturated soil, $S_r = 1$

To determine the degree of saturation, it is necessary to determine void ratio and water content independently.

Soil densities provide a measure of the weight of material related to the amount of space it occupies. Several densities may be defined.

$$\text{Dry density, } \gamma_d = \frac{\text{weight of solids}}{\text{total volume}} = \frac{G\gamma_w}{1+e} \qquad 1.10$$

$$\text{Bulk density, } \gamma_b = \frac{\text{total weight of soil}}{\text{total volume}}$$

$$= \frac{\text{weight of solids + weight of water}}{\text{total volume}}$$

$$= \frac{G\gamma_w + S_r e \gamma_w}{1+e}$$

$$= \left\{\frac{G + S_r e}{1+e}\right\}\gamma_w \qquad 1.11$$

The ratio of the two densities is

$$\frac{\gamma_b}{\gamma_d} = 1 + \frac{S_r e}{G} \qquad 1.12$$

Since, $S_r e = w G_s$

$$\gamma_b = (1 + w)\gamma_d \tag{1.13}$$

It is necessary to determine bulk density and water content to compute the void ratio and degree of saturation which in turn define the physical state of the soil.

1.8.1 Effective Stress Principle

Although soils are regarded as ideal continua, in reality, they are particulate materials and water in void spaces is under pressure. The principle of effective stress determines the effect of pore water pressure on the behaviour of soil with a given total stress. It was stated by Terzaghi (1936) as

> The stress in any point of a section through a soil mass can be computed from the total principle stresses, σ_1, σ_2 and σ_3 which act at this point. If thevoids of the soil are filled with water under a stress, u, it acts with equal intensity in all directions. It is called neutral stress. The balance, $\sigma_1' = \sigma_1 - u$, $\sigma_2' = \sigma_2 - u$ and $\sigma_3' = \sigma_3 - u$ represents an excess over the neutral stress and are the effective principle stress components.

Thus, the fundamental Terzaghi's effective stress equation is

$$\sigma' = \sigma - u \tag{1.14}$$

The importance of effective stress considerations in the soil behaviour has been stated by Terzaghi as

> "All measurable effects of change of stress, such as compression, deformation and mobilization of shearing resistance are exclusively due to changes in the effective stress".

1.8.2 Compressibility and consolidation

The compressibility of soils arises from the relatively large percentage of voids in them. The stresses encountered in most engineering works are far too small to produce significant changes in solid volume. Hence the volume change is almost the result of the changes occurring in void volume. For sands, the void ratio will not change very much upto stress levels at which grains begin to crush (1500 kPa) whereas for clays void ratio changes of 100 percent or more are possible for applied effective stresses of engineering interest. Clays and sands differ in their behaviour when stresses are released.

How fast the soils are compressed depends upon the rate of flow of the fluids present in the voids of the soils and its proximity to the drainage boundary. This process is called consolidation which is a stress-strain-time phenomenon. Deformation may continue for months, years, or even decades. This is the fundamental difference between the compression of granular materials and consolidation of clays. Compression of granular medium occurs almost instantly due to its high permeability whereas consolidation is a time dependent process in fine grained soils. On the other hand during compaction, compression of the loose soil under compactive effort takes place essentially due to the expulsion of air.

1.8.3 Shear strength

The ability of particulate materials such as soils, to support an imposed loading or to support itself is governed by the shear strength of the material. As such, the assessment of the shearing resistance of soils is of primary importance in the analysis and design of stability problems.

If we were to deposit a granular soil by pouring it from a single point above the ground it would assume a conical shape. With additional material poured on the top of mound, the slope for a while might appear to be steeper, but eventually the material slides down to the slope equal to the angle of repose. This angle corresponds to the angle of internal friction of the granular material in its loosest state. Sand dunes are an example from nature for the angle of repose.

Apart from the above simple picture of shear strength of particulate media, let us consider a soil system subjected to external loading. Both the initial mean principle stress and shear stress values change. At this stage, an important distinction can be made between the behaviour of soils, a particulate media and metal, a highly bonded system. When a metal is subjected to shearing stresses and it deforms in a manner similar to soils, the volume of the material or density remains constant during the shearing process if the hydrostatic (mean) stress remains unchanged. On the contrary, in the shearing of soils, a volume change manifests even if the mean stress remains constant. Consequently, the shear strength of the soil is pressure dependent i.e., increases with increase in confining pressure unlike in metals.

Further, it would be of interest to consider the response of saturated fine grained soils during shearing. Typically, engineering loads are applied much faster than the rate at which the pore water can escape from the pores, thus increasing the pore water pressures. If the loading is such that failure does not occur then the pore water pressures dissipate decreasing the void ratio with consequent increases in shear strength of the soil. The primary difference in the shear behaviour of sands and clays is due to differences in the time required for the volume changes to occur.

The rate of loading might be such that shear failure is imminent. Since the pore water cannot carry any shear stress, all the applied shear stress has to be carried by the soil structure. The soil skeleton would withstand imposed shearing stresses only if the rate of shear stress application is such that the skeleton acquires the ability to mobilize adequate internal resistance by compatible volume changes.

It is not always true that shearing causes positive pore water pressures or volume reduction. In a dense granular soil the material tends to expand during shear since grains have to climb up over each other to permit distortion to occur. Analogously, but with a different mechanism, shearing of highly overconsolidated clays also results in volume increase. On the other hand, loose sands and normally consolidated clays contract as the shearing deformations breakdown the soil structure which withstood the hydrostatic stresses without shear. Thus not only the shearing stresses but also the hydrostatic stresses do work on the material during the application of shearing stresses to cause volume reductions. During shearing of an unstable material that expands in volume, energy must be spent both for the shearing and the dilatational components.

1.8.4 Permeability

Particulate media such as soils, in equilibrium, have void spaces mostly continuous, permitting flow of fluids. Flow of an incompressible fluid in porous media is controlled by three basic relationships, viz., conservation of mass, energy and linear momentum. The rate of flow is influenced by hydraulic gradient, i defined as energy or head loss per unit length ($i = h/l$). Flow can be described as laminar for fluid flow in parallel layers without mixing. It can also be turbulent with random velocity fluctuations resulting in mixing of fluids and internal energy dissipation. There can also be intermediate or transition states between these two.

As an outgrowth of experiments associated with the design of rapid sand filters for Dijon city in France, Henry Darcy developed an empirical relation describing flow of water through sands (Darcy, 1856). Based on continuity considerations he enunciated that rate of flow, v in clean sands is proportional to hydraulic gradient, i

$$v \propto i, \quad v = ki \tag{1.15}$$

The proportionality coefficient k, is called Darcy's coefficient of permeability or simply permeability . This has units of velocity since i is dimensionless.

From the theoretical relationships for flow through capillary tubes developed by Hagen and Poiseulle during 1840 and the more recent hydraulic radius models of Kozeny-Carman (Kozeny, 1927 and Carman, 1956), the permeability of particulate media can be examined. According to Kozeny-Carman equation for cohesionless soils such as sand, the relation between permeability and void ratio is of the form:

$$k = c_1 \frac{e^3}{1 + e} \tag{1.16}$$

The shapes of the voids and the zig-zag flow paths through the soil pores, called tortuosity, are taken care of by the constant c_1. For fine grained soils with appreciable clay fraction, Kozeny-Carman equation is less successful. Despite the extensive investigations to date, unfortunately, there are only a few physical models, to predict permeability using the basic physical and insitu properties of soils.

It is interesting to note that there is no other engineering property of soil, infact of any construction material, which is as variable as permeability. The range of permeability is over ten billion times for any soil or rock whereas the variation of bulk density of the material is in the range of fifty times and that of strength is of the order of one tenth of a million (Fig. 1.4) (Cedergren, 1967).

These mechanical behaviour of a given soil are characterised and quantified in terms of appropriate parameters such as coefficient of compressibility, coefficient of consolidation, deformation modulus, angle of shearing resistance, cohesion, coefficient of permeability, etc which can be used in methods of analysis to get the responses of a geotechnical structure under imposed loadings and environments. Testing methods have been evolved to determine these parameters both at

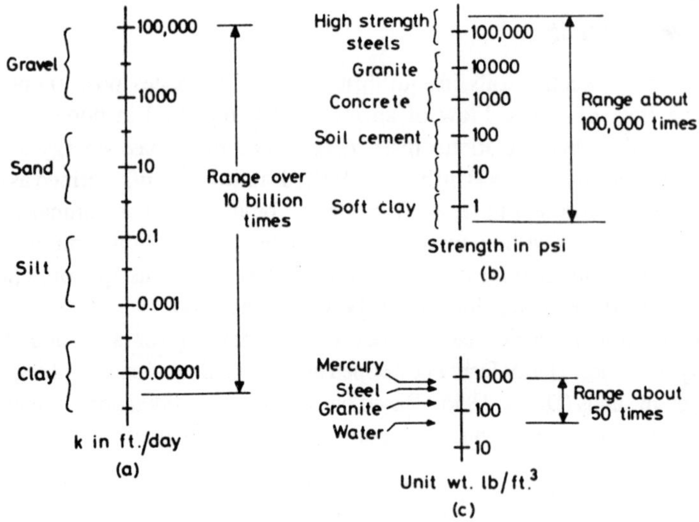

Fig. 1.4 Variation of permeability compared with other engineering properties (log scale). (a) Permeability (b) Strength (c) Unit weight (Cedergren, 1967).

laboratory scale on undisturbed and representative samples and at actual field scale. A brief description of the most widely used testing methods for determining various soil properties will be presented in the next chapter.

2

SOIL PROPERTIES AND THEIR DETERMINATION

2.1 Introduction

Soils being particulate materials, with void ratios most often than not, greater than one, are highly compressible. The shear strength which is synonymous with the strength of soils is dependent on the compression of soils as the stresses become effective. These two properties are interdependent upon the permeability characteristics of soils. For any rational approach to solve stability, settlement and flow problems in soil engineering, strength, compressibility and permeability characteristics are to be determined and expressed in terms of appropriate parameters. Two broad approaches pursued to determine the above engineering properties are laboratory and in-situ testing. The developments, so far taken place in each of these approaches, are so vast, that it would be of an unmanageable task even to briefly cover the same. However a very brief description of the well established methods has been given, culminating at each stage with a coverage to more recent methods. In the ensuing discussions the emphasis is on the method of determination rather than why soils behave differently under the same stimuli. Before elucidation of the methods of testing relative merits and demerits of laboratory and field testing are highlighted.

The determination of engineering properties by laboratory testing offers the following specific advantages over field testing.

(1) Close control over test conditions, including boundary conditions and time, can be attained.

(2) Control can be exercised over the choice of undisturbed or representative material to be tested.

(3) Tests can be carried out under simulated variable field conditions for example, environmental factors such as temperature and humidity can be simulated and varied apart from stress conditions.

(4) Tests can be carried out on the reconstituted (destructured or remoulded) material under different stress conditions to understand basic mechanisms of material behaviour.

Where soils are too difficult to sample or likely to be severely disturbed during sampling, in-situ methods can be prefered for engineering property evaluation in spite of high cost. The tests have a wide range from the relatively simple penetration and vane tests to more sophisticated pressuremeter, dilatometer, and bore hole shear tests.

A major problem with the laboratory testing is one of scale, shape and size of samples. All in-situ soils are likely to contain varied degrees of macro-fabric features such as laminations, joints, fissures, and bedding planes. Where little or insignificant macro-features exist in soil, laboratory tests are reasonable to assess the engineering behaviour provided sample disturbance is not a major problem. At the other extreme, the laboratory tests would still yield realistic parameters provided intensity of geomechanical features are so close that laboratory samples themselves contain features representative of the in-situ material. Between the above two extreme situations, sample size becomes a dominant factor. Engineering behaviour of soils are markedly dependent on stress history and the stress path.

In-situ testing is a recognized and growing activity to assess the mechanical properties right at the stage of subsurface investigation itself. At present the vane and cone penetration tests are usually done for in-place strength determinations of soft and sensitive to stiff clays. Dynamic testing methods are extensively used for mechanical property assessment of sands, gravels, as well as very dense glacial tills. Permeability testing of soils is well established and a wide range of techniques are available to encompass different ground conditions. For example it is possible to carry out in-situ permeability tests in soft clays by continuous sounding device (Torstensson, 1975). In each category of mechanical property assessment many in-situ tests are currently available. Each individual test can be very useful, provided its limitations and range of applicability are recognized and not overlooked.

Many techniques for in-situ property evaluation are available, and each method has its own characteristic features and are appropriate under specific field conditions. In general the cost of in-situ testing is relatively high in comparison to laboratory testing. This is partly due to several factors such as capital investment, mobilization, running and maintenance.

2.2 General Considerations

An important aim of engineering investigations is to establish the ability of soil to support the structure with due considerations to the resulting movements. In a broad sense, the principal parameters to be obtained in these investigations are compressibility, strength and permeability.

2.3 Compressibility and Consolidation — Laboratory methods

In a broad sense all soils undergo certain deformations resulting in a changed particle arrangement to acquire the ability to withstand external loading. Changes in the micro- and macro- structure are reflected either in the distortion of the material or volume change due to decreases in void spaces between particulate constituents. Broadly, deformations can be classified as those resulting in volume change, distortion without volume change or a combination of both. The volume change due to a unit load is compressibility and its time rate is consolidation. During compression practically there is no decrease in the volume of solid particulate constituents. In the case of granular media where interstitial fluid does not play a significant role in time lag, the compressibility is due to displacements between particles combined with particle breakage.

2.3.1 Compression testing methods

When soil layers covering a large area are loaded vertically the compression can be regarded as one dimensional. Laboratory consolidometer test conforms to this condition. Figure 2.1 is a schematic diagram of two types of consolidometers. In the floating ring test the compression takes place from both the faces of the soil

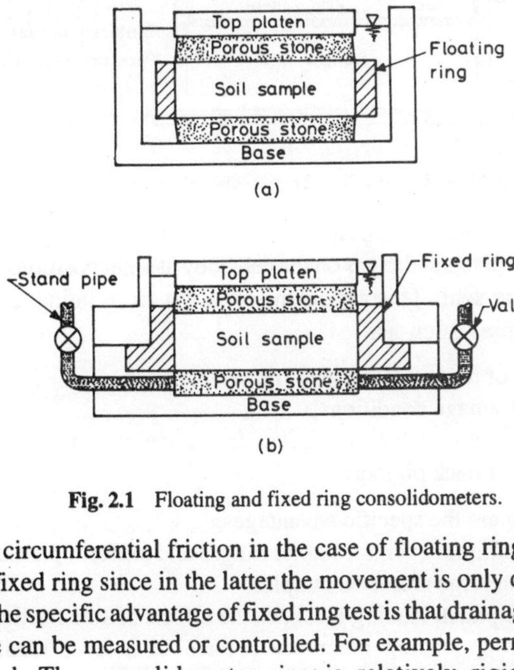

Fig. 2.1 Floating and fixed ring consolidometers.

sample. The circumferential friction in the case of floating ring is comparatively less than in fixed ring since in the latter the movement is only downward relative to the ring. The specific advantage of fixed ring test is that drainage from the bottom porous stone can be measured or controlled. For example, permeability tests can be conducted. The consolidometer ring is relatively rigid to resist laterai deformations upon loading in the range of engineering interest. On the top and bottom of the sample, sintered corundum porous stones are used to permit drainage during consolidation. Side friction can be minimized significantly by switching over to ceramic or teflon lined rings or by the application of silicon grease while conventional rings are used.

Laboratory compression tests are seldom required on coarse grained soils. One reason for this is the practical difficulty associated with obtaining undisturbed samples of granular deposits. Deformations of coarse grained soils occur almost at a rate commensurate with the rate of stress change. Since the rate of settlement of these soils is rapid, they do not often pose problems of post construction settlements. Volume change characteristics are often obtained indirectly from an inplace density determination, relative density tests or through other correlations.

A consolidometer in which the total stress is applied by means of hydraulic pressure acting across a convoluted rubber bellowfram has been developed by

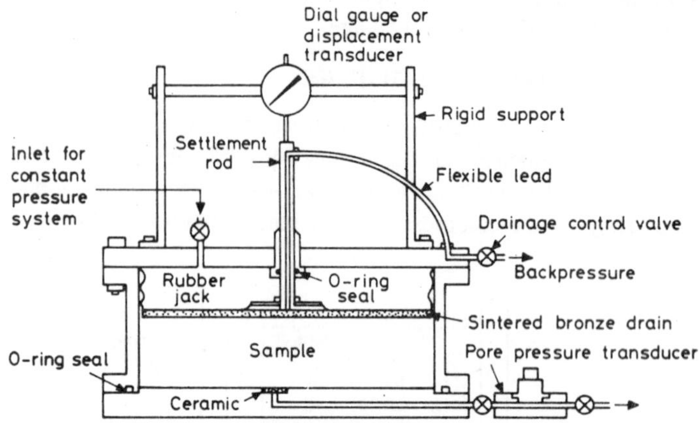

Fig. 2.2 Rowe's consolidation cell (after Rowe and Barden, 1966).

Rowe and Barden (1966). The basic form of the cell is shown in Fig. 2.2. The base and cover are bolted to the flanges on the cell body at eight positions with an O-ring forming adequate seal. This arrangement, apart from being a closed system, has several features such as;

(i) application of uniform load onto the sample,
(ii) control of drainage conditions,
(iii) monitoring of initial and subsequent pore water pressure changes,
(iv) application of back pressure.

The following are the specific advantages:

(i) Loading is simpler even for higher ranges and uniform pressure application is possible to large diameter samples.
(ii) The flexible platen loading localizes the effects of side friction.
(iii) Impact during loading otherwise present in a lever loading system is eliminated.

This hydraulic consolidometer has proved satisfactory for a variety of purposes, viz.;

(a) vertical consolidation of remoulded, compacted or undisturbed samples of saturated and partly saturated clays, both with and without a back pressure and with various sample diameters and thicknesses,
(b) inward and outward radial water flow during consolidation of undisturbed and remoulded samples of layered or laminated clays to study their consolidation behaviour,
(c) vertical and radial flow permeability tests.

In all types of consolidometers discussed above it is not possible to determine the horizontal stresses developed at various stress levels and during consolidation at any stress increment. A cell with three part split ring has been developed by

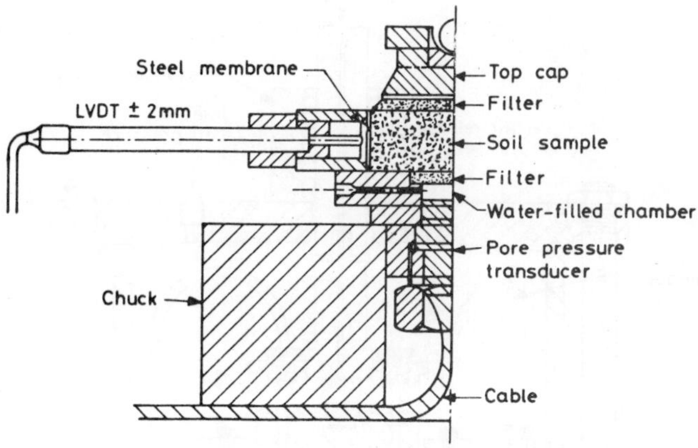

Fig. 2.3 Cross-sectional details of split ring oedometer (after Senneset, 1989).

Senneset (1989). Figure 2.3 shows the cross sectional details of this oedometer. In each part of the ring a sensitive LVDT records the horizontal stresses developed during the test.

Controlled gradient consolidation tests : In the conventional consolidation test the overall consolidation which is occurring during the test is an average of consolidation over different points with varying range of effective stresses. For the element in the field to which the test data is applied, the effective stress is the same throughout the element at any time during consolidation. To overcome this limitation controlled gradient consolidation test has been developed (Lowe et al., 1969).

The controlled gradient test is performed on a thin cylindrical specimen identical to that used in the conventional test. Drainage is in the axial direction and only towards top face of the sample. At the bottom face a constant small hydrostatic excess pressure is maintained while gradually applying axial load to the specimen to achieve a constant pore pressure distribution. As the specimen consolidates the pacer arrangement automatically functions to readjust loading to achieve constant gradient. Another specific advantage of this test is that loading procedure can be interrupted thereby enabling to monitor primary and secondary compressions.

Consolidation at constant rate of strain (CRS Test) : By conventional procedures, consolidation tests on fine grained soils require considerable time. To overcome this limitation, the constant rate of strain consolidation test has been developed by Smith and Wahls (1969). In this test the soil sample is loaded continuously rather than incrementally at any desired rate of strain. Subsequently a general purpose consolidometer has been developed by Wissa et al., (1971). Figure 2.4 is a schematic diagram of the same. The base plate of the cell contains the pore pressure measuring system. The loading system uses a rolling diaphragm seal and runs in ball bushings. Vertical deformations of the test specimen during loading are determined by the movement of cell piston both with a dial gauge and

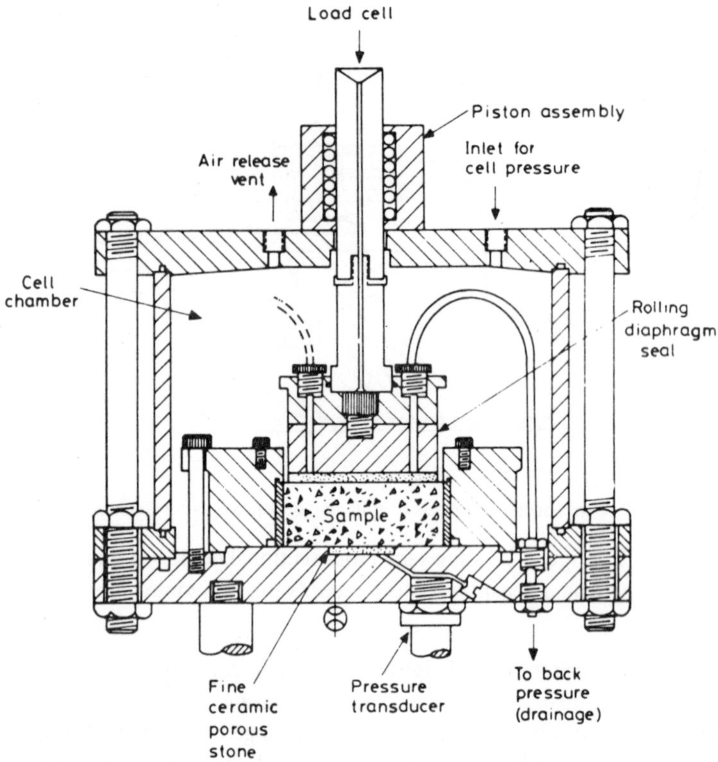

Fig. 2.4 Details of consolidometer for test under constant rate of strain (after Wissa et al., 1971).

an electrical displacement transducer. The load, applied externally to the piston, is monitored with a load cell. The back pressure and the cell pressure at the impervious surface are measured with pressure transducers. Since electronic sensing devices are used, the test can be fully automated using a digital computer for data recording, analaysis and graphical display of the results.

Constant rate of strain loading devices have been introduced into the testing programme at Norwegian Geotechnical Institute since 1984 (Sandbackken, et al., 1986). Such devices have also been successfully used as routine tests in Sweden for over several years (Larsson and Sallfors, 1986). These tests provide continuous relations of stress - strain - strain rate and, permeability versus void ratio and thus coefficients of consolidation versus either stress or void ratio.

With the development of digital controller system (Menzies, 1988) the potential use of hydraulically operated consolidometer, controlled gradient and constant rate of strain consolidation tests can very well become routine tests with the same level of popularity as the conventional test with considerable built in flexibility. The digital controller is a microprocessor hydraulic actuator for the precise regulation of liquid pressure and liquid volume change. Pressure is detected by means of an integral solid state pressure transducer. Control algorithms are built into programmable memory to cause the controller to seek a target pressure or step to a target volume change. The change is detected by counting the steps of the stepping motor. This digital controller has a volume capacity of 200 cc and a pressure range of 0-2000 kPa and is a general purpose constant pressure source, a volume change gauge and a pore pressure measuring system.

2.3.2 Presentation and analysis of compression data

Compression test data are presented in several modes depending upon the preferences and experiences of individuals. Because compression is due to changes in the void spaces, change in void ratio in descending scale versus increase in loading stress is one common method of presentation. The coefficient of compressibility, a_v is the decrease in void ratio per unit increase of stress. It is therefore equal to the slope of the compression curve. Thus,

$$a_v = \frac{de}{d\sigma_v} \qquad 2.1$$

which constantly changes with stress increment. Hence it is difficult to use a_v in settlement analysis. The coefficient of volume compressibility, m_v is the corresponding decrease in unit volume and is given by

$$m_v = \frac{a_v}{(1 + e)} \qquad 2.2$$

Both these coefficients are dimensionally inverse of stress (cm^2/kg or ft^2/ton). When semi log coordinates are used, $e - \log \sigma_v$ plot approximates to a straight line. In the case of constant rate of strain test the rate of change of effective stress at any point and of the pore pressure distribution are constant. In terms of total stress change

$$m_v = \frac{\Delta\varepsilon}{\Delta\sigma_v} = r\frac{\Delta t}{\Delta\sigma_v} \qquad 2.3$$

in which $\Delta\,\sigma_v$ is the change in measured total stress over an increment of time t and r is the average strain rate.

Another method of presentation of data is unit change in sample thickness (or strain) versus logarithm of stress. This yields the same results as in the earlier presentation, but has the specific advantage of minimizing the amount of computation associated with reducing the test data. With this method no conversion to void ratio is necessary. The compression data for the sample are converted into strain by dividing total compression by original sample thickness and this strain is plotted against logarithm of stress.

2.3.3 Stress history effects

The overburden stress at a point in a soil mass is simply the weight of the material above it. The effective overburden is the stress from this soil less the pore water pressure. Soils which at no time in their history have been subjected to stress greater than the existing overburden are regarded as normally consolidated. On the other hand there are situations wherein the removal of overburden takes place. Erosion and glaciation are some of the practical situations causing removal of overburden. Such deposits are regarded as overconsolidated in which the pressure due to present overburden is less than the past effective stress level. The $e - \log \sigma_v$ compression paths of such soils are different from those of normally consolidated. The compression is less upto past maximum stress level, beyond which the compressions are marked and follows the path of normally consolidated state. The stress level at which transition takes place is the preconsolidation stress. In characterizing soil behaviour the relation between preconsolidation stress and existing overburden stress has an important bearing on the settlement analysis. It provides a means to assess as to how, due to "memory" effects, the soil will respond under additional loads from structures or ground water lowering. Geostatic stresses are not the only means of inducement of stress history effects. State of stress of in-situ soils is the resultant effect of various factors such as time and environment apart from stress history effects. Several factors responsible for causing the effect of preconsolidation in soils are listed in Table 2.1.

Table 2.1 Possible causes for preconsolidation of soils

Causes	*Mechanisms*	*Remarks*
A. Stress dependent		
1. Change in total stress due to: (a) removal of overburden (b) past structures (c) glaciation	changes in effective stresses and the consequent stress release effects	commonly encountered in practice
2. Change in pore pressure due to: (a) water table elevation including artesian conditions (b) deep pumping		
B. Environmental factors		
(a) wetting and drying cycles and desiccation due to plant life (b) desiccation due to freezing (c) time dependent deformation due to secondary compression and strain rate of loading	capillary stresses and the consequent changes in soil fabric	formation of drying crusts overlying relatively weaker stratum
C. Quasi-preconsolidation due to physico-chemical reasons		
(a) use of chemical admixtures (b) in-situ chemical bonding either soft and sensitive or stiff cemented	cementation due to precipitation, chemical reactions imparting non-particulate behaviour	yield stress reflects cementation bond strength of soft and sensitive

Several procedures are in vogue to estimate the value of preconsolidation stress, σ_p'. Of these Casagrande's graphical construction method is the earliest one (Casagrande, 1936). The procedure is as follows (Fig. 2.5).

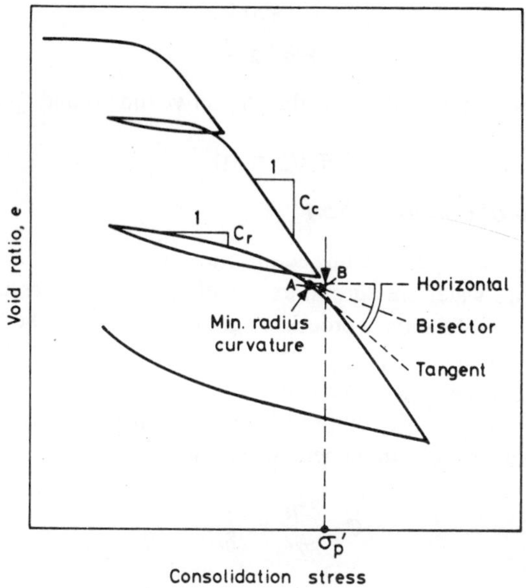

Fig. 2.5 Laboratory compression and recompression curve with geometrical construction to determine preconsolidation stress.

(i) By observation locate the point A of minimum radius in the $e - \log \sigma'$ curve.
(ii) Draw two lines, one horizontal and the other tangential to the curve at this point.
(iii) Bisect the angle between these two lines.
(iv) Project backwards the straight line portion of the laboratory curve to meet the bisector at a point.

The stress corresponding to the meeting point is the preconsolidation stress (point *B* in Fig. 2.5).

Jose et al. (1989) proposed a simple log-log method for determination of preconsolidation stress. In this method both the stress σ' and the void ratio, *e* are in the log scale. The $\log e - \log \sigma'$ plot results in two distinct straight lines, the intersection of which is regarded as the preconsolidation stress. This method is direct and free from any judgment errors in the location of maximum curvature point required in Casangrande method.

Referring to Fig. 2.5, the slope of the curve at stress levels less than σ'_p is the recompression index, C_r. The slope of the curve beyond σ'_p is the compression index, C_c. The magnitude of C_c can be computed from $e - \log \sigma'$ data from the relation

$$C_c = \frac{de}{d\,(\log_{10}\sigma')} \qquad 2.4$$

In the case of constant strain rate test the strain index C is computed from the relation

$$C = \frac{r\,\Delta t}{\log \sigma_2/\sigma_1} \qquad 2.5$$

in which σ_1 and σ_2 are total stresses at the respective time t_1 and t_2 and

$$\Delta t = (t_2 - t_1)$$

2.3.4 Time rate of consolidation

In the process of the clay layer reaching equilibrium under an external load or under its own weight, pore water pressure in excess of hydrostatic pressure dissipates with time. This is consolidation, a process of gradual transfer of the load from the pore water to the particulate skeleton, with a concurrent increase in effective stress. Terzaghi (1925) advanced one dimensional consolidation theory considering the volume of water flowing out of a compressible soil. The partial differential equation for one dimensional consolidation phenomena is given as

$$C_v \frac{\delta^2 u}{\delta Z^2} = \frac{\delta u}{\delta t} \qquad 2.6$$

where $C_v = \dfrac{k}{\gamma_w}\left\{\dfrac{(1 + e_0)}{a_v}\right\}$ is the coefficient of consolidation and contains material properties that govern the consolidation process.

a_v is the coefficient of volume compressibility which is the change in void ratio to the change in effective stress and

k is the coefficient of permeability.

2.3.5 Determination of Coefficient of consolidation

The closed form solution to the partial differential equation (Eqn. 2.6) advanced by Terzaghi (1925) in terms of the Fourier series expansion is of the form

$$\Delta u = \Delta\sigma \sum_{n=0}^{\infty} f_1(Z)\, f_2(T) \qquad 2.7$$

where $f(Z)$ is a function of non dimensional geometrical parameter, Z/H and

T is the nondimensional time factor related to coefficient of consolidation, C_v by

$$T = C_v\left\{\frac{t}{H^2}\right\} \qquad 2.8$$

The progress of consolidation after time t at any depth Z in the consolidating layer can be related to the void ratio, e at that time and the initial and final void ratios. This relationship is called consolidation ratio and it is expressed as

$$U_Z = \frac{e_1 - e}{e_1 - e_2} \tag{2.9}$$

The ratio in terms of pore pressures would be

$$U_Z = \left\{\frac{u_i - u}{u_i}\right\} = \left\{1 - \frac{u}{u_i}\right\} \tag{2.10}$$

From the above relationships, it is evident that U_Z often expressed percent consolidation is zero at the start of loading and it gradually increases to 100 percent as void ratio decreases from e_1 to e_2. The degree or percent consolidation, U_Z represents the conditions at a point in the consolidating layer. Equation 2.7 in terms of U_Z can be expressed as

$$U_Z = 1 - \sum_{n=0}^{\infty} f_1(Z)\, f_2(T) \tag{2.11}$$

Figure 2.6 shows the relationship between average degree of consolidation, U and the time factor, T. The shapes of theoretical curves U versus T plotted by

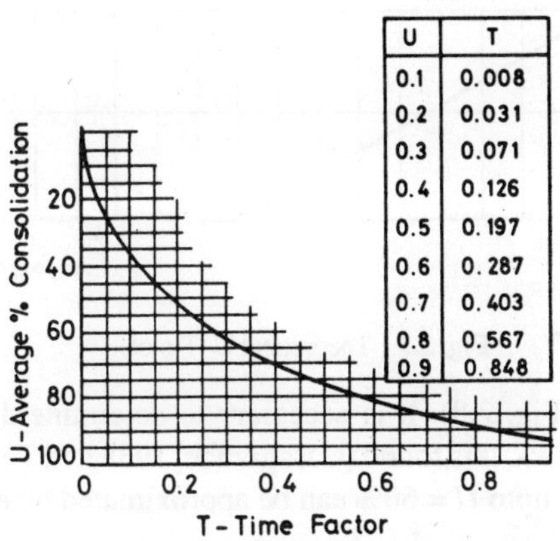

Fig. 2.6 U-T relation for uniform stress distribution in clay.

different modes are shown in Fig. 2.7. The curves of compression versus time for a given load increment are similar to the theoretical U versus T curves. By taking advantage of this fact, C_v can be determined by curve fitting methods. Casagrande's logarithm of time fitting method and Taylor's square root of time method are the most commonly used fitting methods.

Casagrande's method : In this method the deformation indicated by dial readings are plotted against the logarithm of time. Casagrande (1938) based on the behaviour of theoretical as well as experimental U versus log T plots, suggested that the dial reading corresponding to hundred percent consolidation, R_{100} could be approximated by the intersection of two (Fig. 2.7 b) tangents to the laboratory

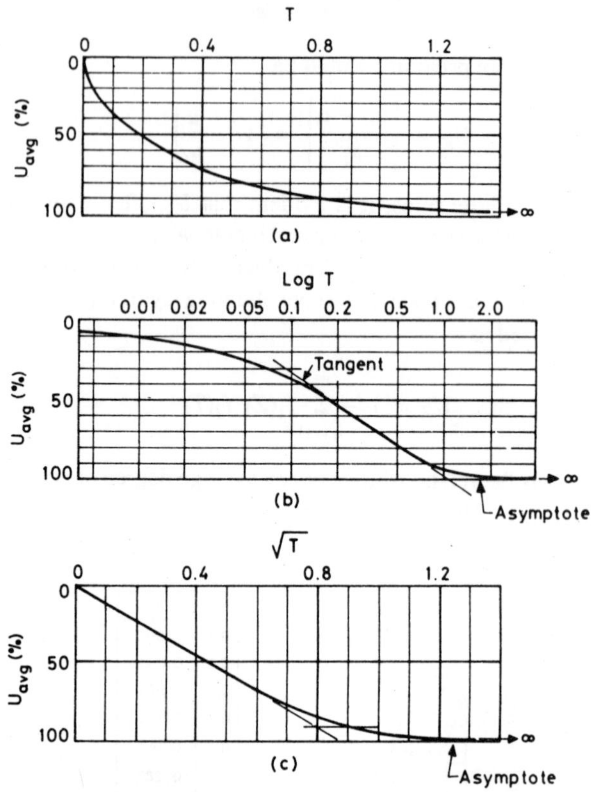

Fig. 2.7 Theoretical U–T plots.

consolidation curve (Fig. 2.8). It is necessary to determine dial reading at zero percent consolidation, R_0. It is found (Casagrande, 1938 and Taylor, 1948) that the T versus U_{ave} relation upto $U = 60\%$ can be approximated by equation;

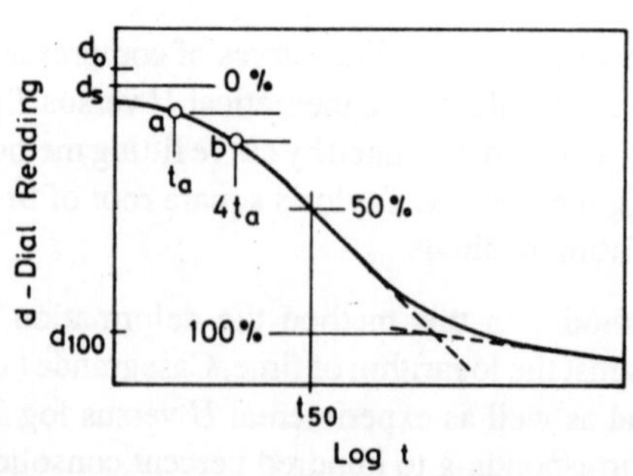

Fig. 2.8 Casagrande's log t fitting method for C_v determination.

$$T = \frac{\pi}{4} U^2 = \frac{\pi}{4} \left\{ \frac{U\%}{100} \right\}^2 \qquad 2.12$$

The first part of consolidation curve is a parabola. After choosing any two time intervals t_a and t_b in the ratio 4 to 1 the corresponding dial readings d_a and d_b respectively are noted. The corrected initial or zero dial reading, d_0 is then located at a distance $(d_a - d_b)$, above point,'a'. Knowing d_{100}, it is now possible to locate d_{50} and t_{50} which is the time for 50 percent consolidation. To evaluate C_v, T_{50} = 0.197 (from equation 2.12) and the drainage path, H which is half the average height of specimen, (in the oedometer test with drainage at top and bottom surfaces) is substituted in the expression

$$C_v = \frac{0.197\, H^2}{t_{50}} \qquad 2.13$$

Taylor's method : Taylor (1948) developed a procedure, based on the similarity between the shapes of theoretical and experimental U versus T plots, to evaluate C_v. Usually a straight line can be drawn through the data points in the initial part of the dial reading versus square root of time plot (Fig. 2.9). The line is

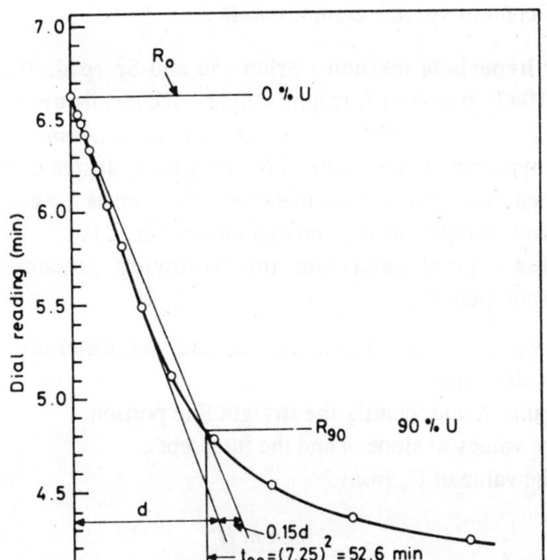

Fig. 2.9 Geometrical construction of Taylor's $\sqrt{t}$ method for C_v determination.

projected backward to zero time to define R_0. Taylor observed that the abscissa of curve at 90 percent consolidation was about 1.15 times the abscissa of the extended straight line. Hence the procedure developed includes drawing a second line from

R_0 with abscissa 1.15 times the corresponding values on the first line. The intersection of this second line and the laboratory curve defines R_{90} and the time for 90 percent consolidation, t_{90}. C_v is determined using the known value of T_{90}(=0.848) as well as the determined t_{90} and average height. That is,

$$C_v = \frac{0.848\, H^2}{t_{90}} \qquad 2.14$$

The specific advantage of the $\sqrt{t}$ fitting method is that t_{90} can be determined without going far beyond t_{90}. This not only reduces considerable testing time but also minimizes the effect of secondary compression on $e - \log \sigma$ curve.

In the case of constant rate of strain test it has been shown by Wissa et al. (1971) that coefficient of consolidation C_v from considerations of linear variation, can be computed from the relation

$$C_v = \frac{r\, H^2}{2\, u_h\, m_v} \qquad 2.15$$

where r = rate of strain
u_h = excess pore water pressure at the bottom
H = sample thickness
and m_v = coefficient of volume compressibility.

Rectangular hyperbola method : Sridharan and Sreepada Rao (1982) and Sridharan et al. (1987), based on Terzaghi's theory of consolidation, observed that the plot of time factor T versus the degree of consolidation, U can be approximated by a rectangular hyperbola in the range of 60-90 percent degree of consolidation. Consequently when T and U are plotted in the form T/U versus T, a curve is obtained upto 60 percent and straight line beyond that range (Fig. 2.10).

Based on this typical behaviour the following procedure has been recommended to compute C_v.

(i) Obtain the values of t and δ from the standard oedometer test, preferably at equal intervals of time.
(ii) Plot t/δ against t, and identify the straight line portion.
(iii) Measure the values of slope m and the intercept c.
(iv) Compute the value of C_v from

$$C_v = \frac{0.24\, m\, H^2}{c} \qquad 2.16$$

The specific advantage of the rectangular hyperbola method is that initial compression need not be known. The method is simple and can be used for all shapes of time compression curves. It is not necessary to record time -compression readings, at equal intervals, to obtain a straight line plot of t/δ versus t. The estimated C_v value from the rectangular hyperbola method generally lies between those obtained from Taylor and Casagrande methods.

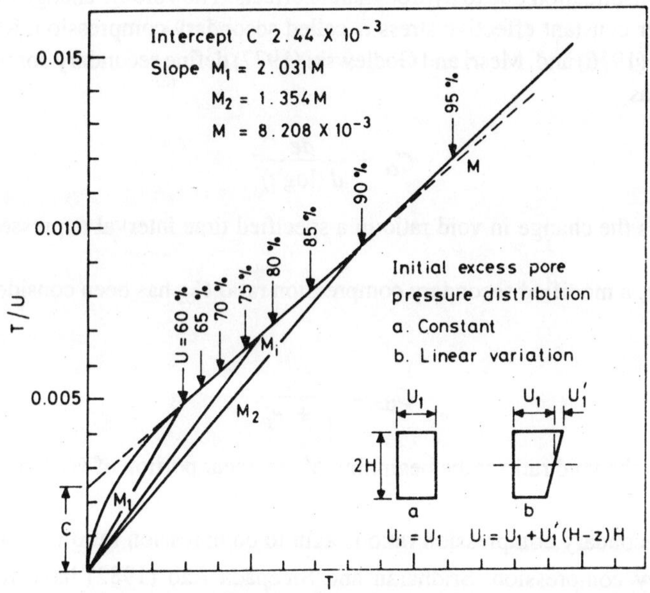

Fig. 2.10 Rectangular hyperbola method to determine C_v (after Sridharan and Rao, 1987).

By further examination of Terzaghi's theoretical *U–T* curve (Pandian et al., 1992) it has been shown that log *U/T* versus log *T* plot exhibits a bilinear behaviour. The intersection point corresponds to a theoretical time factor of 0.793 and a degree of consolidation of 88.5 percent. Using this property, the following equation for the coefficient of consolidation has been proposed,

$$C_v = \frac{0.793\, d^2}{t_{88.5}} \qquad 2.17$$

where, time t corresponds to 88.5 percent consolidation which can be obtained from the log δ/t versus log t plot.

2.3.6 Time dependent deformation

To distinguish it from the deformations caused by loadings and by changes in environmental factors, the time dependent deformation is defined to reflect the deformation under sustained loading only. This is generally referred to as creep. In the case of fine grained soils the compressions with time under constant loading beyond primary consolidation are regarded as secondary compressions. Secondary compression or delayed consolidation has been recognised as a significant phenomenon in connection with foundation settlement in clayey soils. Such settlements have been found to be of significant magnitude in the total settlement (Leonards and Altschaeffl, 1964; Zeevert, 1973).

Terzaghi's classical theory of one dimensional consolidation considers only primary consolidation due to hydrodynamic effects. The volume change that takes place under constant effective stress is called secondary compression. Raymond and Wahls (1976) and, Mesri and Godlewski (1977) define secondary compression index, C_α as

$$C_\alpha = \frac{de}{d(\log t)} \tag{2.18}$$

where de is the change in void ratio in a specified time interval expressed in log scale,

Further, a modified secondary compression ratio $C_{\alpha t}$ has been considered and defined as

$$C_{\alpha t} = \frac{C_\alpha}{1 + e_p} \tag{2.19}$$

where e_p is the void ratio at the beginning of the linear portion of void ratio versus log t plot.

This secondary compression ratio is akin to compression ratio C_c' considered for primary compression. Sridharan and Sreepada Rao (1982) have redefined secondary compression ratio, C_s as the ratio of secondary compression ΔS per unit value of log t to the thickness of the sample h_f at the end of compression for that load increment, i.e.,

$$C_s = \left\{\frac{\Delta S}{\log t}\right\} \frac{1}{h_f} \tag{2.20}$$

The advantages of this definition are that it takes into account the thickness of the sample and C_s is nondimensional.

The secondary compression coefficients can be determined from the slope of the straight line portion of compression versus log time plot beyond the primary consolidation. Usually ΔS is determined over one log cycle of time. Corresponding changes in void ratio and thicknesses of sample are noted for computations. Further a simple two constant hyperbolic equation has been proposed to characterise compression time curves in the laboratory consolidation tests to predict ultimate compression (Narasimha Rao and Kodandaramaswamy, 1982). The functional form of the hyperbolic relation can be expressed as

$$\frac{t}{C} = a + bt \tag{2.21}$$

in which C = compression

t = time

a and b are constants for the sample under compression.

As t approaches infinity, ultimate value of compression

$$\lim_{t \to \infty} C_u = \frac{1}{b} \qquad 2.22$$

That is, the reciprocal of slope gives the ultimate compression.

It has been found that C_α/C_c is fairly constant for many normally consolidated clays and hence forms a working hypothesis for estimating the secondary compression coefficient. A value of 0.05 is acceptable for the ratio of C_α/C_c for preliminary calculations.

2.4 Compressibility and Consolidation - In-situ methods

There are many situations wherein load - deformation characteristics of soils in-situ are required to be assessed during investigation stage itself for the analysis and design of structures. Pressuremeters are the first in the series of load-deformation measuring devices for soils in the field.

2.4.1 Pressuremeter

The basic idea of pressuremeter is relatively simple and involves expanding a cylindrical flexible membrane against a hole drilled in substratum. It is an instrument for in-situ measurements of load - deformation parameters of soil and rock. Probe and volumeter are the two main components (Fig. 2.11).

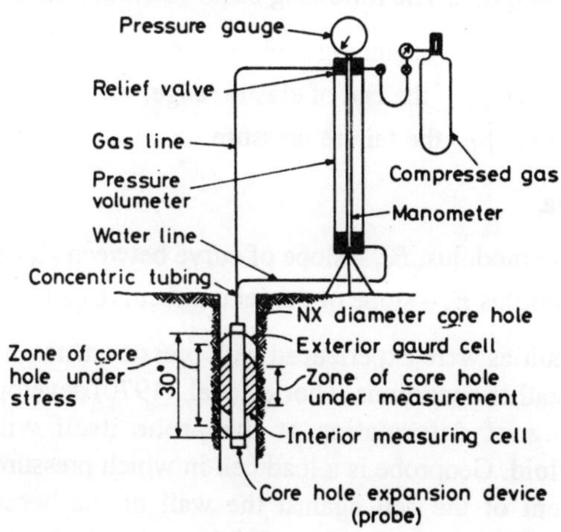

Fig. 2.11 Details of pressuremeter test equipment.

Probe: A cylindrical metal body is surrounded by two flexible rubber membranes. The interior membrane forms the measuring cell and the exterior provides guard cells at the two ends of the probe.

Volumeter: An arrangement located at the surface to apply gas pressure on the water in the manometer tubes which inflates the cells and develops radial stresses depending upon the lateral constraints of the material into which the probe is inserted. In addition to application of increments of pressure to the guard cell, the volumeter - manometer enables to record the radial expansion under each pressure increment by measurement of water flow into the probe.

The procedure involves lowering of the unit into a predrilled bore hole and activating the guard cells by gas pressure while the measuring cell is pressurized with water and maintained at a slightly higher pressure than the guard cells. This ensures the contact between measuring cell and the bore hole wall. The role of guard cells is to reduce end effects on the measuring cell and to provide an essentially two dimensional cylindrical stress condition. Figure 2.12 is a typical

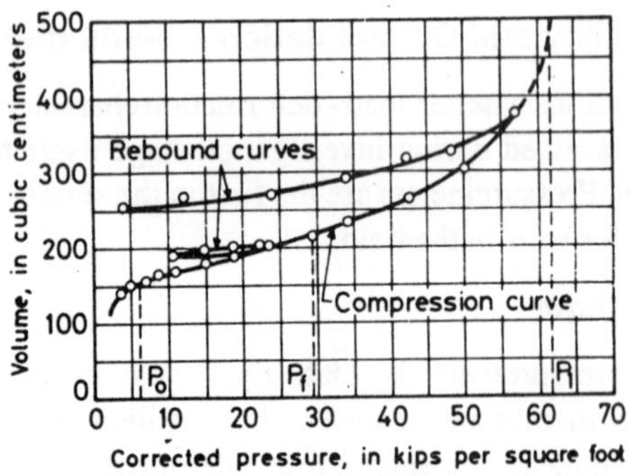

Fig. 2.12 Typical plot of pressuremeter test results.

pressure-volume diagram. The following basic quantities are measured.

(1) Initial pressure, P_o – beginning of elastic stress range after seating is ensured.

(2) Creep pressure, p_f – the end of elastic range.

(3) Limit pressure, p_l – the failure pressure.

From this data,

(1) compression modulus, E_m – slope of curve between P_o and P_f

(2) rebound modulus E_r – slope of the rebound curve can be determined.

Certain difficulties were experienced with pressuremeters while water is used in the subzero weather conditions. Thorley et al. (1970) developed a geoprobe for direct monitoring of deformation at the probe itself without use of any incompressible fluid. Geoprobe is a load cell in which pressure is applied by gas and the movement of the cell against the wall of the borehole is monitored electronically by a linear potentiometer of high resolution. Another significant step in the development of the series of pressuremeters is the introduction of Camkometers (Wroth and Hughes, 1973). These are self boring pressuremeters capable of being introduced into the ground with the bearest minimum of mechanical disturbance. In this respect, the test carried out is quite different from that done in a prebored exact hole where the boring tool has to be withdrawn before a test can be conducted.

The test is carried out by load application to the soil by gas pressure acting on the inside of an elastic membrane encasing the instrument. The instrument measures the pressure, the resulting deflection and pore water pressure using electrical transducers. All signals transmitted to the surface are processed and recorded by an electronic system. The test itself can be conducted in two ways either at constant rate of change of stress or at constant strain rate. The later method is reliable and versatile and hence frequently preferred.

Pressuremeter test results can be used to estimate the settlement of a shallow foundation at the design stage itself. The method differs from the classical method based on parameters obtained from one dimensional consolidation test. This is primarily due to the different stress path followed in the pressuremeter test. Still it is possible to determine parameters such as preconsolidation pressure σ_p, coefficient of volume compressibility, m_v and coefficient of consolidation, c_v from pressuremeter test.

For a pressuremeter inserted without disturbance into clay, the initial cavity pressure is equal to the preconsolidation pressure. According to Calhoon, in the Menard method of settlement analysis, $1/m_v$ is equal to 4.5 E_m (Baguelin et al., 1978). Further a method has been suggested by Baguelin (1973) to estimate the coefficient of horizontal consolidation, C_{vH} by performing holding test. In these tests the pressuremeter having been expanded, the total pressure in the cavity or deformation of the cavity is held constant. If the diameter of the cavity is held constant relaxation is observed by the fall of the measured excess pore pressure and of the total cavity pressure. On the other hand when total pressure is held constant; relaxation is observed in the measured excess pore pressure and the continuing increase in cavity diameter. (Clarke et al., 1979). Coefficient of horizontal consolidation, C_{vH} is computed based on changes in measured pore water pressures.

2.4.2 Flat plate dilatometer

This is an in-situ testing device developed in Italy by Marchetti (1980). It is a penetration device that can laterally expand upon inflation. The test therefore combines many of the features contained in the cone penetration test and pressure meter test. The flat plate dilatometer (Fig. 2.13) consists of a stainless steel blade

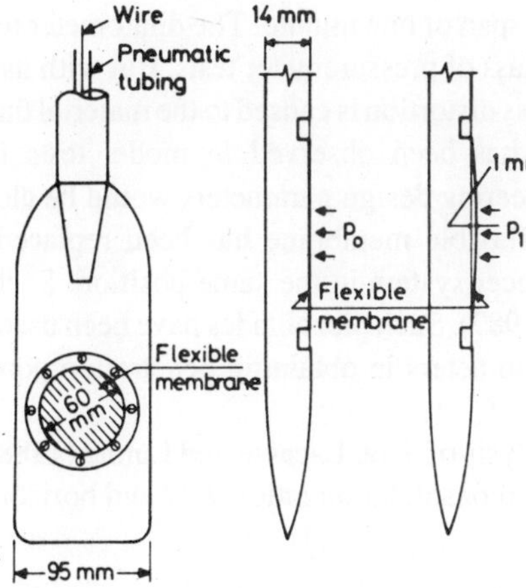

Fig. 2.13 Illustration of flat plate dilatometer (after Marchetti, 1980).

with a flat circular expandable membrane on one side of the flat plate, a short distance behind the sharpened tip. When at rest the external surface of the membrane is flush with the surrounding flat surface of the blade. The drill rods connected to dilatometer carry an air and electrical line from the surface to the position of the membrane. The blade is jacketed into the ground using penetrometer rig to a desired depth, preferably by quasi-static penetration. The membrane can then be inflated by pressurised gas from the control unit by nylon tubing. Beneath the membrane is a measuring device which turns a buzzer off in the control box when the membrane starts to lift off the sensing device and turns the same buzzer on again after 1mm deflection of the membrane is attained.

By use of control unit with a pressure regulator, a gauge and audio signals, the operator determines;

(i) lift off pressure, p_o required to move the center of membrane (reading *A*)
(ii) pressure, p_1 required to move the center of membrane by 1 mm into the soil (reading *B*) and
(iii) pressure at which diaphragm establishes recontact with DMT blade (reading *C*)

The results are presented as diagram of p_0 and p_1 versus depth. Material index, I_D, Horizontal stress index, K_D and Dilatometer modulus, E_D are computed from the above data using the following relations.

$$I_D = (p_1 - p_o)/(p_o - u_o) \qquad 2.23$$

$$K_D = p_o - u_o/\sigma_v \qquad 2.24$$

$$E_D = 38.2\,(p_1 - p_o) \qquad 2.25$$

Where u_o is the static equilibrium pore pressure and σ_v is overburden pressure.

The *C* reading which generally reflects the pore water pressure acting on the membrane can be obtained by installing a simple flow control valve coupled with sensitive pressure gauge in the pressure vent line (Lutteneger and Kabir, 1988). The deflection *C*-reading provides a measure of external pressure forcing the diaphragm to recontact the DMT blade. In general all three pressure readings *A, B* and *C* are obtained in a span of one minute. The dilatometer test belongs basically and practically to the class of pressuremeter tests. But with its flat plate shape and its sharp cutting edge less distortion is caused to the material than by a conventional conical tip. This fact has been observed in model tests in sand. Hence the correlations with engineering design parameters would be closer to experimental findings. Dilatometer flexible membrane has been replaced by flush mounted porous stone or transducer system in the same position. Such devices are called piezoblades (Boghrat, 1987). Such piezoblades have been used separately adjacent to use of flat plate dilatometers in obtaining penetration pore pressure and their dissipation with time.

From a detailed analysis of data, Lacasse and Lunne (1988) have established a correlation between overconsolidation ratio, *OCR* and horizontal stress index, K_D of the form.

$$OCR = 0.225\, K_D^m \qquad 2.26$$

with m value being in the range of 1.35 and 1.67. The values of 'm' are lower for clays of high plasticity and vice versa. Flat dilatometer tests carried out in the recent deposits of Singapore and their analysis (Chang, 1988) have revealed that overconsolidation ratio profile is similar to the pore pressure ratio profile deduced from piezocone test. Pore pressure ratio is the ratio of difference between total pore pressure and in-situ water pressure $(u - u_o)$ to total overburden pressure, σ_{vo} i.e. $(u - u_o)/\sigma_{vo}$.

It is possible to obtain data from flat dilatometer test to assess the consolidation characteristics of in-situ soils. The procedure has been presented by Robertson et al (1988). It is necessary to conduct closing pressure test for this purpose. Closing pressure represents the pore water pressure on dilatometer membrane. The C-reading (after deflation) is a result of external pressure forcing the diaphragm to recontact the DMT blade. This pressure reading, corrected for membrane stiffness closely represents the pore water pressure acting on the membrane.

Flat plate dilatometer can also be used to monitor dissipation rates of penetration pore water pressures. A-reading with time can be determined at any depth of penetration. The contact pressure, A-reading, when corrected for membrane stiffness is the lift off pressure, p_o. The difference between the p_o and penetration pore water pressure u is the effective stress acting on the membrane. Soon after penetration, in soft clays, effective stress σ_{Hi} is low. As the pore pressure dissipate with time the effective stress increases. Hence it should be possible to estimate the coefficient of consolidation, C_h using a standard dilatometer test. The time factor T_D can be computed using the relation

$$T_D = C_h t/R^2 \qquad 2.27$$

where T_D is dilatometer time factor

t is time for percent dissipation

and R is equivalent radius of the blade

The theory of cylindrical cavity expansion in an ideal elasto-plastic material has been used to calculate the coefficient of consolidation, C_h. The dilatometer blade has been modelled as a vertical cylinder with an equivalent radius calculated on the basis of cross-sectional area of the blade. Radial drainage on horizontal planes has been assumed.

To estimate C_h from the flat plate dilatometer test the recommended procedure (Robertson et al., 1988) is as follows:

(i) At any specified level of penetration A, B and C readings are recorded by inflating and slowly deflating the dilatometer membrane.
(ii) The above step is repeated with the time elapsed since the stop of penetration is monitored.
(iii) C-readings (pressure) versus log time is plotted and verified that the final C-readings approximately equal the static equilibrium pore pressure u_o.
(iv) From the time t_{50} and the corresponding R_{50}, C_h is computed from equation 2.27.

Dilatometer modulus: This characterizes the stress displacement curve during 1mm expansion and is related to the stiffness of the soil. In silt and clays undrained conditions can be assumed whereas in clean sands practically complete drainage takes place while 1mm expansion is attained. The dilatometer membrane expansion to an extent of 1 mm in the test obliterates the effects of stress and strain history and hence dilatometer modulus does not correspond to undisturbed modulus in all its entirety. Engineers are, more often, in need of a deformation parameter for estimation of settlement. Hence attempts have been made to correlate dilatometer modulus, E_D with oedometric constrained modulus, M at the effective in-situ overburden stress, σ_{vo} (Marchetti 1980). More specifically the correlation is for the local reload modulus. As such in normally consolidated soils, the dilatometer test does not provide information about the reload modulus and in overconsolidated soil this test will not provide information of the virgin compression modulus. Further dilatometer modulus has been correlated with constrained modulus obtained from oedometer tests. Irrespective of whether the expansion occurred under drained or undrained condition such correlations have been attempted. The dilatometer test in clays and silts is being conducted under undrained conditions while constrained modulus, M is derived from an oedometer test under drained state. However, as evidenced by recent published results (Lacasse and Lunne, 1982 and Lacasse, 1985) the dilatometer provides reasonable estimates of the constrained modulus in soft clays, over consolidated clays and loose sands.

Davidson and Boghrat (1983) suggested that in highly overconsolidated clays the value of modulus obtained from unconsolidated undrained triaxial compression tests could be related directly to the dilatometer modulus, E_D using a factor of 1.4. Recently the studies by Robertson et al., (1988) reveal a factor of 10 for clays and 2 for sands. All these indicate that by a simple dilatometer modulus, E_D the elastic modulus can be assessed.

2.5 Shear Strength and Deformation - Laboratory methods

In a broad sense strength is regarded as the limiting value of shear stress under which the material fails . As discussed earlier shear strength assessment is of prime importance in solving stability problems. If the imposed stresses are anisotropic, there would a tendency within the particulate system to develop shearing resistance to counter the imposed shearing stresses.

2.5.1 Direct shear test

The test, in principle, is quite simple. The apparatus for performing direct shear is made up of a rectangular box consisting of lower and upper halves. The sample to be tested is compressed under a specific normal load and the upper half of the apparatus is then moved laterally by a shear force, forcing the sample to shear along the horizontal plane between the two halves of the apparatus. The normal load is maintained constant during the test. The mobilization of shearing resistance keeps on increasing until the sample is failed. Shearing can be accomplished either by controlled rate of strain or stress. The possible rotation of the loading platen of direct shear box can be circumvented by resorting to the modification effected by

Wernick (1977). The primary modifications to the conventional apparatus are shown in the Fig. 2.14. The roller bearing arrangement ensures that the loading

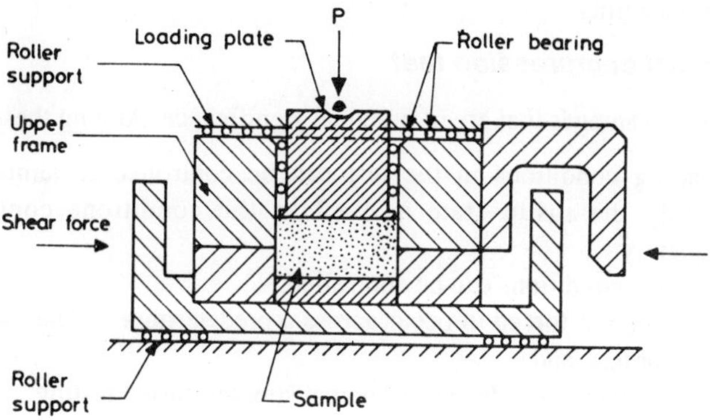

Fig. 2.14 Modified direct shear box (after Wernick, 1977).

platen moves in the vertical direction only thereby maintaining the normality of loading. Usually a record of the magnitude of shearing force at various lateral movement levels and the resulting vertical movements is maintained for obtaining shear stress-strain displacement plots. Figure 2.15 shows a typical Mohr-Coulomb plot from such data. Principal stresses are not directly obtained in the direct shear test. Instead they can be obtained from the Mohr-Coulomb failure envelope.

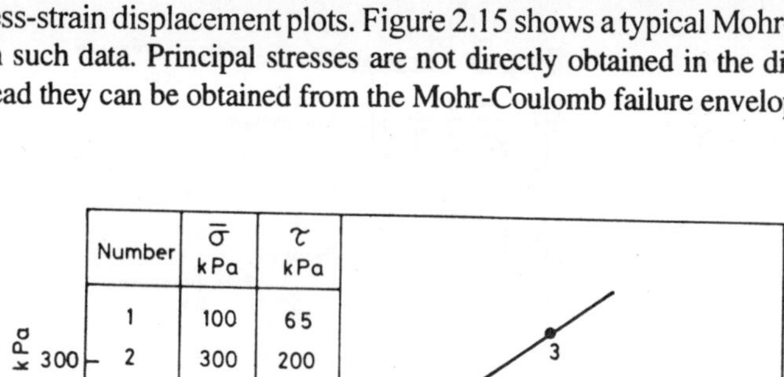

Fig. 2.15 Typical plot of direct shear test data.

Primarily, the direct shear test is inexpensive, fast and simple especially for granular materials. Inability to control drainage in the case of fine grained soils is a disadvantage. Consequently the test is suitable only under drained conditions.Since failure takes place on a pre-identified plane, it need not be the weakest plane. Other limitations are that there are stress concentrations at the sample boundaries leading to non-uniform stress conditions within the sample and uncontrolled rotation of principal stresses. Several modifications to conventional direct shear test can be traced in the literature (Matthews 1988) which have improved applicability.

2.5.2 Triaxial compression test

It has several advantages that are of practical significance. Among these are

(a) the loading conditions in the test can be controlled to simulate more accurately the insitu state as well as the conditions contemplated subsequently,
(b) the drainage conditions can be controlled,
(c) the failure is not forced to occur across a predetermined plane, as that in a direct shear test, and
(d) there is no rotation of plane of principal stresses during testing.

The basic test facility is illustrated in Fig. 2.16. The triaxial test methods described by Bishop and Henkel (1962) are still followed in routine work. In recent

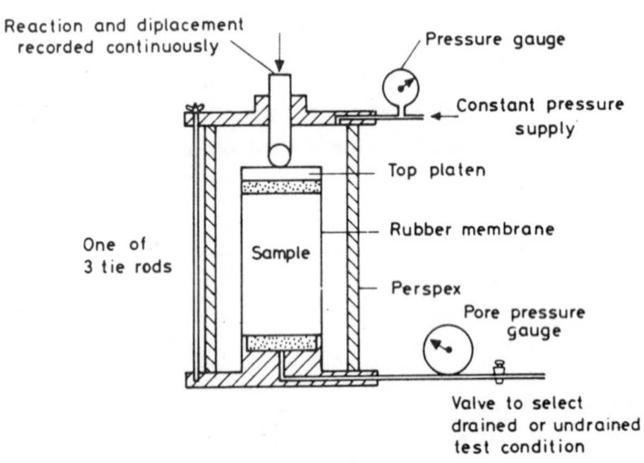

Fig. 2.16 Salient details of triaxial compression test.

reevaluation of conventional triaxial methods (Baldi et al., 1988), sources of measurement errors have been clearly identified (Fig. 2.17). By exercising proper care in testing, such errors can be avoided.

Following a detailed analysis of various inhomogeneities that develop in a cylindrical soil specimen tested in triaxial compression, Hettler and Vardoulakis (1984) suggested an improved version of triaxial setup. Figure 2.18 illustrates the details of units, such as:

(1) lubricated and enlarged end platens to minimise end restraints and
(2) a guided upper plate to minimise load eccentricity.

Critical appraisal of earlier findings permit to make the following recommendations.

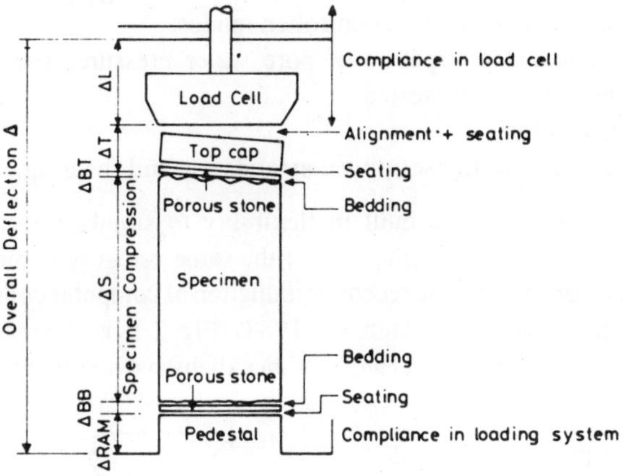

Fig. 2.17 Illustration of sources of errors in triaxial test (after Baldi et al., 1988).

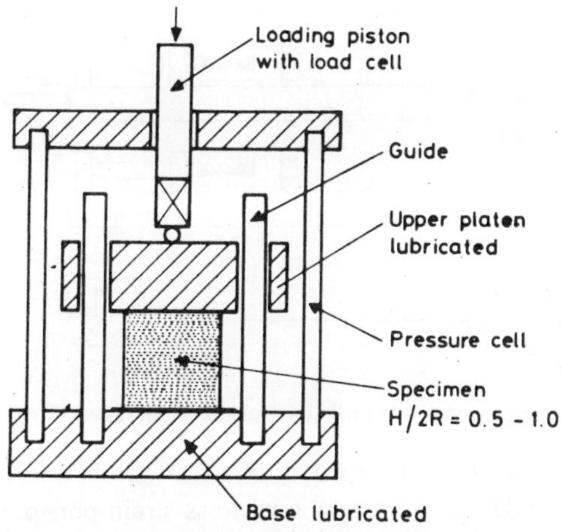

Fig. 2.18 Improved triaxial apparatus (after Hettler and Vardoulakis, 1984).

(1) Use of enlarged lubricated end platens are particularly desirable in research investigations, where the emphasis is on studying basic soil behavior as far as possible free from the effects imposed by the test apparatus.

(2) The enlarged lubricated end platens, being simple and inexpensive to adapt, are readily applicable to routine commercial testing where their use can permit considerable reductions in testing times in tests with pore pressure measurements as well as minimizing other common sources of error. It would also be possible to effectively use the available length of the undisturbed samples.

Triaxial test facilities need to possess considerable flexibility to simulate stress and deformation conditions likely to be encountered in various practical problems. Basically, it should be possible to monitor the following test parameters.

(1) Cell pressure - development and maintaining constant during testing.
(2) Axial load - monotonically increasing or decreasing by either strain - rate controlled or load increment controlled modes.
(3) Development and dissipation of pore water pressures particularly while particulate materials are tested.
(4) Axial deformation.
(5) Volume change during isotropic consolidation and shearing.

Triaxial test has far more built in flexibility to simulate various total and effective stress paths than any other test of the same category commonly used in practice. This is more so with the recent introduction of computer controlled triaxial system with digital controllers (Menzies, 1988) (Fig. 2.19). The reader is referred to an excellent treatise by Head (1986) for an exhaustive treatment.

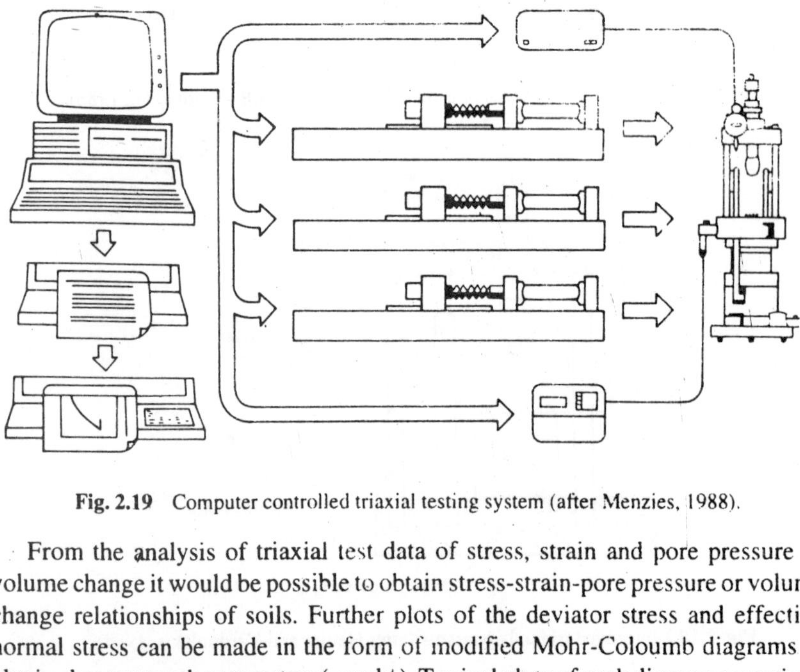

Fig. 2.19 Computer controlled triaxial testing system (after Menzies, 1988).

From the analysis of triaxial test data of stress, strain and pore pressure or volume change it would be possible to obtain stress-strain-pore pressure or volume change relationships of soils. Further plots of the deviator stress and effective normal stress can be made in the form of modified Mohr-Coloumb diagrams to obtain shear strength parameters (c and ϕ). Typical plots of such diagrams are given in Figs. 2.20 and 2.21.

2.5.3 Sample preparation

In the case of soils, cylindrical samples are preferred since they can be directly prepared by extrusion from sampling tubes, trimmed from block samples. It is important that soil samples are subjected to least disturbance while handling and preparation of test specimens. Apart from undisturbed samples, laboratory preparation of remoulded saturated samples can be accomplished by consolidation of soils from their liquid limit states to desired levels of consolidation stress using the large consolidation cell developed by Rowe and Barden (1966). The consolidation system developed by Sheeran et al., (1971) consists of stainless steel

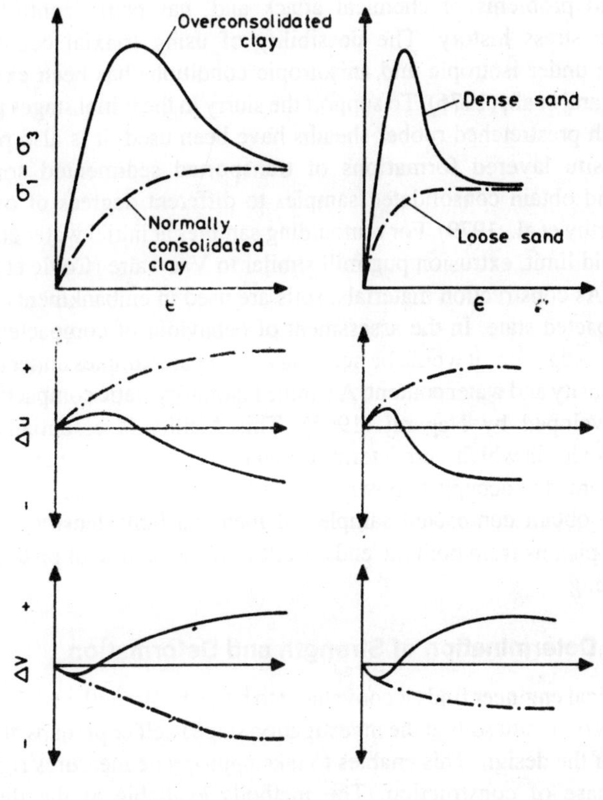

Fig. 2.20 **Schematic representation of stress-strain-pore pressure/volume change response of soils in triaxial test.**

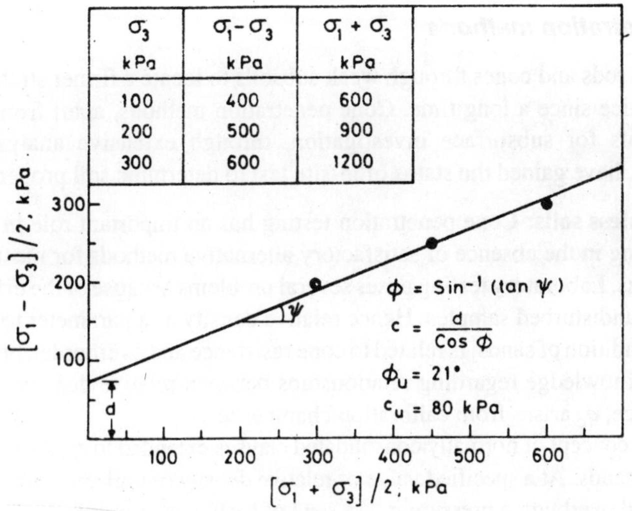

Fig. 2.21 Modified Mohr-Coulomb representation of triaxial compression test results.

or brass consolidometer, the control panel and load pacer. This facility averts corrosion and problems of chemical attack and, has better control over the consolidation stress history. The possibility of using triaxial cell for slurry consolidation under isotropic and anisotropic conditions has been examined in detail by Nagaraj et al., (1976). To support the slurry in the initial stages perforated cylinders with prestretched rubber sheaths have been used. It is also possible to simulate in-situ layered formations of transported sedimented soils in the laboratory and obtain consolidated samples to different degrees of overburden (Krishna Murthy et al., 1979). For remoulding samples at initial water contents far less than liquid limit, extrusion pug mill similar to Vac - aire (Coyle et al., 1968) can be used. As construction materials, soils are used in embankments and dams in their compacted state. In the assessment of behaviour of compacted soils for laboratory investigations it would be necessary to prepare samples under controlled predefined density and water content. A simple laboratory static compaction device has been developed by Nagaraj (1963). This facility is essentially a static compaction device in which a predetermined quantity of conditioned premixed soil and water is forced to occupy a known volume from both the ends of a cylindrical tube so as to obtain compacted samples of identical bulk density. The lateral movement of pistons from both the ends is effected by rotation of piston rods in a screwed bushing.

2.6 In-situ Determination of Strength and Deformation

The geotechnical engineer finds it convenient to have the strength and deformation characteristics of in-situ soils at the investigation stage itself or perhaps during very early stages of the design. This enables to take appropriate measures right during the initial phase of construction. The methods available at the disposal of geotechnical engineers are so many that it would not be possible even to provide a brief description of the same. The discussions here pertain to penetration methods, vane and bore hole shear testing methods.

2.6.1 Penetration methods

Probing with rods and cones through weak subsoils to locate a firmer stratum has been in practice since a long time. Cone penetration methods, apart from being effective tools for subsurface investigation, through extensive analysis and development, have gained the status of in-situ test to determine soil properties.

Cohesionless soils: Cone penetration testing has an important role in in-situ strength testing in the absence of satisfactory alternative methods for most of the field situations. Laboratory testing poses several problems because of the difficulty in obtaining undisturbed samples. Hence relative density as a parameter to reflect the in-situ condition of sands, is related to cone resistance and overburden pressure. Most of the knowledge regarding relationships between relative density, D_r and cone resistance, q_c arises from calibration chamber tests.

The same concept of normally consolidated clays is extended to characterize the behaviour of sands. At a specified value of relative density of uniform, uncemented clean sand and overburden pressure σ_{vo}, a sand of high compressibility has a lower

q_c than a sand of low compressibility (Jamiolkowski et al., 1985). The word normally consolidated can be used only to mean that the influence of the present overburden pressure can directly be taken into account in the cone resistance mobilized. Sands are regarded as overconsolidated if they deviate from the above enunciated behaviour. No unique relationship exists between D_r and q_c through σ'_{vo}.

It is possible to estimate the peak effective angle of shearing resistance, ϕ' with relative density, D_r as a parameter. A more direct correlation between q_c and ϕ' is shown in Fig. 2.22 (Durgunoglu and Mitchell, 1975). This relation has been deduced from bearing capacity theory, ignoring the effects of soil compressibility employing a soil-to-cone friction angle equal to $\phi'/2$ and a lateral earth pressure coefficient $K_o = 1 - \sin \phi'$. Robertson and Campanella (1983) support the above approach on the basis of analysis of data from a number of researchers. For overconsolidated sands Fig. 2.22 may slightly overestimate the friction angle.

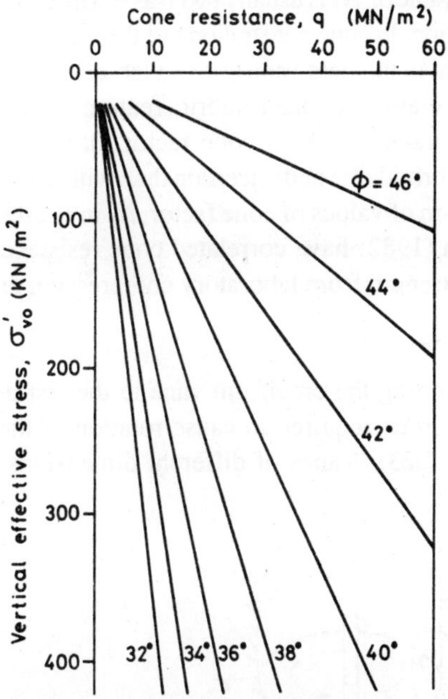

Fig. 2.22 Relationship between angle of shearing resistance and cone resistance for uncemented normally consolidated quartz sand (after Durgunoglu and Mitchell, 1975).

Cohesive Soils: Penetration with rods and cones through relatively soft soil has been in vogue from a long time. Cone penetration methods, apart from being effective tools for subsurface investigation, through extensive analysis and development have gained the status of *in-situ* test to determine soil properties.

In contrast to cohesionless soils cone penetration testing is of less importance in cohesive soils. In normally consolidated and lightly overconsolidated clays, soil parameters can be obtained from several other methods such as in-situ vane tests and laboratory tests on undisturbed samples. Despite this, cone penetration tests have the advantage of rapid coverage of large area.

The relationship between cone resistance, q_c and undrained shear strength of clays, c_u can be expressed as

$$q_c = N_k c_u + \sigma_{vo} \qquad 2.28$$

where σ_{vo} is the total overburden.

N_k is cone factor which is analogous to bearing capacity factor N_c.

The contribution of total overburden pressure, σ_{vo} has been interpreted as either the in-situ vertical total stress σ_{vo} or the in-situ horizontal stress, σ_{ho} or in-situ octahedral stress σ_{oct}. The factor N_k is not a constant but affected by shape of penetrometer tip, penetration rate, strength anisotropy and macrofabric. Except in highly sensitive clays, the value of N_k is usually taken as 9. The cone resistance in clays varies with rate of penetration. In cone penetration test the shearing rate is of the order of 100 times faster than that in the field vane test or a laboratory compression test.

In overconsolidated clays the macrofabric features such as fissures and discontinuities have a marked effect on cone factor making the assessment of undrained shear strength formidable and uncertain than with normally consolidated clays. For the determination of values of cone factor, N_k in overconsolidated clays, Marsland and Quaterman (1982) have correlated cone resistance with both field vane shear strength and strength from laboratory compression tests.

2.6.2 Vane shear tests

The test comprises of inserting the cruciform vane to the test depth followed by the measurement of the torque required to cause rotation of the vane and hence shearing of the soil (Fig. 2.23). Vanes of different dimensions are suggested to

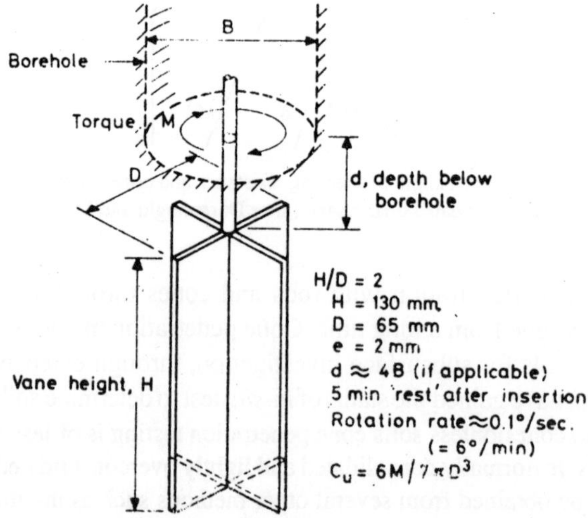

Fig. 2.23 General details of field vane.

allow larger (diameter and/or width) vanes for softer clays, so that torque measuring system can, for greater accuracy, operate over comparatively limited ranges. A vane 100 mm long by 50 mm wide by 1.6 mm thick has been considered to be of the smallest size for the accurate determination of shear strength in the range of 50 to 75 kPa (Andersen, 1981). Since the most widely used dimensions appear to be H = 130 mm and D = 65 mm for shear strength values upto 50 kPa, these would seem to be most appropriate for standardization (Chandler, 1988). Other dimension on which there is general agreement is the blade thickness of 2 mm and the area ratio (the ratio of volume of soil displaced by the vane to the soil volume swept by the rotated vane) to be less than 12 percent. The test results yield peak and remolded torque values in equilibrium with the mobilized shear strength on the cylindrical failure surface induced in the soil.

Let S_V be undrained shear strength on the vertical surface and S_H that on the horizontal surface. Then the torque, T is given by

$$T = \pi D H S_V D/2 + 2 S_H \int_0^{D/2} 2 \pi r^2 \, dr$$

With the assumption of uniform shear strength over the horizontal failure surface and upon integration of the above equation results in

$$T = \pi D^2 H S_V/2 + D^3 S_H/6$$

For an isotropic case $S_V = S_H$, then

$$T = \{S_V \pi D^2 H/2\} \{1 + D/3H\}$$

For the standard vane of $H = 2D$

$$S_V = (6T/7 \pi D^3) = 0.86 \, T/\pi D^3 \qquad 2.29$$

Based on recent studies on vane testing (Sridharan and Madhav, 1964; Donald et al., 1977; Menzies and Merrifield, 1980; Wiesel, 1973; Torstensson, 1977; Chandler, 1988 and Mahmoud, 1988) it can be summarised that:

(i) Distribution of shear stress on the vertical faces of the blades can be assumed to be uniform and non-uniform on the top and bottom surfaces. Since c_u values are likely to be conservative due to this, a factor of 0.91 is suggested in place of 0.86 in the relation given by equation 2.29.

(ii) The measured strengths decrease with increasing time to failure in the case of highly plastic clays i.e., the strength is strain dependent. A different trend has been observed in the case of low plasticity clays.

It has been generally agreed upon to adapt a rate of rotation of 6°/min to measure undrained shear strength in uniform soft clays. Development of an instrumented vane with pore pressure transducer is advocated to monitor the pore water pressures during the suggested rest period of 5 minutes after insertion and different rates of rotation.

For general purposes it is sufficient to assume isotropic strength conditions ($S_H = S_V$). The effects of anisotropy on vane strength can be determined by using vanes of different *H/D* ratios and deduce anisotropy ratio. Aas (1965 and 1967) and Silvestri and Aubertin (1988) have investigated strength anisotropy in clays by using vanes of different sizes, and configuration with *H/D* ratios from 0.5 to 4.0. With S_H being different from S_V, terms in the equation for torque (Eqn. 2.29) can be rearranged as,

$$2T/\pi D^2 H = S_V + S_H (D/3H) \qquad 2.30$$

This implies that in a graphical plot in which vertical and horizontal axes represent $2T/\pi D^2 H$ and $D/3H$ respectively, Eqn. 2.30 describes a straight line intersecting the vertical axis at a value of S_V and having a slope equal to S_H. The intersection of the straight line extrapolated with the negative ($D/3H$) axis yields directly the magnitude of anisotropic strength ratio (S_V/S_H) (See Fig. 2.24).

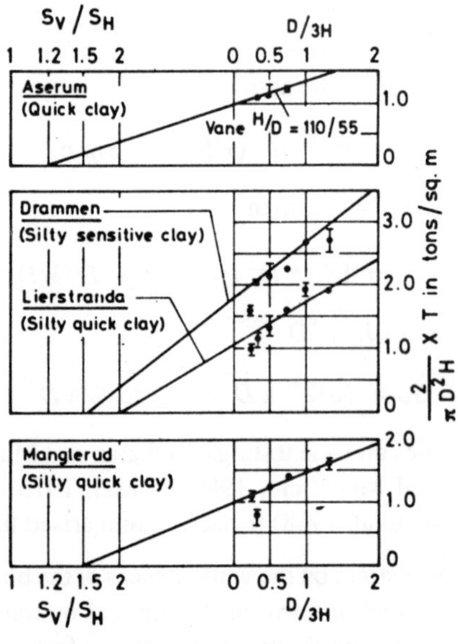

Fig. 2.24 Illustration of method to determine anisotropic strength ratios of clays (after Aas, 1965).

In the simplest form of testing, inspection vanes developed by Norwegian Geotechnical Institute have been used for the determination of the undrained shear strength in trenches and excavations (Andresen and Sollie, 1966). Fig. 2.25 shows the salient features. When the handle is rotated the spring deforms and the upper part rotates with respect to the lower part. The annular scale mounted below the handle reads the torque at various stages. The maximum shear strength that can be measured is 200 kPa. For the use of vane at different depths, the vane and rods are inserted into the soil through a cased borehole. To minimize the effects of borehole

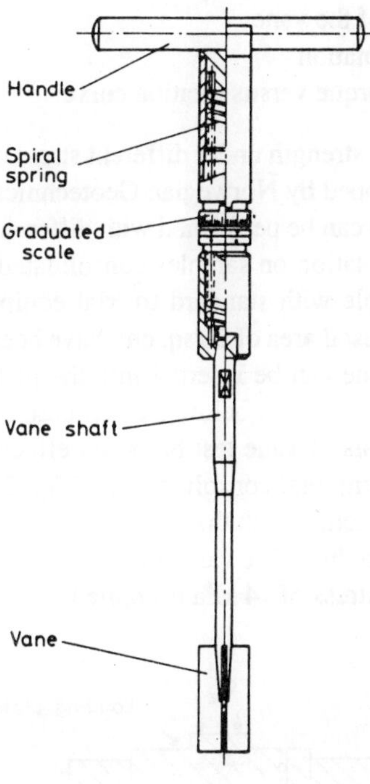

Fig. 2.25 Details of field inspection vane (after Andresen and Sollie, 1966).

disturbance the test is carried out at a depth of four times the diameter of bore hole from the bottom of bore hole. With greater penetration, friction on the torque rod increases. This can be minimized by providing a lubricated sleeve over the length embedded in the clay. For economy, vane testing can also be carried out without using a borehole. Sleeved torque rods with protective shoes is one type, the other being the use of unsleeved torque rods with slip coupling. Further, the state of-the-art of offshore vane testing has improved greatly in the last decade. Advanced electromechanical systems have operated reliably and efficiently under water of more than 1000 m deep, independent of motion of vessels providing digital recording and processing of data (Young et al., 1988). The vane is usually rotated within the soil about five minutes after insertion with monitoring of the angle or torque. Rotation of vane is continued until a peak shear strength (torque) is obtained. The remolded strength is then obtained by continuing rotation of the vane to cause remolding of the cylindrical failure surface until a constant reading is obtained. Madhav and Ramakrishna (1977), with an assumption that at small rotations the pressure over each vertical strip of the blade is normal and constant with depth but is a function of distance from the center, have derived the following equations for the undrained modulus, E_s of the clay from vane shear test data.

$$E_s = \frac{T}{\theta} \frac{1}{I_\theta D^3} \tag{2.31}$$

where T is the torque applied
D is the diameter of the vane
θ is the angle of rotation
I_θ is the slope of torque versus rotation curve.

To determine the vane strength under different stress conditions, Vane triaxial apparatus has been developed by Norwegian Geotechnical Institute (Kenney and Landva, 1965). Vane tests can be performed with different sizes and shapes of the vane at various rates of rotation on samples consolidated and/ or stressed to any set of conditions obtainable with standard triaxial equipment. Standard Geonor triaxial cells having a pedestal area of 20 sq. cm, have been used to construct vane triaxial apparatus. The vane can be inserted into the pedestal either from above before the sample is placed or from below the cell base by temporarily removing the metal plug. For analysis of vane test based on effective stress, Karube et al. (1988) modified the conventional consolidometer (Fig. 2.26). The specimens are consolidated in the ring together with the vane in it, using stress history similar to the cylinder shear test procedure. The vane rod connected to a torque meter is used to shear after the vertical stress of 74 kPa is applied.

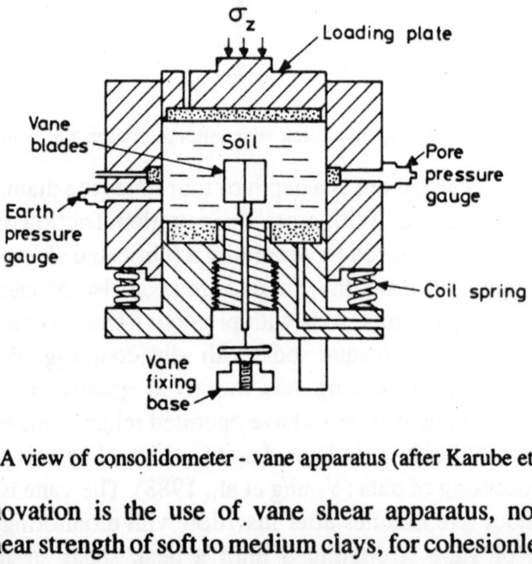

Fig. 2.26 A view of consolidometer - vane apparatus (after Karube et al., 1988).

Another innovation is the use of vane shear apparatus, normally used to determine the shear strength of soft to medium clays, for cohesionless soils (Barros and Barros, 1989). An analytical approach based on the solution for the thick wall cylinder in a triaxial stress field has been proposed for the evaluation of the angle of internal friction. The test results indicated that a definite relation exists between void ratio, e and the failure shear moment M. The predicted internal friction angle ϕ obtained from the analytical approach agreed very well with the laboratory direct shear tests on sands. The differences observed for the sand investigated was no more than 2.5° over a range of 33° to 49°. Further investigations on sands with different granulometric characteristics and under simulated field conditions are suggested to reconfirm the applicability of vane tests on sands.

The combination of vane and cone devices into one device known as Vane - cone apparatus has been developed by Youssef (1989) (Fig. 2.27). The use of this

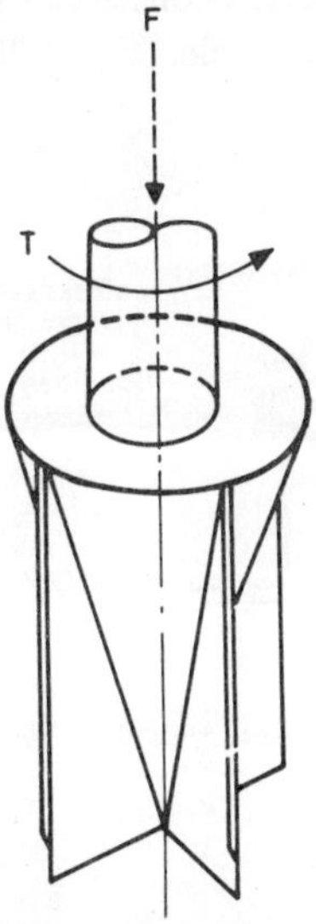

Fig. 2.27 Vane-cone penetrometer (after Youssef, 1989).

apparatus permits deforming the soil vertically and tangentially. This enables to determine vertical and tangential shear resistances in soil. The following torque penetration relation has been developed from the considerations of radial and vertical pressure distribution along the failure surface.

$$T = \alpha c + \beta F \tan \phi \qquad 2.32$$

where T is applied torque
F is the total penetration force
c and ϕ are the soil parameters under undrained condition
α and β are penetrometer shape constants.
$\alpha = \{\pi D^2 H/2\} \{1 + D/8H\}$
$\beta = H \{1 + D/4H\}$
D is the diameter of the vane cone device
H is the height of vane cone device

2.6.3 Borehole shear tests

The borehole shear device is an apparatus that can be used at any point in a borehole to determine the in-situ shear strength parameters of soils quickly and economically than by conventional laboratory tests. This test, developed by Handy and Fox (1967), has received considerable attention and popularity among practicing geotechnical engineers (Schmertmann, 1975; Wineland, 1975; Demartinecourt and Bauer, 1983; Bauer et al., 1980; ASTM Subcommittee D18.02-1987 and others).

The test involves placement of a bilaterally expandable probe in a predrilled borehole, expanding the probe against the wall of the borehole to engage the soil and causing a shear failure in the soil by pulling the probe axially along the borehole. The borehole shear device consists of basically three parts, the shear head, the pulling device and the console. (Fig. 2.28).

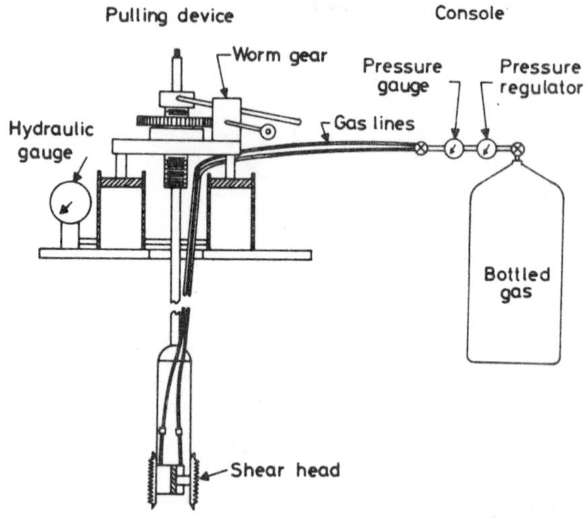

Fig. 2.28 General details of borehole shear device.

The shear probe consists of lengthwise split opposite quarters of a cylinder interconnected to a gas-operated piston to expand against or contract from the surface of borehole. When expanded against the wall of the bore, it is mainly to engage the soil and to apply normal pressure on the soil. It is possible to preset the expansion pressure of the probe. The pulling device consists of an aluminum plate which rests on the ground surface. A worm gear and screw arrangement enables to pull the rod and in turn the shear probe. Two hydraulic cylinders with hydraulic pressure enable to measure the vertical shear force. The console consists of a bottled gas supply, a pressure regulator and a pressure gauge. Plastic tubing is provided to apply and regulate the gas pressure to the gas operated piston in the shear probe.

The test is carried out in two separate phases.

(i) Consolidation stage involves application and maintenance of probe expansion pressure, under which the clay will consolidate. Five minutes (sands) to fifteen minutes (clays) waiting time has been suggested (Schmertmann, 1975). This would induce almost full drainage in sand in the vicinity of shear plates and only partial in case of clays. Handy (1975) indicates that drainage time is greatly reduced due to decreased drainage paths.

(ii) After the completion of consolidation, the shear test is run by turning the crank on the pulling device at the rate of 2 revolutions per second which results in a vertical translation of 0.05 mm/sec. The shearing resistance increases as the crank is rotated until shear occurs and then remains constant or falls off. Unless pore pressures mobilized during shearing are monitored, the test cannot be categorized as drained or partially drained. The maximum pressure reached by the hydraulic gauge on the pulling device along with the gas pressure is read from the console. Knowing the gauge pressures and the area of the shearing plates, the normal and shear stresses can be computed. The values of normal and shear stresses thus determined constitute a point on the Mohr envelope.

The shear force is released by turning the worm-gear handle counter clockwise until the base plate reading returns to initial value. The test is repeated with five increments of normal pressure consisting of both consolidation and shear phases. This would enable to have five points on the Mohr envelope to perform linear regression analysis to obtain strength parameters. The entire testing programme generally takes about an hour.

Several modifications have been effected to enhance the applicability of conventional test to cover a wide range of in- situ soils and to overcome certain difficulties in the conduct of test (Demartinecourt and Bauer, 1983). These modifications include pore water pressure and load transducers, a motorized pulling mechanism and different types of shear plates. Figure 2.29 illustrates the details of modified borehole shear device. In order to monitor pore water pressures

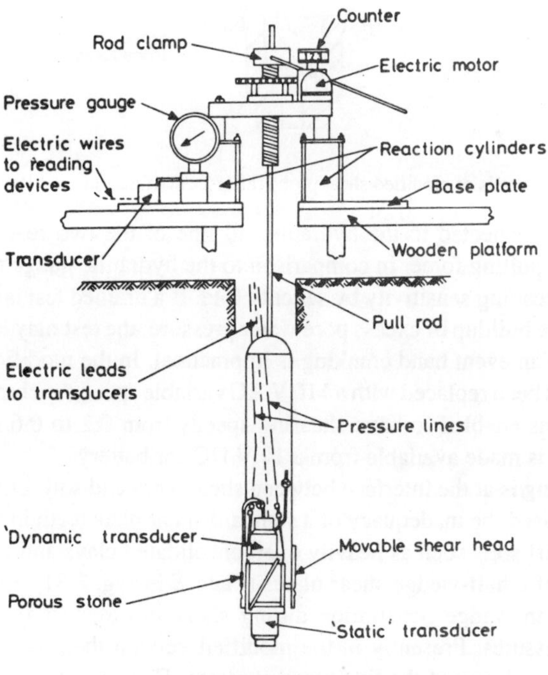

Fig. 2.29 Schematic details of modified borehole shear device (after Demartinecourt and Bauer, 1983).

during consolidation and shear phase two pore water pressure transducers are incorporated into the shear probe. The details of this modified shear probe are shown in Fig. 2.30. The gauge measuring the pulling force is too coarse and insensitive with soft clays or loose sands. A pressure transducer with a range of 0

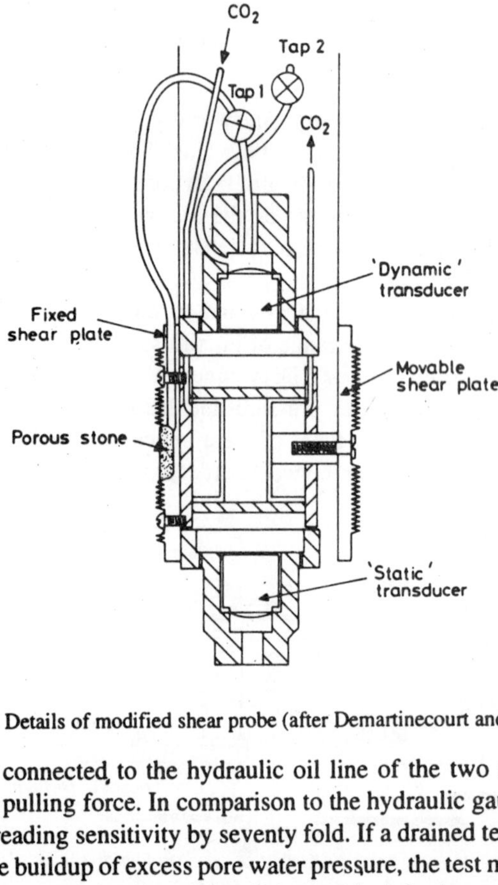

Fig. 2.30 Details of modified shear probe (after Demartinecourt and Bauer, 1983).

to 690 kPa is connected to the hydraulic oil line of the two reaction cylinders measuring the pulling force. In comparison to the hydraulic gauge the transducer increased the reading sensitivity by seventy fold. If a drained test is contemplated by avoiding the buildup of excess pore water pressure, the test may last for several hours. In such an event hand cranking is impractical. In the modified version the hand crank has been replaced with a 110 V AC variable speed synchronised electric motor. This has enabled to have shearing speeds from 0.2 to 0.6 mm/min. The power needed is made available from a 12 V DC car battery.

The shearing is at the interface between shear teeth and soil. Lutenegger et al. (1978) recognised the inadequacy of a standard shear plate teeth in building up of soil cake in stiff soils such as heavily overconsolidated clays and glacial tills. At that stage itself a half-wedge shear plates (Plate *E* in Fig. 2.31) were adopted to allow a slide in wedge penetration during shear in order to avoid the use of preseating pressures. Presently in the modified version there is a possibility of interchanging with any of the five types of plates (Fig. 2.31) with different depth and spacings of teeth.

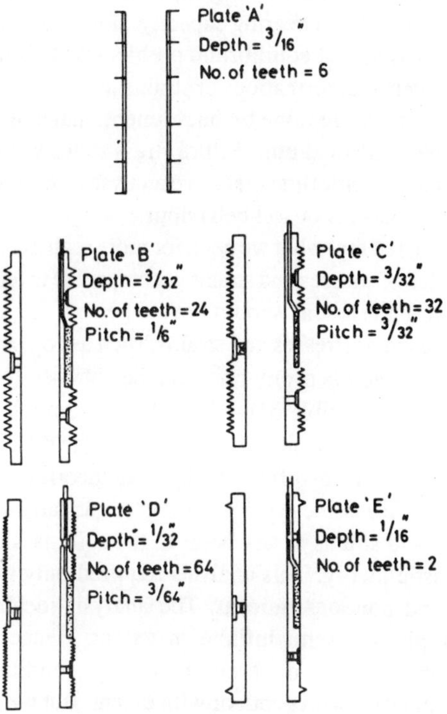

Fig. 2.31 Shear plates with different depth and teeth spacing (after Demartinecourt and Bauer, 1983).

The potential use of borehole shear test has been markedly enhanced with the incorporation of the above modifications. Still certain aspects of the test merit examination, particularly regarding effects of borehole disturbance, drainage conditions and shear rates. Despite this, the equipment has proven to be rugged, versatile and efficient to use. Some of the specific advantages of this test recognized even prior to the above modifications are that (Handy, 1975);

(i) the test data can be plotted in the site itself during the test enabling immediate repetition if necessary,
(ii) the test can be conducted at the same layer in situations where three triaxial samples from the same layer are not possible to obtain,
(iii) the tests provide relatively accurate information economically almost at one tenth of time required for conventional triaxial tests.

2.7 Constitutive Modelling of Soil Behaviour

As we have seen, the parameters c and ϕ which can be used to compute the magnitude of shearing resistance available at failure and the parameters C_c and E_s for computations of the deformations or strains in the soil for particular type of loading can be determined by a vide range of laboratory and in-situ tests. Conventionally, practical problems in geotechnical engineering are being solved with this information. Stability analysis in bearing capacity, slope stability or earth pressure problems are analyzed as a limit equilibrium problem with failure c and ϕ values. No consideration is given to deformations or strains in these analyses. On the other hand settlement calculations are done by one-dimentional consolidation analysis, treating the soil as an elastic medium. Failure stress states were never considered in these analyses. For quite some time these two analyses were carried out as though they were two different aspects of soil behaviour.

It has been now realized that what we need for more realistic analysis and design is a complete knowledge of stresses and strains at compatible levels right upto failure. And also the stress-strain relations should have a general formulation containing all the 6 components each of stresses and strains, such as $(\sigma_{ij}) = (D_{ijkl})\ (\varepsilon_{kl})$ so that the deformation in any direction at any point can be obtained for any combination of loading. This would be possible only if the stress-strain relations are based on certain laws of continuum mechanics. This has led to the development of several constitutive relations based on different physical theories of elasticity, elasto plastic, visco plastic etc. Boussinesq (1885) had given as early as 1885 closed form solutions for stresses and strains at any point in the soil mass for applied surface loading based on elastic theory. This analysis required only two soil parameters elastic modulus, E and poissons ratio, μ. The analysis idealized the soil to be homogeneous, isotropic and semi infinite in extent. Hence such closed form solutions are not possible for different boundary conditions and for non-homogeneous soil properties with depth or with extent. But we now have methods to solve such nonhomogeneous soil problems by finite element methods. But still, the elastic solutions are not satisfactory except may be for heavily overconsolidated or dense soils at low stress levels. Otherwise soil is generally known to exhibit highly non linear and irreversible strains which can be seen from Fig. 2.32 for compression and shear loadings. Such non linear behaviour can be reflected by using elastic constants E and μ to be variable functions of stresses or strains. But the irrecoverable response cannot be described by elastic models.

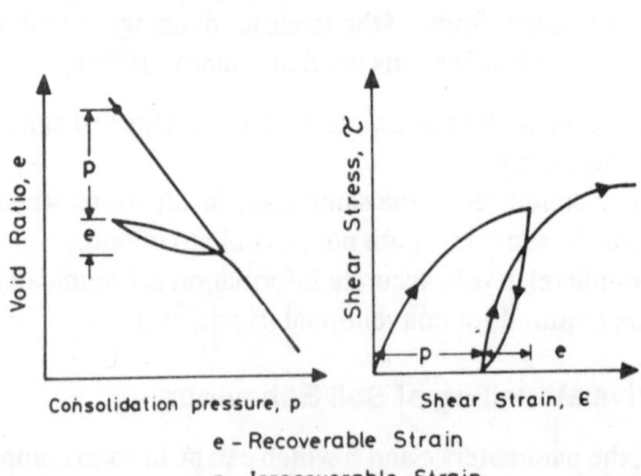

Fig. 2.32 Recoverable and irrecoverable components of clay deformation during compression and shear.

More recently elasto-plastic models have been developed, originally started in sixties, with several refinements and extensions being effected subsequently. Today these form very powerful tools for obtaining the complete stress-strain behaviour of a given soil for any type of loading, knowing only its response under very simple loading paths. For example, with just the data of consolidation test and an undrained shear test, it is possible to predict the stress-strain behavior for shear tests under any other confining stress level, drained or undrained, plane strain or triaxial case, normally or overconsolidated state, compression or extension etc. These models have been developed from consideration of plastic work done or energy dissipated. Since they are developed to be applicable for a general 3-dimensional problem, to include the work done by all the components of stress and strain, the mathematical relations are written in terms of invariants i.e., the octahedral normal stress or mean principle stress $p = (\sigma'_1 + \sigma'_2 + \sigma'_3)/3$ and the octahedral shear stress = $2/3\ [(\sigma_1 - \sigma_2)^2 + (\sigma_2 - \sigma_3)^2 + (\sigma_3 - \sigma_1)^2]^{1/2}$ (which reduces to $q = (\sigma_1 - \sigma_3)$ for triaxial state of stress) and correspondingly strain invariants, volumetric strains, $\varepsilon_v = (\varepsilon_1 + \varepsilon_2 + \varepsilon_3)/3$ and octahedral shear strain $\varepsilon_s = 2/3\ [(\varepsilon_1 - \varepsilon_2)^2 + (\varepsilon_2 - \varepsilon_3)^2 + (\varepsilon_3 - \varepsilon_1)^2]^{1/2}$ ($\varepsilon_s = 2/3\ (\varepsilon_1 - \varepsilon_3)$ for triaxial case).

The basic soil parameters required for working with any of these elasto-plastic models are the elastic and plastic bulk and shear modulii and the energy dissipating factor, M. The determination of these parameters do not require any additional testing methods, but can be obtained from the conventional tests themselves. It is shown that the energy dissipation factor M is the ratio of q to p at ultimate state of shearing and is related to the friction angle of the soil. The elastic and plastic bulk modulii can be derived from the slopes C_c and C_r of the compression and rebound paths in a conventional consolidation test. The plastic shear modulus is not an independent parameter, but is related to the bulk modulus once the yield function is defined as per the specific model. The detailed interpretation of these parameters and their use in constitutive modelling of the soils will be dealth with in chapter 5.

2.8 Permeability - Laboratory methods

Permeability is the property of a porous material that governs the rate of flow through the material. Soils and rocks are basically pervious to different degrees dependent on the pore structure. Permeability is one of the most important engineering properties associated with these materials. The range of values encountered in practice can vary over ten billion.

Laboratory methods of determination of permeability characteristics are appropriate in the case of reconstituted consolidated soils, compacted soils and undisturbed samples. The standard method is to have the sample confined and monitor the flow into the sample. Consolidation cells work well for undisturbed samples whereas compaction molds can be used directly for compacted soils. Figure 2.33 and 2.34 are schematic diagrams of a consolidation cell and compaction mould permeameters.

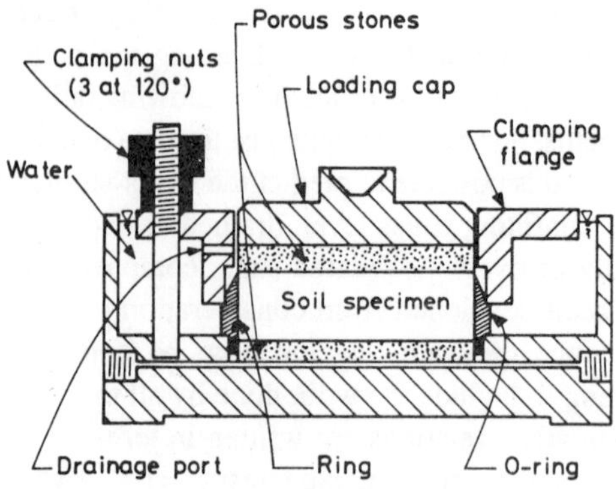

Fig. 2.33 Details of consolidation cell permeameter.

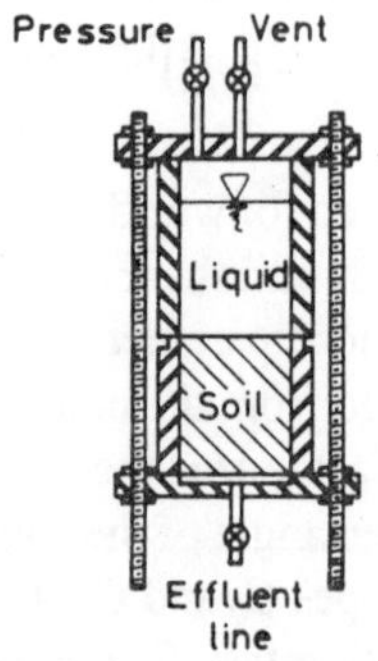

Fig. 2.34 Details of compaction mould permeameter.

While use of compaction mold permeameter is simple, the disadvantages associated with such devices are the difficulties in monitoring the changes in the degree of saturation, measurement of shrinkage or swelling that might take place either due to change in the degree of saturation or due to leachate movement and lack of control over the stress applied to the soil. The details of the double ring compaction mold permeameter (Anderson et al., 1985) are shown in Fig. 2.35. In both the above mentioned methods, being rigid wall types, there is the possibility for spurious leakage of permeating liquid along the sides of the test specimen. To a certain extent this can be detected in the compaction mold permeameter by comparison of flows into outer collection ring and into the inner ring. To circumvent the disadvantages of rigid wall permeameter, flexible wall permeameter has been developed. The samples are confined with a flexible membrane in a chamber so that the pressure applied to the outside of the membrane presses it against the sample conforming to its exact shape and size at all the stages of the tests. This eliminates the flow along the sides of the sample. Figure 2.36 shows a typical triaxial flexible wall permeameter (Carpenter and Stephenson 1985, Uppot

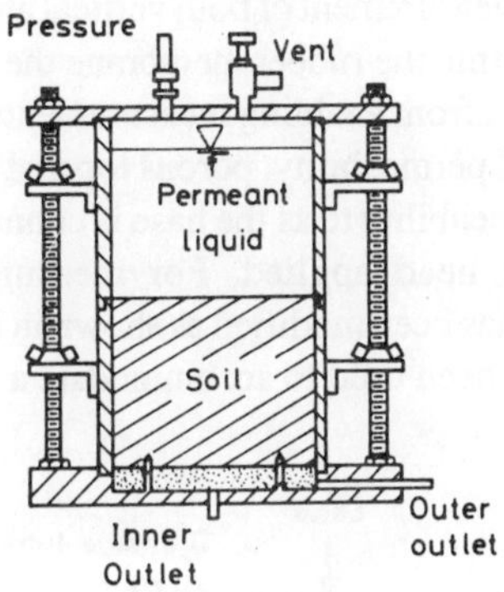

Fig. 2.35 Double-ring compaction mold permeameter (after Anderson et al., 1985).

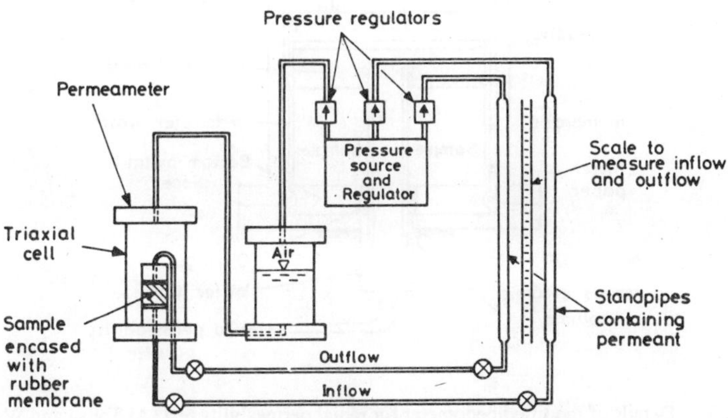

Fig. 2.36 Details of flexible wall permeability test (after Carpenter and Stephenson, 1985).

and Stephenson, 1989). The specific modifications of this setup over the earlier designs are:

(i) independent measurement of volume change of sample during testing
(ii) chemical resistance to the organic solvents to be used, and
(iii) facility of safe and easy handling of toxic permeants.

Other built in facilities of triaxial flexible wall permeameters are in the application of back pressure and measurement as well as the control of external stress state on the sample.

Flow through the sample can be caused by constant head or falling head. In the constant head method the hydraulic gradient $i = h/l$ is maintained constant and the total volume of flow, Q is measured during time interval, t. Then from Darcy's equation,

$$k = \frac{QL}{hAT} \qquad 2.33$$

More recently Al-Tabbaa and Wood (1987) have reported the details of the modified oedometer for the measurement of both vertical and radial permeabilities. Since the clay is enclosed within the rubber membrane the flow along the sides of the sample is eliminated apart from enabling measurements of horizontal stresses. For measurements of vertical permeability, porous top and bottom plates are used. To perform falling head permeability tests the base is connected to a 5 mm burette and a difference of water head applied. For measurements of horizontal permeability the oedometer has been modified as shown in Fig. 2.37. Impermeable top and bottom platens have been used to accommodate a central porous ceramic

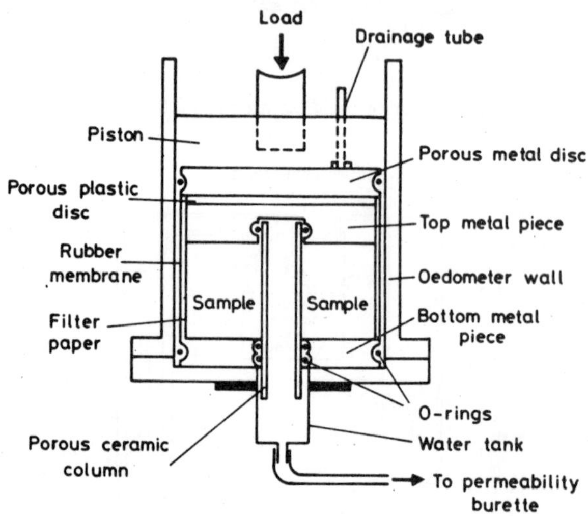

Fig. 2.37 Details of modified oedometer for radial permeability test (Al Tabaa and Wood, 1987).

column 18 mm in diameter which locates in the center of the top platen and is able to slide through the center of the bottom platen as compression of clay proceeds. The filter paper provided on the inside of the rubber membrane around the sample is in hydraulic contact with the porous plastic disc placed above the top of impermeable metal platen.

At the end of consolidation cycle, the porous column is connected to the 5mm burette and a head difference is caused across the sample to actuate radial flow towards filter paper drainage layer and move towards the drainage tube in the top porous metal piece. In this way the falling head horizontal permeability test can be conducted.

2.9 In-Situ Permeability Tests

A knowledge of the in-situ permeability and its possible degree of variation are needed to examine:

(i) ground water seepage and its influence on the stability,

(ii) erosion and piping in earth dams,
(iii) construction dewatering, and
(iv) competency of soil liners to meet the regulations of environmental agencies.

Laboratory devices can handle only relatively small specimens of these materials. On the other hand, in-situ tests offer an opportunity to test larger and hence more representative volumes of material under boundary conditions as encountered in- situ. It has long been known that geomechanical features (Rowe, 1972; Sowers, 1981 and others) have a dominant influence on permeability. For example, uniform deposits of clays, more often than not contain thin partings of silts and fine sands, either as laminations within the mass or randomly dispersed. As a consequence of this macrostructure, in-situ permeabilities of such deposits from field observations are generally upto several orders of magnitude more than those determined from laboratory tests. Such discrepancies, to a lesser extent, are also prevalent in the case of sands and coarse silts.

2.9.1 Well tests

This simplest field test is performed by drilling a hole below water table without the use of casing, down to an impermeable layer. In the absence of an impervious layer the hole is advanced to a depth below the water table to at least ten times the diameter. The water is pumped out and allowed to refill. Then the rate of rise of water within the hole is observed. Figure 2.38 illustrates the details of the test. The

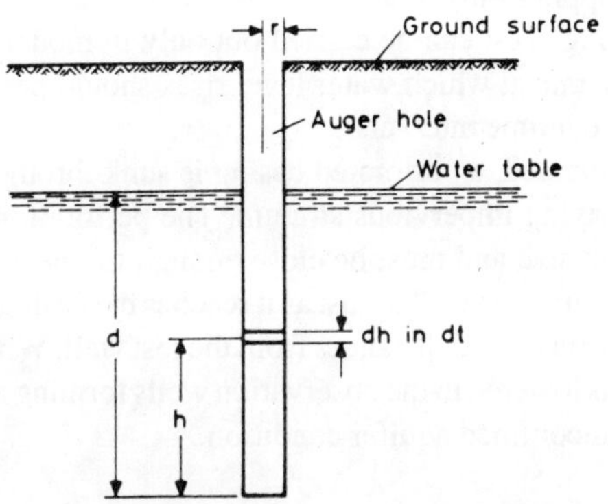

Fig. 2.38 Auger hole method for assessment of in-situ permeability

formula for determining the permeability coefficient suggested by Spangler as early as 1953 (Spangler, 1953) is as follows:

$$k = 0.617 \frac{r}{Sd} (dh/dt) \tag{2.34}$$

where h is the depth of water in the hole at the time, t
(dh/dt) is the rate of rise of water level at depth h,
r is the radius of hole

d is the depth of hole below water table
S is coefficient dependent on ratios of *h/d* and *r/d* (vide Fig. 2.39).

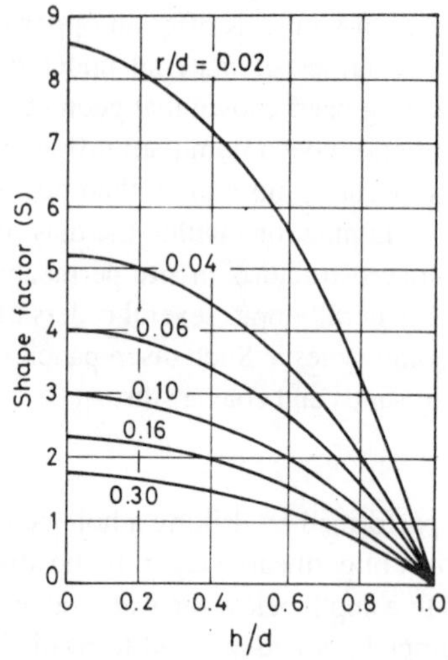

Fig. 2.39 Shape factors for use with auger method (after Spangler, 1953).

This approach applies only to an incompressible soil and no draw down takes place during pumping. Test can be carried out only in moderately pervious soils because of the flow rate at which water level rises should permit observations of water level at different time intervals.

In a well point method, a perforated casing is sunk through the water bearing stratum to an underlying impervious stratum. The perforation in the well casing must be of sufficient size and must be close enough to one another to permit the ground water to flow into the well as fast as it reaches the casing. Observation wells are positioned at various radial distances from the test well. When water is pumped out, the water level is lowered in the observation wells forming a cone of depression (Fig. 2.40), for an unconfined aquifer condition.

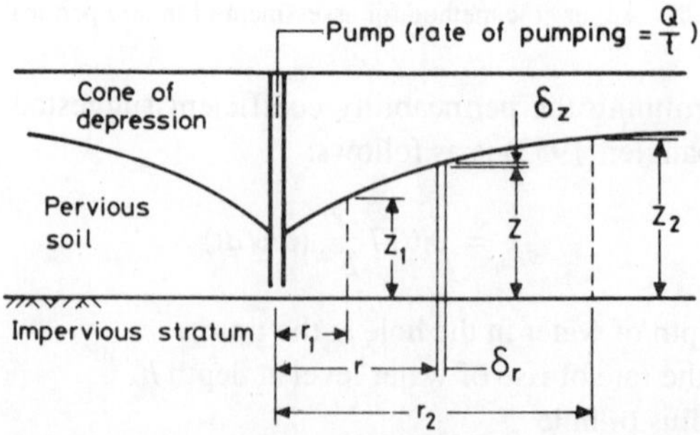

Fig. 2.40 Illustration of principle of well point method.

The rate of flow Q/t is the rate at which water is being pumped out. The surface area of the soil element at distance r from the well point is equal to the surface area of a cylinder of radius r and height z i.e., $A = 2\pi rz$. The hydraulic gradient at the point is the rate of change of head per unit length or $i = dz/dr$. Then from Darcy's law

$$\left\{\frac{Q}{t}\right\} = k\,A\,i = k\,2\,\pi rz\left\{\frac{dz}{dr}\right\}$$

or

$$\frac{1}{r}\,dr = k\left\{\frac{2\,\pi}{Q/t}\right\}z\,dz$$

Integrating between the limits r_2 to r_1 and Z_2 to Z_1

$$\log_e\left\{\frac{r_2}{r_1}\right\} = k\,\frac{2\,\pi}{Q/t}\left\{\frac{(Z_2^2 - Z_1^2)}{2}\right\}$$

$$k = \left[\frac{2.303\,Q/t}{\pi\,(Z_2^2 - Z_1^2)}\right] \times \log_{10}\left\{\frac{r_2}{r_1}\right\} \qquad 2.35$$

Hence by measuring the height of ground water at two points at distance r_1 and r_2 from the well points, the coefficient of permeability can be determined.

Another field situation is to have confined aquifer with test and observation wells. The water in the observation wells rises above the level of the aquifer due to artesian pressures (Fig. 2.41). The two cases that might arise due to pumping are

(i) the water level rise in the test well might remain above the roof level of aquifer at steady flow conditions ($Z > d$),

(ii) the water level in the test well might fall below the roof level of aquifer ($Z < d$).

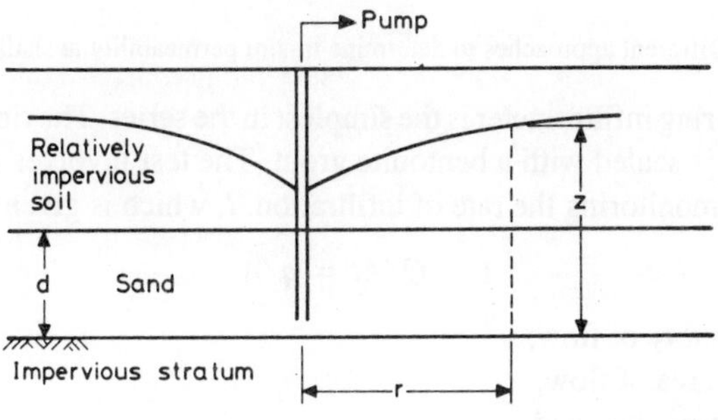

Fig. 2.41 Well point method under confined aquifer condition.

The following relationships can be derived from the same principles detailed above : for case (i)

$$k = 2.303 \left\{ \frac{Q/t}{2\,\pi d\,(Z_2 - Z_1)} \right\} \times \log_{10} \left\{ \frac{r_2}{r_1} \right\} \qquad 2.36$$

and for the case (ii)

$$k = 2.303 \left\{ \frac{Q/t}{\pi\,(2Hd - d^2)} \right\} \times \log_{10} \left\{ \frac{R}{r} \right\} \qquad 2.37$$

where H is the depth of water table from the impervious boundary at the radius R from the center of the test well within which the cone of depression is formed.

Apart from the above tests many in-situ tests such as bore hole, packer and porous probe tests, have been developed during the recent past. For detailed discussions about these tests the reader is referred to Nagaraj (1993).

2.9.2 Infiltrometers

Infiltrometers are used to measure infiltration rate and/or permeability of soils in-situ. Infiltration is the rate at which a given volume of water crosses the air-soil interface into a unit area per unit time. Figure 2.42 schematically indicates the types of infiltrometers available for in-situ tests.

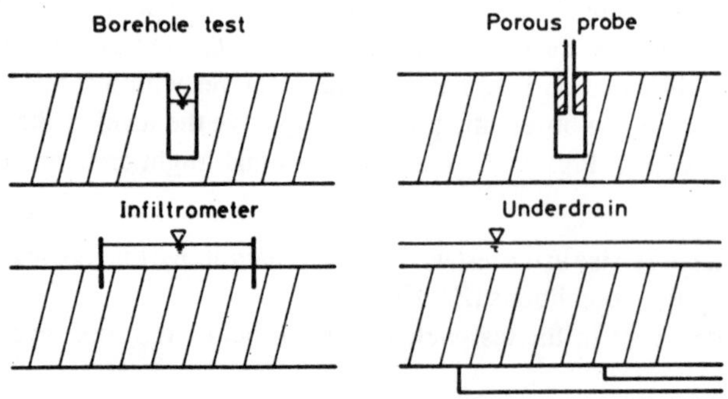

Fig. 2.42 Different approaches to determine in-situ permeability at shallow depths.

Open single ring infiltrometer is the simplest in the series. The ring is embedded in a trench that is sealed with a bentonite grout. The test involves filling the ring with water and monitoring the rate of infiltration, I, which is given by

$$I = Q/At = q/A \qquad 2.38$$

where Q is quantity of flow,
A is the area of flow,
t is elapsed time and
q is rate of flow.

This mode of computing the infiltration rate is based on the assumption that the soil is homogeneous and uniformly soaked behind the wetting front. There is no leakage through the seal and evaporation losses have been accounted for. The rate of infiltration has to be sufficiently large for direct measurement.

Hydraulic conductivity is calculated from the relation

$$k = I/i = IL_f/(H + L_f) \tag{2.39}$$

where I is the rate of infiltration
h is depth of ponded water
L_f is depth of the wetting front.

For one dimensional flow conditions L_f would be within the depth to which the ring is installed. It is also possible to assess the depth of infiltration using the volume of permeant, porosity, dry density, degree of saturation and area of the soil.

To overcome evaporation losses and to have the flexibility of handling low rate of infiltration either by variable or a constant head test, the ring is sealed at the top with a flexiglass lid with provision for fixing a graduated cylinder or constant pressure system along with an arrangement to monitor the volume of inflow water. Figure 2.43 shows the single-ring infiltrometer sealed at the top with graduated cylinder for variable head test. The rate of infiltration and hydraulic conductivity can be computed from equation 2.38 and 2.39 respectively.

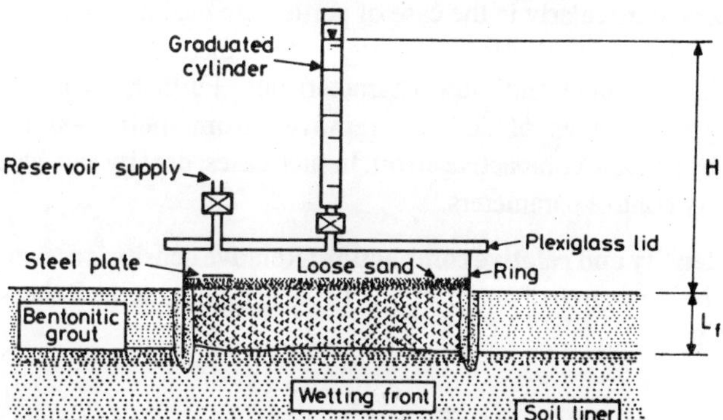

Fig. 2.43 Details of sealed single ring infiltrometer (after Daniel, 1989).

The following are the specific advantages of open and closed single ring infiltrometers.

(i) Easy to install at moderate cost
(ii) Adequately large infiltrometers can be used to test large volume of soil
(iii) Can measure low hydraulic conductivity in the vertical direction down to $10^{-8} - 10^{-9}$ cm/sec.

The reader is referred to the field data reported by Daniel (1989) and Fernuik and Hang (1990) to appreciate the versatality and applicability of the wide spectrum of infiltrometers in environmental geotechnique.

2.10 Inferential Testing

In a broad sense material testing could be either inferential or property characterization in nature. Laboratory testing in which mechanical properties are not directly measured but are inferred through empirical correlations form inferential testing programme. Consistency limits, colloidal activity and swelling potential form a few inferential test parameters for soils. In addition, tests to determine density, moisture content, degree of saturation provide information about the in-situ state of soils. These data at times are used as inferential parameters to assess the probable range of mechanical properties. In all cases, any correlation between inferential parameters and mechanical properties would be of proven value so far as there is a clear understanding of the basic mechanisms of inferential tests with respect to mechanical properties they reflect. However, the behaviour of materials under changing stress, time and environment etc. cannot be easily assessed with the information obtained only by inferential tests.

2.10.1 In-situ and placement state parameters

Soils are basically multicomponent and multiphase systems. The word "phase" is of a Greek origin meaning appearance. It is the homogeneous part of a system separated from other parts by abrupt transitions. Each physically or chemically different, homogeneous and mechanically separable part of the system constitutes a distinct phase. As cited earlier, density and water content form two prime material state parameters. Particularly in the case of particulate media such as fine grained soils, these state parameters reflect, to a large extent, their stability with respect to strength, volume change and flow characteristics. Further, as a construction material, large quantities of soil are removed from their in-situ state and recompacted by suitable compactive effort. In such cases, density and water content form the quality control parameters.

Relative density and relative compaction: Relative density is a term generally used to describe the degree of compaction of coarse grained soils. Relative density D_r is defined as

$$D_r = \frac{e_{max} - e}{e_{max} - e_{min}} \quad 2.40$$

where e_{max} is maximum possible void ratio,

e_{min} is minimum possible void ratio

and e is void ratio in natural state of granular material.

Using the relation $\gamma_d = G\,\gamma_w/(1+e)$ equation 2.40 can also be expressed interms of the dry unit weights as

$$D_r = \frac{\gamma_d - \gamma_{d\,(min)}}{\gamma_{d\,(max)} - \gamma_{d\,(min)}} \times \frac{\gamma_{d\,(max)}}{\gamma_d} \quad 2.41$$

Relative density is generally expressed as percentage.

Another common method of representation and evaluation of compacted fills is in terms of relative compaction, R_c defined as

$$R_c = \frac{\gamma_d}{\gamma_{d\,(\max)}} \tag{2.42}$$

The terms relative compaction and relative density can cause confusion. Hence it is necessary to understand the simple practical correlation between them. Both terms range from 0 to 100%. Zero percent relative compaction refers to zero dry density which has no meaning. Whereas zero percent relative density refers to a possible state.

2.10.2 Water content determination

As cited earlier, the simplest method to determine moisture content is to drive away the moisture from the material by heating and determine the change in weight and weight of dry material. The major classification of methods of determining water content are;

Direct: (i) Electrical Oven Drying
(ii) Microwave Oven Drying

Indirect: Chemical, electrical resistance, nuclear methods.

Of these nuclear methods which are relatively new to the profession are discussed in detail to bring about their potential to determine the *in-situ* density and water content of soils.

Direct Methods

Direct drying: For engineering purposes, determination of moisture content generally requires drying the material at a controlled constant temperature (about 105°C) until the weight of material reaches a constant value. Careful control of oven temperature usually within 5°C variation is needed to obtain reproducible results. This long accepted standard procedure is slow with large samples taking time far beyond 24 hours. For the tests commonly performed in the laboratory, this delay is usually acceptable. However, for tests commonly performed in the field compaction control of earth works faster techniques are desirable. Some of the faster methods of drying are stove-top drying, infrared lamp drying and the more recent method of microwave oven drying.

Microwave oven drying: The use of temperature controlled microwave ovens for drying is gaining acceptance. Microwaves are electromagnetic radiations that have frequencies between 10^3 to 10^6 MHz with the wave lengths in the range of 1 mm to 1m. In other words, it is an ultra high frequency radio wave. It can be used as a new energy source for heating and desiccating dielectric materials. The response of the material to electromagnetic fields depend on the degree of polarity of the material. Highly polar materials, with high dielectric constant, in the presence of an alternating electric field experience continued reorientation of polar molecules. This continued molecular motion causes the material to heat up. Microwave heating reduces the time needed for heating since it is generated inside the material. This is in contrast to conventional heating through convection. Temperature regulations can be accomplished by inserting a temperature probe in the material being heated. The oven is programmed to automatically shut off and on intermittently when either the desired temperature is reached or to be maintained during the period of test.

The possible use of temperature controlled microwave oven drying for the rapid determination of moisture content of clay soils and shale of widely varying plasticity characteristics has been examined by Charlie et al. (1982). A high degree of agreement with the values obtained from conventional temperature controlled drying has been noticed. Drying times were two to seven times faster. It has been possible to achieve even 13 times faster drying by employing full microwave power without any temperature control. But this results in slightly higher moisture contents compared to conventional drying.

Nuclear methods

Normally, density and moisture content measurements are done by direct tests. Reports of such conventional tests usually strike a background of familiarity to all concerned. There are a number of engineering situations such as critical moisture contents below subgrades and compaction control, the spatial distribution of these parameters need be determined over long periods of time without causing any disturbance to the system and structure. It should also be possible to estimate moisture content inspite of moisture being in the form of ice or vapour. Direct simple methods of measurements cannot fulfill all these requirements. Hence considerable research efforts have been made with experimentation and field trails in the development of facilities for the measurements of density and moisture content based on nuclear radiations.

In each collision or scattering process between neutrons which have been scattered, hydrogen atoms would lose most of their energy and would return as slow moving electrons. This characteristic observed for hydrogen atoms enabled to develop a method to measure concentration and distribution of moisture. By counting the number of slow neutrons at or near the source, the corresponding hydrogen atoms can be assessed and hence the moisture content. It is also a fact that this scattering and slowing down process is practically independent of whether or not the hydrogen is bound chemically and is in vapour, liquid or solid state.

When gamma rays emitted by a source such as radium, pass through a material, these rays are scattered as a consequence of their interaction with the electrons in the atoms of the material. The scattering is stronger as the number of electrons in a given volume of the material increase which inturn would be roughly proportional to the bulk density of the material. This principle is explored to develop density meters.

A device for measuring density should have: (i) gamma ray source and (ii) a detector mounted at a well defined distance from the source and separated by a lead shield to prevent direct gamma rays from the source reaching the detector. The probe for measuring moisture content requires: (i) a source of fast neutrons (ii) a detector for slow neutrons rigidly placed near the source and (iii) an electronic device for counting and recording the number of slow neutrons reaching the detector per unit time. Recently Delft Soil Mechanics Laboratory has developed a nuclear density device to measure, directly during penetration tests, the in-situ density of saturated soils. A radioactive source and radiation detector were built into a cone shaped probe of 44 mm dia. with an effective area of 15 cm^2 (Fig. 2.44) (Tjelta et al., 1987). This probe can be penetrated with standard cone penetration testing methods and equipment. While the cone is penetrated the soil is irradiated

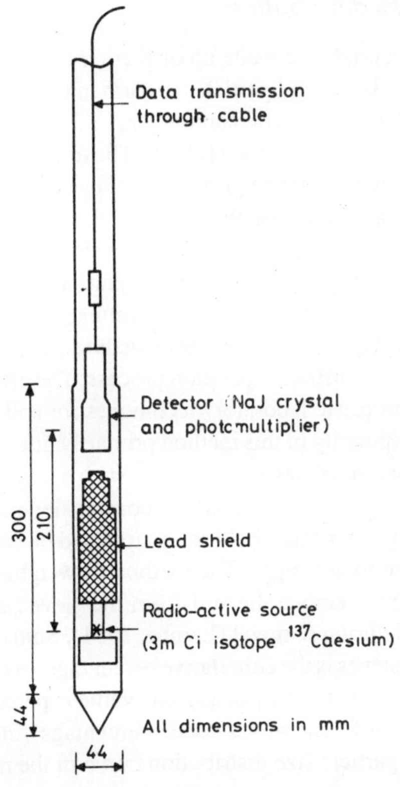

Fig. 2.44 Details of nuclear in-situ density cone (after Tjelta et al., 1987).

with gamma photons emitted near the tip. The bulk density of the soil is determined by comparing the energy level of emitted photons from the source and the detected photons entering the detectors after irradiation of the soil mass. During penetration the in-situ density is continuously recorded.

Applications: The advantages of the nuclear methods are the speed of measurements, nondestructiveness and repeatability of test at the same location. In order to reduce the cost of compaction control work, to avoid many problems of personnel scheduling on large compaction control jobs and to speed up field moisture and density determinations, nuclear methods have been fully exploited.

Surface nuclear density moisture content probes have been extensively used for compaction control of base and subbase course for highways, bridge approaches, air field pavements, building sites and backfills. On the other hand, depth density and moisture probes have enabled evaluation and correction of construction failures such as highways and airfield strips, soft spots, moisture migration due to climatic conditions, improper backfills and embankments.

Despite the potential of nuclear method being apparent, it has not gained popularity as one would otherwise expect. This is mainly due to limited awareness of the methodology among practicing engineers.

2.10.3 Particle size distribution

Soils as particulate materials are made up of particles of differing size and shape over a very wide range. In general particles in earth materials can vary in size from 10 ÅU as the smallest unit to several centimeters in thickness when rock material is considered. This is a range of one to a billion. The tremendous magnitude of this range in earth materials can better be grasped by considering the analogy when the smallest particle is compared to a moth ball or child's marble and the size of earth corresponds to rock material, the ratio of which is also close to a billion. There is no other engineering material with this range of particle size.

In general, particle size distribution is determined by a combined process of mechanical and sedimentation methods. Sieve analysis, a mechanical method is an important method as an industrial seperation process. The efforts to standardize as well as to develop appropriate laboratory techniques for soils are attributed to the strongly entrenched popularity of this method primarily due to its simplicity. Sieve analysis essentially consists of nesting sieves, in such a way that each sieve has an opening approximately one half of that of the coarser sieve above the nest.

After ten minutes of horizontal shaking soil retained on each sieve is expressed as percent in relation to total weight. The method of weighing the sieve plus soil rather than attempting to remove the soil from the sieve has been suggested to prevent the loss of soil during removal (Lambe, 1951). Sum of all the percentages retained on all coarser sieves is the cumulative percentage retained on the particular sieve size. The difference between 100 and this value is percentage finer than any size corresponding to different sieves used. Percentage finer is plotted against particle size to obtain particle size distribution curve in the range covered by this mechanical method.

Grain size separation below 75 μm size is neither practical nor accurate by the conventional methods of dry or wet mechanical sieveing. Sedimentation methods have been employed for grain size determination in the subsieve range. These methods are based on the Stoke's equation for the terminal velocity of a falling sphere. Of the methods in vogue hydrometer method is extensively used. It is essentially a measure of the density of suspension with time. A hydrometer calibrated in the density range of 0.995 to 1.030 is used for this purpose. For detailed description of this method the reader is referred to the text by Lambe (1951) and to Nagaraj (1993) for the critical review of this method of particle size analysis.

Based on the principle of Edlefsen and Cole's specific gravity balance (Edlefsen and Cole, 1936) the plummet balance has been designed by Marshall (1956). Figure 2.45 shows the components of plummet balance.

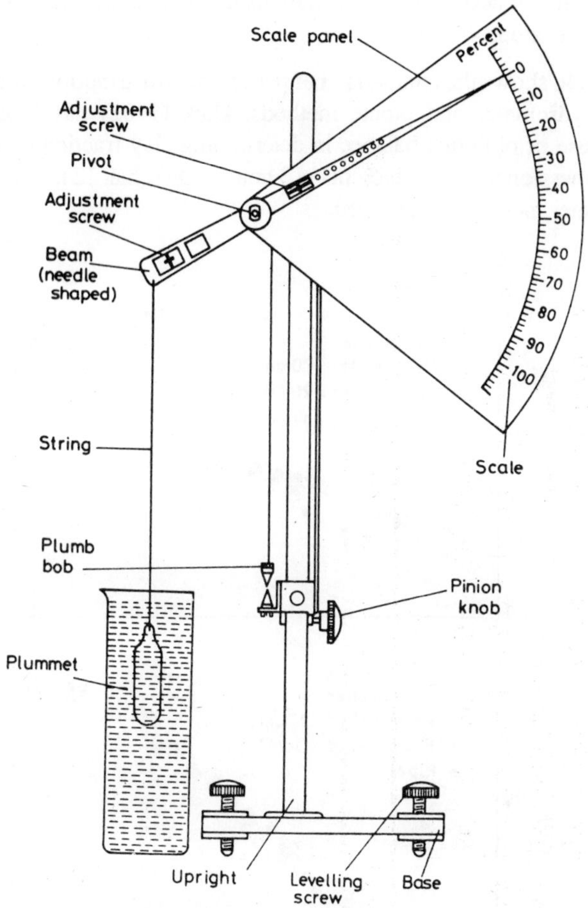

Fig. 2.45 Schematic details of plummet balance.

More recently, Nagaraj and Sivapullaiah (1981) have brought out the potential use of plummet balance as a direct reading device in the subsieve particle size analysis. In brief, the method consists of lowering the plummet to a predetermined depth in the soil suspension. As the particle settles down, density reduces and the plummet raises altering the depth of immersion. With the help of a rack and pinion arrangement, the depth of immersion is reset to original level. The pointer, which has been earlier calibrated to zero and hundred reading, now indicates directly percentage of particle diameter, *d* finer than the value obtained by the following relation

$$d = K\,(Z_e/t)^{1/2} \qquad 2.43$$

where Z_e is effective depth of immersion,

t is time in minutes when pointer readings are taken,
K is a factor accounting for variations in temperature and viscosity of the suspension.

Figure 2.46 shows the comparison of particle size distribution data obtained by plummet, hydrometer and pipete methods. Dick (1982) has brought out the significant use of plummet balance in determining clay fraction by reducing the depth of immersion to as low as 5 cms in 3 hours rather than 12 hours needed when the hydrometer method is employed.

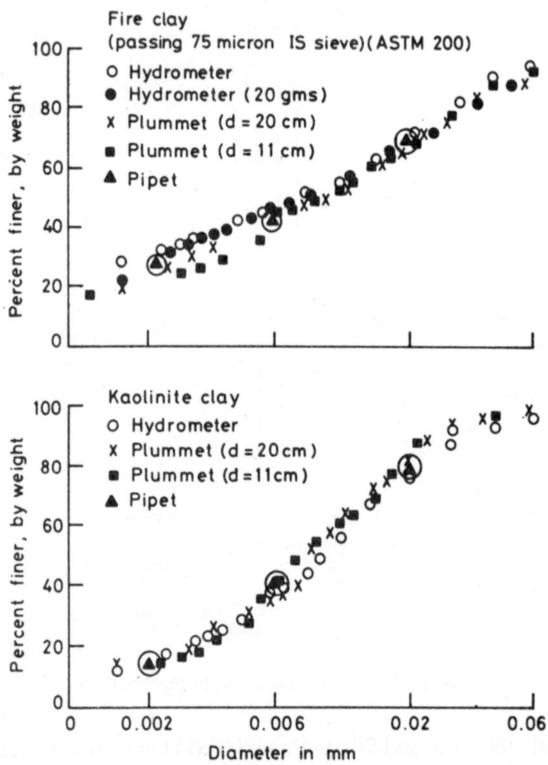

Fig. 2.46 Comparison of different methods of particle size distribution (after Nagaraj and Sivapullaiah, 1981).

2.10.4 Consistency limits of soils

It has been demonstrated by Casagrande (1932) that in the case of coarse grained soils, gradation of particles is adequate for characterization of specific properties, whereas for fine grained soils additional parameters are required to account for physico-chemcial interactions between soil and water. Consistency limits of soils satisfy this requirement. It is essentially the manifestation of physico-chemical forces of cohesion and adhesion acting within the soil at various moisture contents. In 1911, the Swedish soil scientist Atterberg (1911) reported an extensive study on the plasticity of soils, dividing the range from solid to the liquid into five stages by setting arbitrary limits. In the progressive transition from liquid state to solid state, the soil undergoes dramatic changes in consistency. Figure 2.47 schematically indicates the change in volume of the soil with changes in water content together

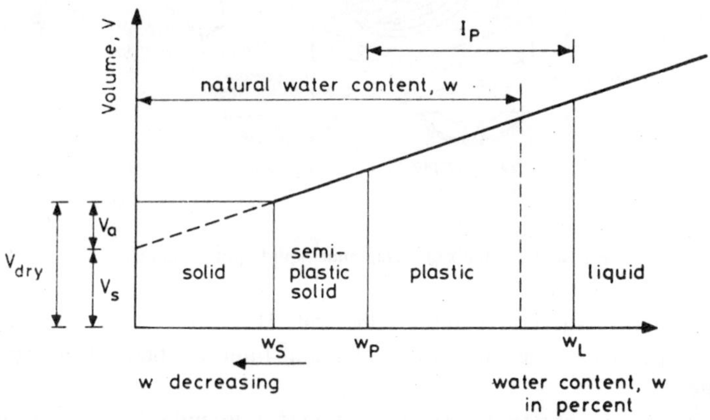

Fig. 2.47 Consistency relationships of clays.

with the limits of different states of engineering significance. All limits are expressed as water contents. Realistically all these transitions in the rheological states are not abrupt but are gradual. The importance of these limits was neither realized in Atterberg's own field of agriculture nor in soil mechanics until Terzaghi (1926) could see its importance and potential as applicable to soil mechanics. More recently, the principles and potentials of consistency limits have been critically reviewed by Nagaraj and Srinivasa Murthy (1988).

Liquid limit

The liquid limit is essentially a measure of a constant value of viscous shearing resistance, as the soil approaches the liquid state and it is expressed in terms of water content as a percentage. Liquid limit of soils can be determined by the following two methods.

Percussion method: This method was developed by Casagrande (1932). The details of the device are illustrated in Fig. 2.48. It essentially consists of a standard cup which can be lifted through a standard height by a cam and dropped on a base of standard material. The soil paste is transferred into the cup by a spatula and finished to get smooth surface without air voids. A groove of standard dimension is cut in the soil using a standard grooving tool. The number of drops required to close the groove by half an inch along the bottom of the groove is recorded. The water content of the sample paste removed from the cup is determined. The procedure is repeated by altering the water contents by spreading the soil paste on a glass plate and working with a spatula. The water contents are plotted against their corresponding logarithm of number of blows within a range of 10 to 40 blows to obtain a linear relationship. The water content corresponding to 25 blows is read

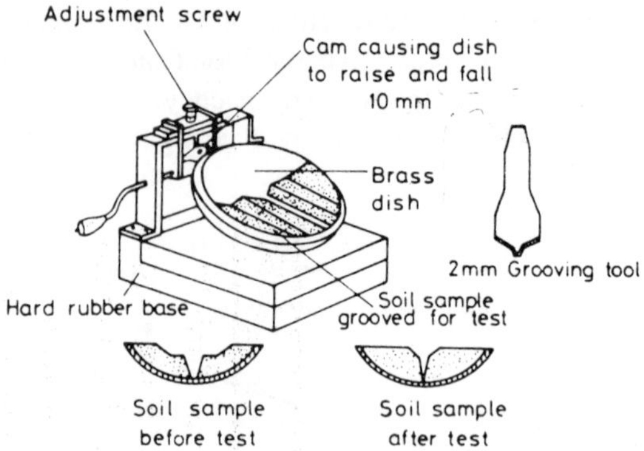

Fig. 2.48 Details of Casagrande liquid limit apparatus.

from the graph as the liquid limit of the soil. This number was suggested by Casagrande to obtain the same value of liquid limit as obtained by Atterberg technique.

After thirty years of usage, a reexamination (Casagrande, 1958) revealed that liquid limit test is akin to dynamic shear test. Kenney (1963) interpreted that liquid limit determination is a dynamic, stress controlled, strain dependent shear test in which maximum strength is not involved. Simultaneously standardization of height of drop and hardness of the base (Norman, 1958) were examined and a new grooving tool to handle low plasticity soils was introduced (Hovayni, 1958). Another factor which has received attention recently is the significance of specimen preparation for consistency limits (Armstrong and Petry, 1986). The moisture distribution upon reconstitution from oven dried stage, over a period of 24 hours yielded liquid limit water contents 3 to 6 percent higher than those not left for reaching equilibrium. A statistically controlled experiment was performed to investigate how well an operator can reproduce the Atterberg limits and to what an extent his experience affects the test results (Liu and Thonburn, 1964). The conclusion was that these tests can be regarded as reproducible from the engineering standpoint. The amount of experience an operator has, was not found to cause variations in experimental determinations (Olson, 1921).

Cone penetrometer method: During the same period when Atterberg developed the method to determine plasticity characteristics of soils, a fall cone test was developed to determine the consistency of soils by the adaption of Swedish Brinell hardness test. Decision to use free falling cone, instead of pressing a cylinder or ball, stemmed from the geometrical similarities of impression being independent of the depth and elimination of difficult decision about rate of loading. A metal cone of certain weight and apex angle is suspended over a horizontally leveled surface with the point barely touching the surface. Then the cone is released for penetration under its own weight. The depth of penetration is related to shear strength. Since there is no fall of the cone through a certain height before it penetrates into the soil, this method can very well be regarded as cone penetrometer method rather than fall cone method. From the consideration of forces mobilized along the cone, the weight of cone, W depth of penetration, d and undrained shear strength, C_u can be functionally related in the form (Hansbo, 1957)

$$C_u = k\frac{W}{d^2} \tag{2.44}$$

where k is the proportionality constant for a given cone angle and rate of shear.

In the cone penetrometer method standardized in BS - 1377 (1975), a cone of apex angle of 30° ± 1° fixed to a vertically sliding shaft is positioned with its tip just touching the surface of the clay. Liquid limit is defined as the water content of the clay at which the cone penetrates 20 mm in 5 ± 0.5 sec. Figure 2.49 illustrates the details of the cone penetrometer.

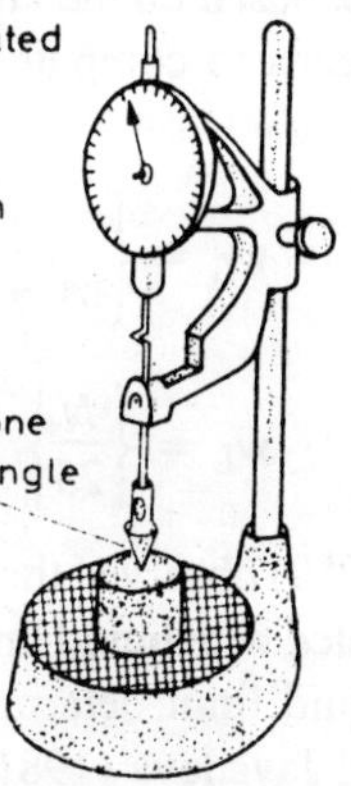

Fig. 2.49 Cone penetrometer details and method to determine liquid limit.

Wroth and Wood (1978) have provided adequate supporting evidence in favour of cone method, as a liquid limit measuring device, instead of the more elaborate percussion device. One of the recent significant findings (Nagaraj et al., 1987), in favour of the cone method, is the possibility of computing the liquid limit water contents almost instantaneously without necessarily waiting for conventional oven drying for water content determination. This is due to the obvious fact that amount of soil paste transferred to the cup is a function of both water content and the liquid limit of the soil. Hence, instead of water content determination, the bulk density of the soil can be determined just prior to the cone penetration. From the bulk density versus penetration plots, the bulk density corresponding to 20 mm penetration can be read. Detailed experimental study has indicated that a degree of saturation of 98 per cent can conveniently be assumed. The errors associated with the assumption of specific gravity of soil in the range of 2.65 to 2.80 are well within the acceptable levels of accuracy of determination. Further, this approach also permits one to independently check the values obtained by conventional water content determinations.

The values of liquid limit water contents determined from percussion and cone methods are not identical for the entire range of liquid limit water contents more particularly in the higher ranges.

One point method: Liquid limit tests do not directly provide water contents at liquid limits. Flow lines are obtained between water content on arithmetic scale and number of drops on the logarithmic scale. Functionally,

$$w = -F \log N + c \qquad 2.45$$

where w is the percent water content
F is flow index being the slope of flow line
N is the number of drops
C is a constant being the intercept at unit drop.

Two sets of formulae, based on the algebraic expression for the characteristic flow lines, provide a means to compute w_L from one point data by percussion method.

$$w_L = \left\{\frac{1}{(A - B \log N)}\right\} \qquad 2.46$$

$$w_L = \left\{\frac{N}{25}\right\}^{\alpha} \qquad 2.47$$

Mohan and Goyal (1959) observed that for values of α in the range from 0.068 to 0.121, the error associated with liquid limit water content is less than one percent. A scientific base for liquid limit determination by one point method has been provided by Nagaraj and Jayadeva (1981). It is derived from the uniqueness of generalized flow lines, which is obtained on the basis of the physico-chemical interactions. Generalised flow line is obtained by normalization of water contents by liquid limit values (Fig. 2.50). This nullifies the effects of specific surface (m^2/gm) and associated physico-chemical factors which otherwise spatially spread the flow lines in the plot of water content versus number of drops. From the data of Sherwood and Ryley (1970) the following two relationships to determine the water contents at liquid limit from cone penetrometer method have been developed (Nagaraj and Jayadeva, 1981)

$$\frac{w_L}{w} = \left\{\frac{1}{0.77 \log D}\right\} \qquad 2.48$$

$$\frac{w_L}{w} = \left\{\frac{1}{0.65 + 0.017\, D}\right\} \qquad 2.49$$

where D is the depth of cone penetration in mm.

The above functional relationships result from normalization of the flow lines of soils in w versus D or w versus $\log D$ space by their respective liquid limit water contents.

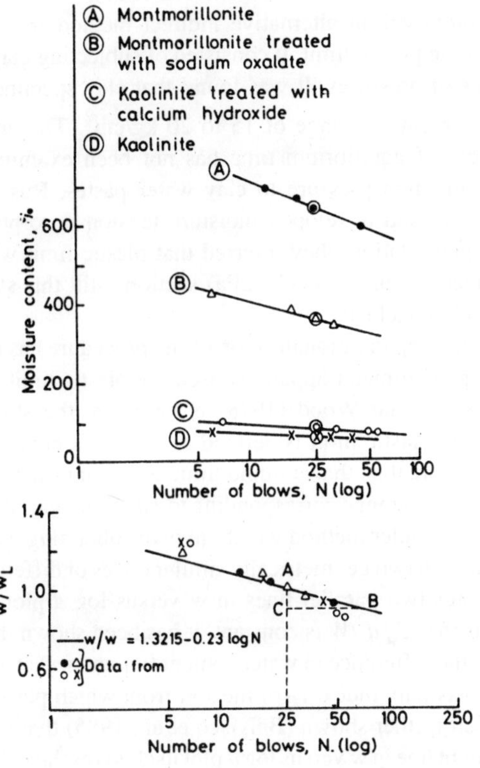

Fig. 2.50 Flow curves and their normalization (after Nagaraj and Jayadeva, 1981).

Plastic limit

Basically, plastic limit of fine grained soils is a reflection of both strength and deformability. As such, most of the methods of determination involve shaping wet clay of different consistencies into balls and threads, and subjecting them to different types of manipulation. The most commonly used method is the Casagrande method, in which wet clay is rolled into threads and the limiting water content is determined at which the threads begin to brake just when rolled to 1/8 inch (3 mm) diameter. According to Kenney (1963) this test is a complicated version of splitting test in which during rolling the soil specimen is subjected to rotational stresses while the diameter is reduced. The test conditions are more strain controlled than stress controlled. Another method, the cube method (Abdun-Nur, 1959) consists of moulding the wet soil into an approximate 3/4 inch (19 mm) size cube and pressing the opposite faces between spatula on nonabsorptive surface till it develops cracks on being formed. The moisture content at this stage is plastic limit. The determination of plastic limit by the above methods is subjected to personal impression and can very much cause considerable discrepancies even though it is possible to eliminate human factor in reproducibility of test results (Ballard and Weeks, 1963; Nuyens and Kockaerts, 1967).

Ekamn (1959) proposed an alternative indirect method free from personal impression to determine plastic limit. It consisted of subjecting clay specimens to different magnitudes of pressures. It was found that clay specimens reach their plastic limit water contents in range of 15 to 20 kg/cm^2. The uncertainty with respect to assessment of equilibrium time has not been examined. Instead of application of consolidation pressure to clay water pastes, Russel and Mickle (1970), applied suction and developed moisture tension desorption curves for different soils. By interpolation, they inferred that plastic limit water content is reached at approximately 162 psi (1112 kPa) suction. Still this study cannot be regarded as completely conclusive.

Another significant step in rationalisation of the procedure has been in the use of liquid limit cone penetrometer apparatus itself for plastic limit determination. The approach by Wroth and Wood (1978) is based on the strength criterion according to which the plastic limit is defined as the water content at which the strength is 100 times more than that at liquid limit. Since, it is difficult to have the device to give a penetration range corresponding to 100 times variation in strength, they have suggested a simpler method which involves obtaining water content, w versus penetration, d with two geometrically similar cones of different weights W_1 and W_2. This generates two parallel lines in w versus log d plot. Based on the dimensional analysis that $C_u\, d^2/W$ is constant, it has been shown that the spacing between the lines i.e the difference in water content (w_1 and w_2) at any penetration is a function of the plasticity index, I_P of the soil from which plastic limit can be computed. It has been further shown (Belvisco et al., 1985) that the slope of the flow line i.e, the straight line in w versus log d plot itself gives the value of plasticity index. Thus the same test enables to determine both liquid limit and plastic limits.

Derived parameters

Although in engineering practice, the consistency limits are often used directly in property correlations, a few other derived parameters have gained acceptance in the formulation of correlations.

Plasticity Index : This is defined as the difference between liquid and plastic limit water contents expressed as a percentage.

$$I_P = w_L - w_P \qquad 2.50$$

It is the range of water content over which the soil remains plastic. In soils containing negligible clay fraction it's value is relatively low, whereas in the case of highly active clay minerals, it's value may even exceed 500. Plasticity index, apart from being used in engineering property correlations, has been extensively used in the classification of fine grained soils.

Relative Consistency and Liquidity Index: Analogous to the relative density of cohesionless soils, the state of fine grained soil was identified by relative consistency of the soil, C_R . This is the ratio of the difference between the liquid limit of the soil and the natural water content to the plasticity index and is expressed as

$$C_R = \frac{w_L - w}{w_L - w_P} \qquad 2.51$$

This parameter was found to be deficient to the extent that in-situ water content more than liquid limit water content could not be accounted for.

The modified form of expression (eq. 2.51) is liquidity index, I_L, which is the ratio of difference between natural water content and plastic limit to the plasticity index, and is expressed as

$$I_L = \frac{w - w_P}{w_L - w_P} \qquad 2.52$$

The value of liquidity index reflects the stress history of in-situ soil. The normally consolidated clays at shallow depths have values closer to unity, while the value decreases and approaches zero for heavily overconsolidated clays. The liquidity index values can be far greater than unity for highly sensitive clays or can be negative for in-situ weathered stiff clays.

Activity Number, A : The plasticity of the soil depends on the type of clay mineral as well as on the amount of clay present. Soil samples taken over an identified geographical area are likely to have a common geological history. Although, there could be variations in the clay fraction in different batches of samples, the nature of the clay mineral is not likely to show much variability.

Skempton (1953) observed that a plot of I_P versus percent clay content, C for soil samples from a given site, with a particular geological formation generates into a straight line passing through the origin. Such lines were distinctly different for sites with different geological formations (Fig. 2.51). Skempton defined activity number, A of soil as the slope of that line:

$$A = \frac{I_P}{C} \qquad 2.53$$

For soils from the same geological origin, with increase in clay content both plasticity index, I_P and percentage finer than 2 μm increase maintaining a constant value of activity number. Presently it has become a practice, to indicate activity number, A along with other consistency limits in the soil investigation reports.

2.11 Empirical Methods to Predict Soil Behaviour

Geotechnical engineers have always attempted to evolve simple methods to predict soil behaviour with minimum input of parameters that are commonly determined in routine investigations. Empirical approaches that have been developed from time to time are presently satisfying this need. Although numerous empirical correlations between consistency limits or their derived parameters and engineering properties are in vogue, only most commonly used methods are cited in the following paragraphs.

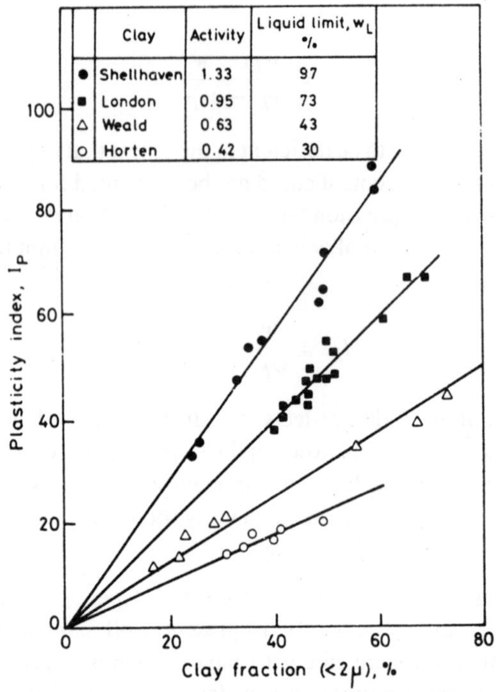

Fig. 2.51 Plastictiy index versus clay fraction (after Skempton, 1953).

Compressibility

A direct method for assessing compressibility is to run laboratory consolidation tests on undisturbed soil samples. Great care and time are required to obtain reliable data where a larger number of samples are to be tested. This also involves considerable cost. Hence there have been several attempts in literature to derive simple empirical relations to correlate compression index with inferential and /or state parameters based on experimental results of iocal soils. Table 2.2 lists a few commonly used relations.

The relations in Table 2.2 can in general be grouped into two categories

(a) in which C_c is related to the liquid limit and
(b) in which C_c is linked with natural water content or in-situ void ratio of soil.

The critical examination of these relationships for their credibility and applicability will be presented in chapter 5.

Shear strength

Empirical relations which are most commonly used for strength predictions are :

(i) c_u/p *vs* I_P relationship (Skempton and Henkel, 1953 and Skempton, 1954)

Table 2.2 Compression Index equations

Equation	Reference	Regions/Conditions of Applicability
$C_c = 0.007\ (w_L - 10)$	Skempton (1944)	Remoulded clays
$C_c = 0.009\ (w_L - 10)$	Terazaghi & Peck (1967)	Normally consolidated, Moderately sensitive
$C_c = 0.01\ w_n$	Kopulla (1981)	Chicago & Alberta clays
$C_c = 0.0115\ w_n$	Bowles (1979)	Organic silt & clays
$C_c = 1.15\ (e - e_o)$	Nishida (1956)	All clays
$C_c = 1.15\ (e - 0.35)$	Nishida (1956)	All clays
$C_c = 0.54\ (e_n - 0.35)$	Nishida (1956)	Natural soils
$C_c = 0.75\ (e_o - 0.50)$	Bowles (1979)	Soils with low plasticity
$C_c = 0.0046\ (w_L - 9)$	Bowles (1979)	Brazillian clays
$C_c = 1.21 + 1.055\ (e_o - 1.87)$	Bowles (1979)	Motley clays from Saopaulo city
$C_c = 0.30\ (e_o - 0.27)$	Hough (1957)	Inorganic silty sandy-silty clay
$C_c = 0.208\ (e_o - 0.0083)$	Bowles (1979)	Chicago clays
$C_c = 0.156\ e_o + 0.0107$	Bowles (1979)	All clays
$C_c = 0.5\ (\gamma_w/\gamma_d^2)^{1.2}$	Oswald (1980)	Soil systems of all complexities and types

Note: C_c – compression index
w_L – liquid limit water content
γ_d – dry density of soil at which C_c is required
e_o – initial or insitu void ratio
w_n – natural water content
γ_w – unit strength of water
S_t – sensitivity of the clay

$$\frac{c_u}{p} = 0.11 + 0.0037\ I_P \qquad 2.54$$

where c_u is the undrained strength
I_P is the plasticity index $(w_L - w_P)$
p is the overburden pressure

(ii) Liquidity index versus remoulded strength : Based on the observation of a definite trend between Liquidity index and remoulded strength of several clays (Skempton and Northey, 1953; Schofield and Wroth, 1968 and Houston and Mitchell, 1969). Wroth and Wood (1979) in their attempt to correlate engineering properties with index properties from critical state concepts, have linearised this relation. However, they did not visualise any scientific basis for such a relation.

The validity of these relations from micro mechanistic considerations, has been examined later in chapter 5.

As already stated, the empirical relations have been developed to predict soil properties, using simple index properties. Why these relations hold good or is their any scientific basis underlying such correlations merits examination. The credibility of such relations and the confidence with which we can use them gets greatly enhanced if we can find the scientific basis for these relations. Either to examine the rationale for the existing empirical relations or alternately to develop new theoretical relations based on established scientific principles, it is essential to have an understanding of the nature of forces operating between different constituents and phases at micro level especially for a multicomponent polyphase system like soil at microlevel. Next chapter presents a brief discussion of the force fields between and arrangement of particles at microlevel which may lead to the possibility of generalising and predicting the behaviour at macrolevel, which is the aim of this book.

3

FUNDAMENTALS OF SOIL BEHAVIOUR

3.1 Introduction

The geotechnical engineer is primarily interested in the strength and deformation behaviour of soils under imposed boundary and loading conditions. These behaviour are influenced by the various forces at work during formation of soils, in addition to those of the present state. Soils are formed from their parent material 'rock' through the weathering processes and the morphological cycle. The effect of chemical processes are reflected with a different crystalline structure of the soil particles than that of the parent material. The chemical processes of differential solution, leaching and deposition of chemicals are highly sensitive to environmental conditions. This induces a multitude of variabilities to the soil. The soil in nature can be described as a multicomponent and polyphase system in a dynamic equilibrium, which is continuously changing due to external loading, deformational constraints and environmental factors. The vagaries of soil in nature are such, the complexities of phase interactions are so great and the factors affecting the soil behaviour are so many, that even for a simple geotechnical engineering problem it may not be possible to completely characterize the soil behaviour by mere testing in either laboratory or field. Even now (1994), our analytical capabilities have far surpassed our understanding of the soil behaviour. Perhaps an understanding of the fundamentals of soil behaviour may reduce this gap.

3.2 A Mechanistic Approach

As typical to other engineering fields of study, the fundamental behaviour of soils can also be studied from the mechanics point of view. The discrete solid particles of soils are not strongly bonded as the crystal structure of solids, and are free to have relative movements which are themselves constrained due to internal friction and are not as free as the molecules of a fluid. Hence the responses of soils cannot be fully characterized either by solid mechanics or by fluid mechanics.

Feda (1982) has described the mechanics of particulate materials as the appropriate mechanics for characterizing the responses of soils. The particulate materials are those which exhibit dilatancy and contractancy and are sensitive to hydrostatic stresses. They are composed of mutually contacting solid particles or structural units dispersed in liquid and/or gaseous phases. The structural unit is that part of the skeleton which moves as a single unit during a deformation process.

3.2.1 Particulate Materials

The following are the specific features of particulate materials.

(1) Existence of mutual contact between the particles/ units of a particulate system, restrains the freedom of movement of individual particles, due to the frictional forces mobilized at these contacts. This results in an ability of the particulate system to be in an anisotropic state of stress.

(2) The deformations in particulate systems are brought out by the relative sliding and rotation of the particles (intergranular deformations) and not so much by the deformations of the individual units themselves (intragranular deformations). The deformations of particulate materials are essentially 'plastic' due to which these materials are often termed as 'dissipative systems'.

(3) The particulate systems undergo volume changes (either dilation and contraction) under the application of deviatoric stress. Their mechanical behaviour are greatly influenced by the mean principal stress.

(4) The particulate materials generally consist of three phases, viz., solid, liquid and gaseous. The mechanical behaviour and their associated parameters of each phase are greatly different from those of the other two. The mechanical behaviour of such a system can only be characterized by resolving the total state of stress into the states of stress of individual phases. This is the well known effective stress principle, applicable to particulate materials. According to this, the effective stress which has its seat in the solid phase, is the difference between the total stress and the isotropic neutral stress of the fluid phases. The mechanical behaviour of the particulate systems are only the functions of this effective stress. For characterizing the behaviour, i.e., for description of complete stress-state relationship, the relationship between the phases, describing the state of the particulate materials, must be defined exactly and its variability with the load must be fully stated. Wheather the above mechanisms proposed for particulate materials can be applied to describe the behaviour of soil systems needs a detailed examination.

For this, from mechanisms point of view, an understanding of the forces at play and their origin are required. A detailed understanding of the nature of solid and liquid phases and the mutual interactions at work is a pre-requisite for the same.

3.3 Nature of Soil Solid Particles

Soils, in nature, consist of particles of all sizes, ranging from colloids (less than 1μm) to boulders and rock masses (several hundred mms). Based on particle size the soil constituents may be grouped as indicated in Table 3.1.

Table 3.1 Range of particle size of soil solid constituents

Group name	*Particle size in* mms
Gravel	greater than 4.75
Sand	0.06 to 4.75
Silt	0.002 to 0.06
Clays	less than 0.002

Generally particles coarser than 0.002 mm are bulky and isometric in shape and exist as individual stable units. This also represents the lower limit of mechanical degradation. The common minerals constituting these particles are silicates (feldspars), oxides (silica and iron), carbonates (calcium and magnesium) and sulphates (calcium).

The clay particles are platy, rod and lath shaped and have specific mineralogical composition defined by the type of clay minerals. Clay minerals are hydrated aluminum silicates and hydrous oxides of aluminum, magnesium and iron, in a crystalline form of relatively complicated structure. The individual crystals look like tiny flakes or plates which consist of many crystal sheets having a repeating unit structure. In fact there are only two fundamental crystal sheets, the tetrahedral or silica, and the octahedral or alumina sheets (Figs. 3.1 and 3.2). The particular

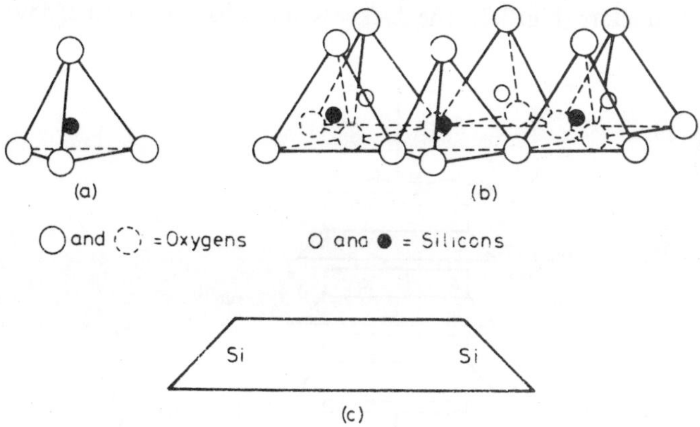

Fig. 3.1 (a) Single silica tetrahedron (b) Isometric diagram of silica sheet (c) Schematic representation of silica sheet.

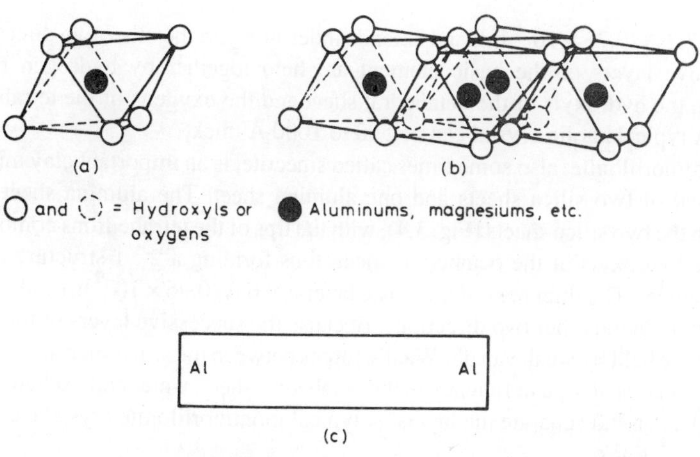

Fig. 3.2 (a) Single octahedral unit (b) Isometric view of octahedral sheet (c) Schematic representation of alumina octhedral sheet.

way in which these sheets are stacked, together with different bonding and different metallic ions in the crystal lattice, constitute the different clay minerals. The clay minerals are divided into three general groups according to their crystalline arrangement, and it is found that roughly similar engineering properties are associated with all the clay minerals belonging to the same group. The three groups are:

(1) Kaolins
(2) Montmorillonites
(3) Illites.

Kaolins consist basically of repeating layer of one tetrahedral sheet and one octahedral sheet. The two sheets are held together in such a way that the oxygen tips of the silica sheet and one of the hydroxyl layers of alumina sheet form a single layer of 1:1 structure (Fig. 3.3). The thickness of this basic element of a two sheet

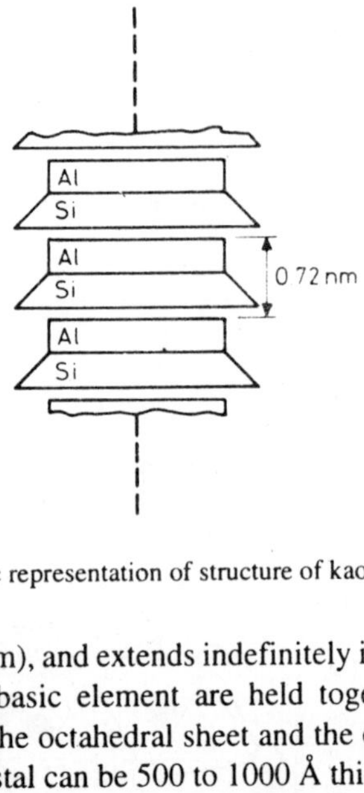

Fig. 3.3 Schematic representation of structure of kaolinite clay mineral.

layer is 7.2 Å (0.72×10^{-9} m), and extends indefinitely in the other two directions. Successive layers of the basic element are held together by hydrogen bonds between the hydroxyls of the octahedral sheet and the oxygens of the tetrahedral sheet. A typical kaolin crystal can be 500 to 1000 Å thick.

Montmorillonite, also some times called smectite, is an important clay mineral composed of two silica sheets and one alumina sheet. The alumina sheet is in between the two silica sheets (Fig. 3.4), with the tips of the tetrahedrons combining with the hydroxyls of the octahedral sheet, thus forming a 2 : 1 structure of the basic element. The thickness of this basic layer is 9.6 Å (0.96×10^{-9} m) and extend indefinitely in the other two directions. Because the successive layers of the basic element are held by weak van der Waal's forces between the silica sheets, and there is a net negative charge deficiency in the octahedral sheet, water and exchangeable ions can enter and separate the layers. A typical montmorillonite crystal can be of 10 to 100 Å thick.

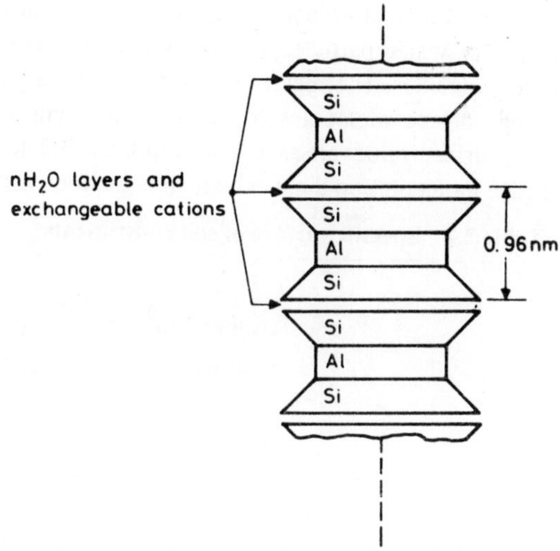

Fig. 3.4 Schematic representation of structure of montmorillonite clay mineral.

Illite is also a mineral of 2 : 1 structure, the basic element layers are bonded together by a strong ionic bond through a potassium ion (Fig. 3.5). A typical illite crystal can be of 100 to 300 Å thick.

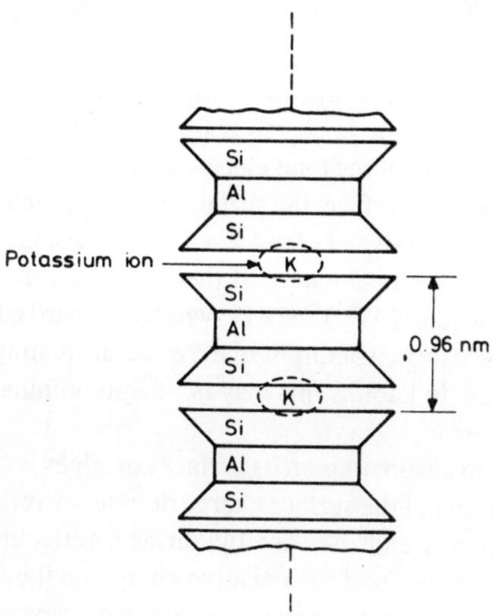

Fig. 3.5 Schematic representation of structure of illite clay mineral.

The direct implication of variation of particle size is their variation in the specific surface which has a direct bearing on the mechanical behaviour. The specific surface is the ratio of the surface area of the particle to its weight or volume.

In silt, sand and larger particles the specific surface is relatively small. Because of this, the surface molecules constitute only a fraction of the total molecules in a particle. In contrast, in clay size particles, a large portion of the total molecules constitute the surface molecules. The specific surface of soil is a good measure of the proportion of the surface molecules to the total molecules in a particle. Typically, the specific surface values are as indicated in Table 3.2. In a latter section the effect of the surface molecules on the mechanical behaviour will be discussed.

Table 3.2 Specific surface of soil constituents

Sands	less than 0.01 m^2/gm
Silts	less than 1 m^2/gm
Kaolins	15 m^2/gm
Illites	80 m^2/gm
Montmorillonites	800 m^2/gm

Every soil particle carries an electrical charge. Although, theoretically, a soil particle can carry either a net negative or positive charge, only net negative charges have been measured. This net negative charge may arise from any one or a combination of the following five factors.

(1) Isomorphic substitution
(2) Surface dissociation of hydroxyl ions
(3) Crystal lattice defects and broken bonds
(4) Adsorption of anions
(5) Presence of organic matter.

Of these five possible causes the first two have been considered as most important for clay particles while in coarse grained soils it may be due to broken bonds. Though the magnitude of the total electrical charge in a soil of unit mass is directly related to the specific surface, the surface activity of sand and other coarse grained soils do not acquire those of clays, even if they are ground to the size of the clays because of non-chemical nature of their surfaces. In most clay mineral structures the aluminum ions (Al^{+3}) may be wholly or partly replaced by other cations of lower valency to form isomorphs of the mineral creating increased charge deficiency on the surface. In kaolins this may not be predominant due to inherent stable structure.

The ability of a clay to adsorb ions on its surfaces or edges is called its exchange capacity which is a function of the surface charge deficiency reflecting the mineral structure and broken bonds and the specific surface, reflecting the size of the particle. Mostly in soils because of net negative charge on the surface it is a case of cation exchange and hence is termed as base exchange capacity. This process is one in which the cation adsorbed on the surface of the soil particle is exchanged for a cation in the pore solution. The base exchange capacity is defined as milliequivalents per 100 gm of dry soil. One milliequivalent is one milligram of hydrogen or the amount of any ion which will combine with or displace one milligram of hydrogen. The typical values of base exchange capacity in meq/100gm of dry soil are 3.8, 40 and 80 for kaolinite, illite and montmorillonite respectively.

Thus the nature of solid particle is not unique. It varies with its size and shape, mineralogical and chemical compositions.

3.4 Nature of Water

In water molecules the centres of positive and negative charges do not coincide and hence they behave like dipoles (Fig. 3.6). The distribution of electrical charges in a water molecule resembles a tetrahedron, with two vertices charged positively

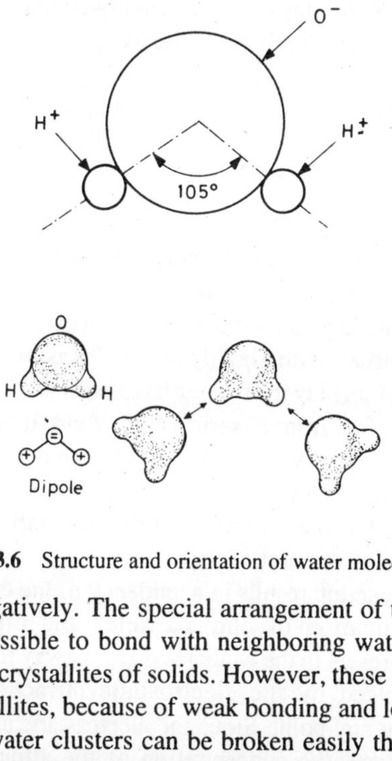

Fig. 3.6 Structure and orientation of water molecules.

and the other two negatively. The special arrangement of the charges in a water molecule makes it possible to bond with neighboring water molecules, to form clusters similar to the crystallites of solids. However, these clusters are not stable, unlike the solid crystallites, because of weak bonding and local thermal effects.

The structure of water clusters can be broken easily through an electrostatic potential of ions or through an electrically charged surface like soil particle. This makes the water molecules to have a relatively immobile structure of oriented dipoles in the immediate vicinity of a charged surface. The original structure of the water crystallite is unaffected only beyond the distance at which the forces conditioning the origin of clusters out balance the influence of electrostatic forces. Thus the structure of water molecule varies from that of a rigid oriented dipole at the charged surface, a distributed and random structure with in the electrostatic field and the crystalline structure of the free water beyond the electrostatic field.

3.5 Solid - Water Interactions

Since the surface of solid particles carry electrical charges and water molecules are dipolar in nature, it is logical to expect electrostatic interactions between them. The quantum of surface charge in a unit mass defines the interaction level. In coarse grained soils, the magnitude of surface charge in a unit mass is relatively low. The presence of water in these soils, does not contribute much to the internal force field except changing their unit weights. An apparent exception to this fact is the case of water present in loose sands subjected to dynamic loadings wherein the mobilized pore pressure due to external constraints of permeability is reflected as a change in the internal force field.

Fine grained soils especially clays are strongly influenced by the presence of water because of their high surface activity. The causes of clay water interactions are:

(a) The water is held to the clay crystal by hydrogen bonding i.e., the hydrogens of the water molecule are attracted to the oxygens or hydroxyls on the surface of the particle.
(b) The water molecules are electrostatically attracted to the surface of the clay crystal.
(c) The hydrated exchangeable cations present in water are electrostatically attracted to the clay surface.

Of these three factors the first factor is the most important one. Now, it seems that clay particles are almost always hydrated in nature by a number of layers of water molecules surrounding each crystal of clay. This water of hydration which is strongly held to the surface with a highly oriented structure is called the adsorbed water. The properties of this layer of water have been shown to be different from those of 'normal water' free from electrical force field. It has been found that the thickness of adsorbed layer of water is almost the same for all clay minerals in a given environment. However, because of variation of specific surface, the volume of adsorbed layer of water in a unit mass of dry soil varies in proportion to the specific surface of the clay mineral.

The clay-water interaction results in a tendency of the counter ions, (resulting from the surface ionization of clay particle) and other dissociated ions (exchangeable cation present in the water) to diffuse away from the surface, which will be counter balanced by the electrostatic attraction. This defines the concentration of ions at any point. Since the electrostatic attraction falls off with distance from the surface, the concentration of the attracted cations will also diminish while the concentration of counter ions increases with distance from the surface (Fig. 3.7). Thus obtained distribution of ion concentration with distance resembles 'atmospheric distribution'. The charged surface and the strongly held cation at the surface together with the relatively mobile counter ions in the medium adjacent to the surface are considered to be two layers, and the whole systems is referred to as the 'diffuse double layer'. The concentration of charges is greater near the surface and falls off with distance.

The mechanism of stress transfer between solid soil particles with the above considerations will be discussed later.

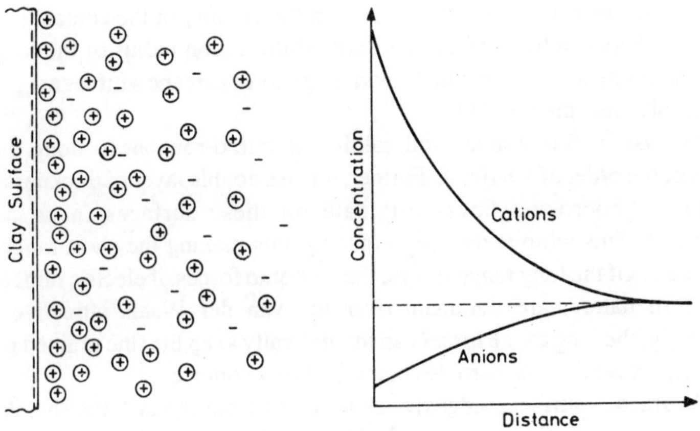

Fig. 3.7 Distribution of ions adjacent to clay particle surface.

3.6 Solid - Solid Interactions

Between two units of matter, there always exist interaction forces of both attractive and repulsive nature. These forces can be grouped as short range and long range forces. Depending on the dominance of one group, the effect of the other group may not be felt. While the long range forces can act over distances of several hundred angstroms, (1 Å = 10^{-8}cms), the action radius of short range forces are only a few angstroms.

The typical short range forces are the homeopolar, the heteropolar and the van der Waals - London attractive forces, the hydrogen bonding and the Born repulsive forces. The typical long range forces are the van der Waals - London attractive forces and coulomb electric repulsive forces.

The short range forces represent the free surface energy brought into being by the interphase boundary i.e., by the failure of crystal lattices on their surface and hence are bound to this surface. The magnitude of short range forces varies as the inverse seventh power of the distance (r^{-7}) between the two solids. The capillary forces in a three phase system which originate from the hydrophylic property of the solid and the surface tension of the water-air interphase are essentially a combination of the short range forces of different phases.

The long range forces originate in particles of colloidal dimensions. A typical example of this are clay particles in an aqueous medium which conditions the dispersion of the particles themselves. The magnitude of long range attractive forces are inversely proportional to the third power of the distance between the particles (r^{-3}) while the repulsive forces are inversely proportional to the second power of the distance (r^{-2}). Hence, of the long range forces, the repulsive forces are dominant. In the case of coarse grained soil solids mostly short range forces are significant. This is because, the microroughness of these solid surfaces attain heights of several thousands of angstroms. Then such surfaces can touch each other only through apexes of microroughness, which are called contact junctions. At these points, only short range forces are at play and the long range forces are in-significant, since at every other point than the vicinity of the contact junctions, the distance between the surfaces far exceeds the action radius of the long range forces. These interactions for which short range forces are the sources are generally termed as physical interactions.

In the case of fine grained soil solids, the micro-roughness along the basal planes is of the order of 10-100 Å. Further, diffuse double layer of thickness greater than that of micro-roughness originate on these surfaces in an aqueous environment. This inhibits the apex contacts thus making the short range forces insignificant. Of the long range forces, the coulomb forces of electric double layer, repulsive in nature, are dominant over the van der Waals attractive forces. Accordingly, the long range forces can theoretically keep the fine grained particles in parallel, preventing the particles being in direct contact.

On the fracture surfaces (edges) of clay particles, the crystal lattice breaks down exposing the ionic lattice, because of which the short-range forces may act there. Since, the layer of octahedrons is the main source of this electric charge, the density of edge charge is greater for 1 : 1 sheet structure minerals like kaolins than for 2 : 1 sheet structure minerals, like montmorillonites. Under ordinary conditions ($pH \approx 7$) the basal planes of clay particles are charged negative and the edges positive, which therefore results in possible edge to face contacts and hence flocculation. This phenomena is more for kaolins than for montmorillonites. However with increase in pH value, the clay particles carry only negative charges even on edges.

Unlike the attractive force, the repulsive force acting between particles strongly depends on the medium, which distinctly affects the thickness of the diffuse double layer. The thickness of this layer decreases with increase in concentration of the electrolyte and the valance of the exchangeable cation. When this layer is very thin, the van der Waal's - London attractive forces out balance the repulsive forces leading to flocculation known as 'salty flocculation'. In this case the particles tend to remain parallel in contrast with the edge to face flocculation. Since in the case of clays, the nature of forces is one of long range which essentially depends on the chemistry of the pore fluid, these interactions can be termed as physico-chemical.

From the above discussions, the long-range forces enable the clay particles to exist in suspensions without mutual contact, even under an external load, which only reduces the distance between the particles to increase the repulsive force between the particles and thus countering the applied load. In such systems, the concepts of particulate mechanics, which are based on effective solid-solid contacts are not applicable.

3.6.1 Physical interactions

The physical interactions between the particles mobilize resistance on the particle surface against deformation through short range forces. These resistances are termed as 'friction', and the points of contact as 'friction bonds'. Since, two particles in their mutual displacement can either slide or roll, sliding as well as rolling friction are to be considered. Generally friction is defined as solid friction between two surfaces with negligible moisture. Since the solid surfaces are generally covered with an adsorbed layer of hydrated ions, or contaminants, friction at the contact bond will be different than the solid friction.

Figure 3.8 shows the fundamental case of a sliding friction. The macroscopic view of the two clean highly polished surfaces appears as in Fig. 3.8(a). The microscopic view of the same would be jagged and very irregular as shown in Fig. 3.8(b) and the corresponding microscopic view of contact is as shown in Fig. 3.8(c).

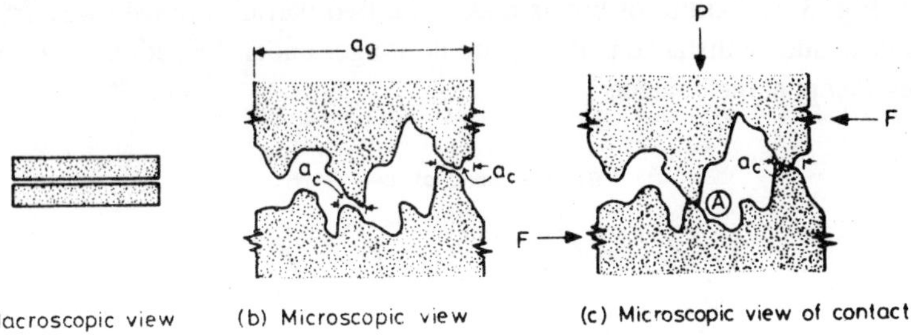

Fig. 3.8 Frictional shearing resistance due to contact junctions.

If the average normal stress on the gross geometrical contact area a_g equals σ_n, then the normal stress in the contact junctions of area a_c is equal to σ_c according to equation:

$$\sigma_c = \sigma_n \, a_g / \Sigma \, a_c \qquad 3.1$$

The effect of the mirco-roughness on the grain surface usually results in $\Sigma \, a_c << a_g$ and σ_c attains the strength of contact junctions, i.e., the contact junctions reach the plastic state.

The shear strength of two contacting surfaces is equal to

$$\tau = \tau_c \, \Sigma \, a_c / a_g \qquad 3.2$$

where τ_c is the shear strength of contact junction. Then combining equations 3.1 and 3.2

$$\tau = (\tau_c / \sigma_c) \, \sigma_n = \mu \, \sigma_n$$

where $\mu = (\tau_c / \sigma_c)$

From the above equation, the friction or the shear strength is independent of gross area a_g and is in direct proportion to the normal stress σ_n. It is found that the size of the contact area, a_c grows not only inversely proportional to σ_c but also on the shear stress τ_c at the contact junction. In soil solid particles, the growth of the contact area is also conditioned by their brittleness.

Bonds produced by internal forces continue to exist even when a sample is not loaded. They have been termed as 'cohesive bond'. There are two types in these bonds.

(1) Cement bonds or brittle bonds
(2) Adhesive bonds or ductile bonds

A number of physico-chemical processes are capable of producing contact between two particles by cementing them together. In such cases, unlike in friction bonds, the effective contact does not depend on the external stress. A contact between two particles produced in this way is essentially no different from the lattice bonds. After breakdown of brittle bonds, the actual area of contact which was equal to gross area, reduces drastically thus changing the brittle bond to friction bond (Fig. 3.9). Failure of bonds makes the two particles move away from one another to such a distance that they are no longer under the action of short range forces except at the contacts.

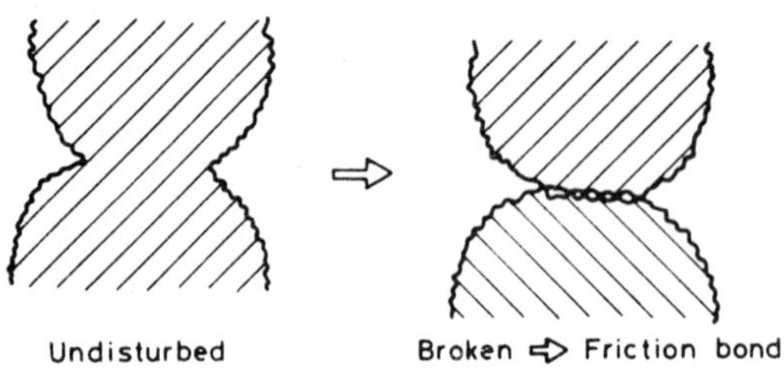

Fig. 3.9 Brittle bonds being transformed to friction bonds upon failure.

The second type of cohesive bond, viz, adhesive bond, does not cease to exist with the relative displacement of the particles even under large deformations. In this, it is similar to friction bonds which also possess the so called ductile structures. Further no contact stress is required for the existence of bonds i.e., with the removal of stress the bonds do not cease to exist. The type of interparticle bond due to either capillary forces or the attractive forces between particles are the examples of this type. Both brittle and ductile cohesive bonds are a result of the action of stronger primary bonds (homeopolar, heteropolar, covalent and hydrogen bonds) and equally of weaker secondary bonds and valance forces.

The cement bonds can be formed due to physico-chemical and environmental changes (like leaching and change in degree of saturation) or due to mechanical processes like deformations and vibrations etc. The adhesive forces are also short range forces and are reversible unlike in brittle bonds. Though generally, they exist in all particles, they are insignificant in larger particles.

In all the above types of bonds (friction, brittle and ductile), it is exclusively the short range forces which are operative. The long range forces (physico-chemical) which are predominant in fine grained soils will be discussed further.

3.6.2 Physico - chemical interactions

Clay particles can repell each other electrostatically over long range, but the process depends on the ion concentration interparticle spacing, and other factors. Similarly there can be attraction of the individual particles due to the tendency of hydrogen bonding, van der Waal's forces and other types of organic and chemical bonding. The interparticle force or potential which depends on the valance and concentration of dissolved ions, temperature and the nature of bonding forces, decreases with increase in distance from the mineral surface.

The individual clay particles interact through their adsorbed water layers and thus the presence of different ions, organic material, different concentrations etc., affect or contribute to the multitude of soil structure found in natural soil deposits.

3.7 Soil Structure and Fabric

In geotechnical engineering the structure of a soil is used to mean both the geometric arrangement of the particles or grains as well as the interparticle forces which may act between them. Soil fabric refers only to the geometrical arrangement of the solid particles. It is believed that the structure strongly governs the engineering behaviour of the soils. Perhaps no other term in geotechnical engineering, in the recent past, has attracted so much of attention as the term structure. It has been defined, quantified, and scaled at micro level and modelled with the fond hope of predicting the soil behaviour. It has helped only in improving our understanding of behavior qualitatively rather than giving parameters which can be directly used in predicting the performance (Lambe, 1973).

3.7.1 Fabric of coarse grained soils

Coarse grained soil particles (larger than 0.01mm), can settle out of a soil-fluid suspension independently of other particles. This will result in what is called as 'single grained structure' (Fig. 3.10).

(a) Loose (b) Dense

Fig. 3.10 Single grained structure of sands.

The deposition media could be water or air. The resulting structure could be loose or dense. Depending on the grain size distribution, shape and arrangement of the particles, a wide range of void ratios is possible. Even a honey comb structure (Fig. 3.11) can result under some depositional conditions. Such a structure is considered as metastable since, they can support static loads due to arching while it is sensitive and collapses when vibrated or dynamically loaded.

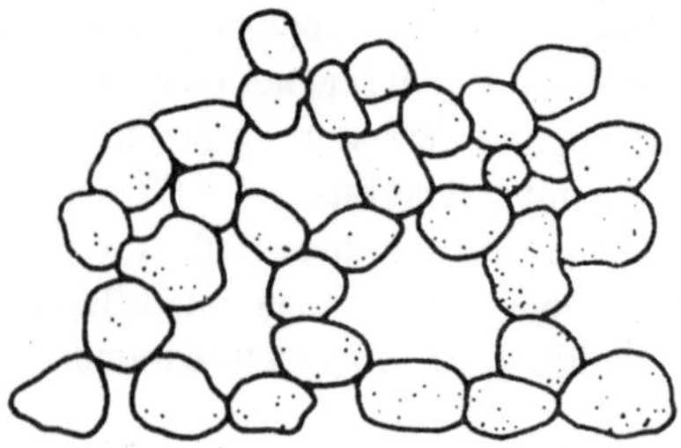

Fig. 3.11 Honeycomb structure in coarse grained soils.

The macroparameters known to define the structure of coarse grained soils are the composition, shape, size and its distribution, of the elementary particles, the relative density and porosity, the pore size and its distribution the water content and degree of saturation. The microparameters are the average number of contacts per particle and their relative frequency and the directions of the contact planes, local variation of porosity etc. When the grains are of equal size (i.e., a monodispersive material) the medium is more porous than the polydispersive medium in which smaller grains can fill the pores between the large particles.

In the case of isometric grains, the number of contacts which is also known as coordination number, is the chief structural parameter. Based on the relative frequency diagram of the number of contacts, one can estimate the local variation of porosity. For monodispersive spherical grains, 12 contacts correspond to the theoretically possible minimum porosity of 25.95% and 6 contacts to the maximum porosity of 47.64%. For sets with random arrangement of grains, for both mono- and poly-dispersive materials many authors (Smith et al., 1929; Bernal and Mason, 1960; Marsal, 1960; and others) have shown that the assymetrical relative frequency curves of a dense medium differ considerably from the symmetrical curves of a loose medium. The latter symmetrical curves are closer to the Gauss-Laplace normal distribution curve. It has also been shown that the arrangement of particles is not regular, due to which all the grains would not have the same number of contacts. The importance of another parameter viz., dip of the contact plane has also been brought out. The conclusion drawn from these studies clearly indicate that the fabric of the granular materials has a distinct geometrical anisotropy which is due to a directional load. This condition manifests itself because, the majority of contact planes orient perpendicular to the direction of loading. The structure of coarse grained soils thus assumes its most stable state for which the dissipation of strain energy is maximum and for which the contact forces are normal to the contact planes. A consequence of this geometrical anisotropy is the mechanical anisotropy.

3.7.2 Structure of fine grained soils

When the size of the structural units (particles) is so small that the physico-chemcial interactions begin to predominate over the weight of the particles, there is no difference in the behaviour of monodispersive and polydispersive materials.

Net force between particles

As brought out in section (3.6.2), there exists both attractive and repulsive forces between colloidal clay particles in dilute aqueous suspensions containing some dissolved electrolyte. The net force at a given separation distance is the algebraic sum of the repulsive and attractive forces acting at this distance. Since, the van der Waal's - London (attraction) force is insensitive to the properties of the separating medium, (Fig. 3.12b) while the repulsive forces are sensitive (Fig. 3.12a) it follows that, the net force of interaction between clay platelets can be varied directly by

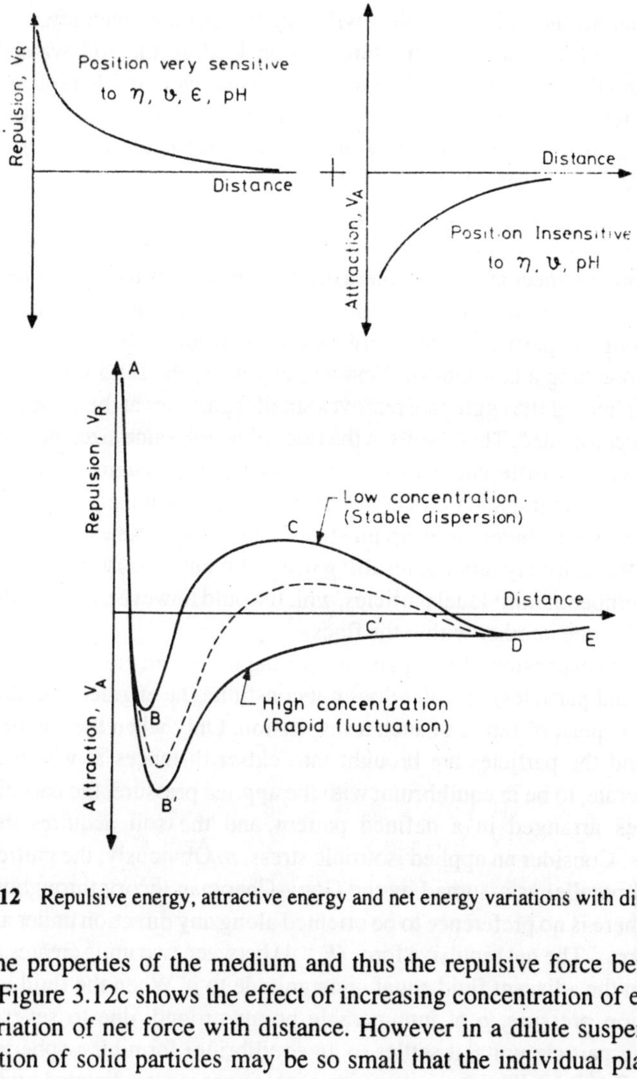

Fig. 3.12 Repulsive energy, attractive energy and net energy variations with distance.

varying the properties of the medium and thus the repulsive force between the platelets. Figure 3.12c shows the effect of increasing concentration of electrolyte on the variation of net force with distance. However in a dilute suspension, the concentration of solid particles may be so small that the individual platelets are separated by distances far greater than the range over which the repulsive or attractive forces operate. Hence the effect of these forces on stability of the suspension is not apparent.

Clay particles of colloidal size in an electrolyte suspension are subjected to chance impacts by water molecules which cause the clay platelets to move randomly (Brownian movements). Such random movements (together with other grosser fluctuations in water currents) will from time to time bring clay particles together, to distances within the range of interparticle forces. Then the subsequent behavior of the particles will depend on the net force between them at the separating distance to which the chance movement has brought them.

If the net force is repulsive, the particles will be kept separate and further random movements will separate them even more. Supposing that all particles consist of the same clay mineral and that the suspension is uniform, one may then assume that all individual platelets will remain separate, even after a long time. This results in a 'dispersed structure'. Upon loading the soil water content is reduced and the external load is resisted by the net interparticle repulsive force.

If the net force is attractive due to higher concentrations, the chance approach of particles in the suspension may bring them further closer. With increase in attraction as the distance diminishes, the chance encounter will lead to coagulation of the two particles. The further behaviour is similar to that of a larger particle. If the particles in the original suspension were of different sizes, the larger particles would probably meet and coagulate with smaller ones as they settle more rapidly through the suspension. Since, each smaller particle encountered would be attracted to the group of particles, the entire assembly would fall at increasing speed, probably reaching a constant size and velocity when the fluid shearing resistance at the periphery of the aggregate removes smaller particles at the same rate at which they are accumulated. This results in the flocs of nearly same size, though the initial particles were of different sizes (i.e., poly dispersive system). The coagulation, flocculation or aggregation process would proceed simultaneously at many points in the suspension. Under these circumstances, the clay particles, in groups or flocs, would settle relatively rapidly, resulting in a sediment of loosely knit assemblages of large number of individual particles, which would however, be virtually in direct contact with one another within the flocs.

A dilute suspension of soil particles (or the basic units i.e., flocs and may not be individual particles) in water does not constitute the engineering material soil. We cannot speak of fabric of such a suspension. Only when the smallest stress is applied and the particles are brought into closer distances at which interaction forces operate, to be in equilibrium with the applied pressure, the particle units get themselves arranged in a defined pattern and the soil acquires engineering properties. Consider an applied isotropic stress, p. Obviously, the particles cannot be placed parallel as assumed by the Gouy-Chapman theory throughout the soil, because there is no preference to be oriented along any direction under an isotropic state of stress. The net repulsive force $(R-A)$ between two units creates an osmotic suction on the adjacent fluid equal in magnitude to p. When the fluid is subjected to a suction pressure, p at innumerable points around, due to several pairs of interacting particle units, it settles to an equilibrium forming a spherical pore as shown in Fig. 3.13. The maximum size of this pore is also dictated by the applied suction from surface tension considerations by the equation $p = 2\,T/R$. A very clear evidence for existence of this sort of a structure is the pore size distribution data

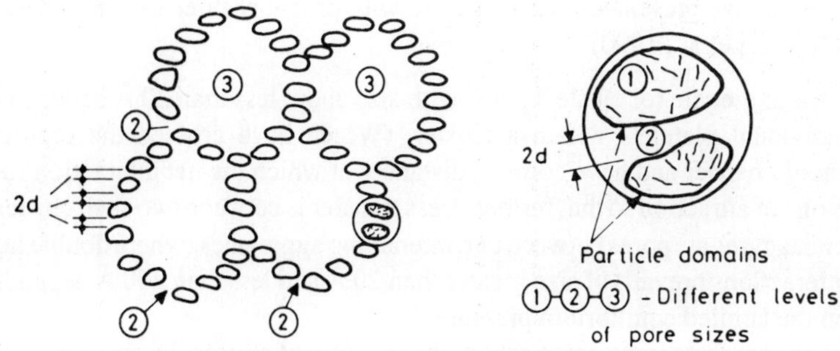

Fig. 3.13 Possible microstructure in fine grained soils.

of several soils (Griffths and Joshi, 1989) using mercury porosimeter. The data shows (Fig. 3.14) that there are pores as large as 10000 Å. It is known that the interparticle forces *R* and *A* cease to exist beyond a distance of about 300 Å and from the pore size distribution data, the volume of pores less than 300 Å is only 5 to 8% of the total void volume, while the rest of the volume, which is more than 90%, is due to pores of larger sizes greater than 300 Å.

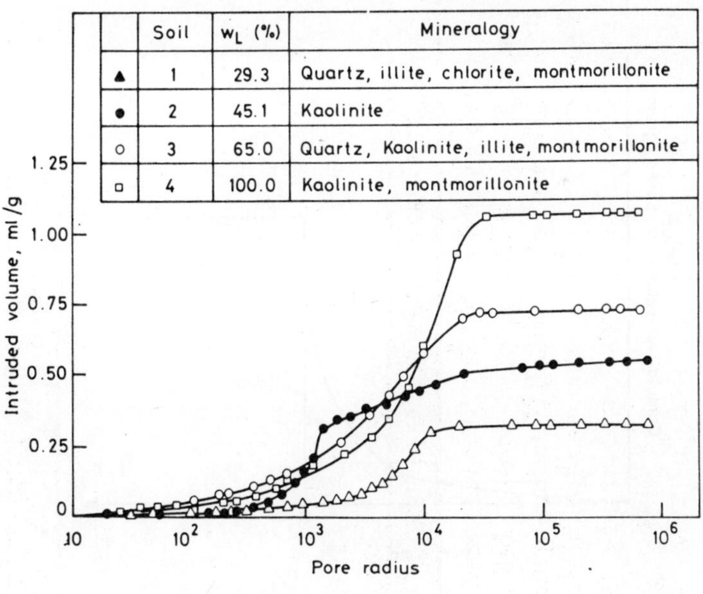

Fig. 3.14 Pore size distribution patterns of clays corresponding to their liquid limit state (after Griffiths and Joshi, 1989).

The existence of pores of size greater than 10000 Å together with double layer interactions between particles being restricted to less than 300 Å, can be explained only by the above presented microstructure with probably three distinct levels of pores. (Nagaraj et al., 1990)

(a) Intra-aggregate (or cluster) pores with size about less than 20 Å between the individual platelets within a cluster. (Which have crossed the repulsive energy barrier and have come to distances at which the net interaction force is one of attraction so that further stress transfer is between two such clusters).
(b) Inter aggregate pores between two interacting aggregates (where double layer interactions prevail) of size greater than 20 Å and less than 200 Å depending on the applied equilibrium pressure.
(c) Inter aggregate large pores held within a group of clusters by surface tension with sizes far greater than 200 Å.

The equilibrium size R of the water pores computed from the usual surface tension equation $p = 2T\cos\theta/R$, for the probable pressure at liquid limit state which is known to be about 6 kPa, works out to be about 50000 Å which is a little higher than the observed values (15000 – 20000 Å). The deviation in the theoretical and measured values may be due to several causes, like, that the mercury porosimeter records the size of the pore entry and not that of the actual pore, the properties of the adsorbed water (T and θ) may be different from that of free water or the p vs R relation may not be completely linear in the range of such small radii.

With these three distinct levels of pores, the ideal pore size distribution curve of a soil should be as shown in the Fig. 3.15. The actual distribution, with all the

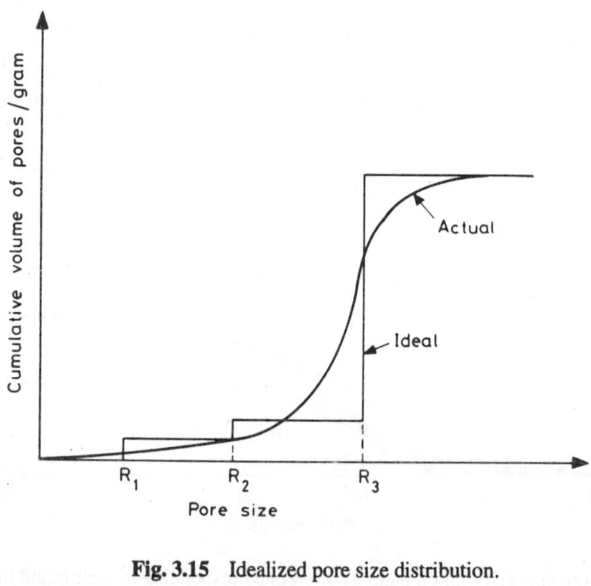

Fig. 3.15 Idealized pore size distribution.

complexities in the natural soils, show nearly the same trend (Fig. 3.16). In this plot, the data for the four soils in Fig. 3.14 is replotted in terms of intruded pore volume per unit volume of the soil instead of per unit weight of dry soil, so that

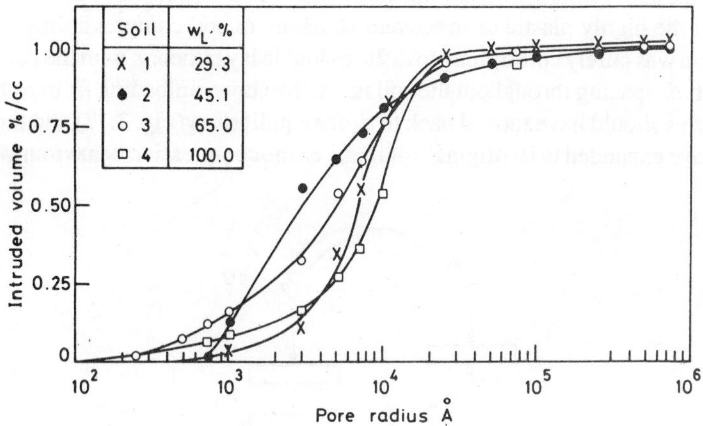

Fig. 3.16 Normalized intruded volume versus pore size.

the total pore volumes for different soils are normalized for comparison. It can be seen that the relative distribution of pores is very much similar for all the soils.

With the above picture of microstructure, it may now be possible to qualitatively explain the mechanical behaviour of finegrained soils under different imposed loading conditions. With increase in load from p_1 to p_2, the particle units come to a closer spacing from d_1 to d_2, so that the net internal repulsive stress is increased from $(R - A)_1$ to $(R - A)_2$ to balance the applied pressure, i.e., $p_2 = (R - A)_2$. Under this increased osmotic suction, there has to be a corresponding reduction in the size of the enclosed water pores from R_1 to R_2 for equilibrium of the water pores. The reduction in the '*d*' spacing between solid units may not be sufficient to accommodate the reduction in the pore size and hence the total perimeter along which the particles are spaced. This may force some of the particle units to coalesce along their basal planes and form larger clusters. A reduction in size from R_1 to R_2 of all the large pores should result in a corresponding volumetric compression. For e.g., if the increase in pressure is such that $p_2 = 2p_1$, the changed radius R_2 will be equal to $R_1/2$ and the total volume of large pores V_2 (which is 95% of the total void volume) should be 1/8 of V_1. Such high magnitudes of volumetric compressions are not observed in reality. This may be because, the grouping of particle units against the repulsive energy barrier requires considerable input of energy. Hence with the available limited energy the soil may equilibrate with larger number of pores of the specified size R_2 so that the reduction in perimeter is manageable with the possible extent of grouping. In any case, the maximum size of the pores will be governed by the osmotic suction, though the volume change is a complex phenomenon involving change in the number of pores and hence cannot be theoretically defined. The pore size distribution of all soils do indicate a systematic reduction in the pore size with increase in pressure (Griffiths and Joshi, 1989).

With the above microstructure and mechanism of stress transfer, we can explain very well the highly plastic or irrecoverable nature of soil compressibility. If the interaction was purely conforming to diffuse-double layer theory, with the particles spaced at 'd' spacing throughout the soil mass, then upon unloading from p_2 to p_1, the particles should have moved back to d_1 for equilibrium (Fig. 3.17) and the soil should have expanded to its original volume, i.e., the compression behaviour would

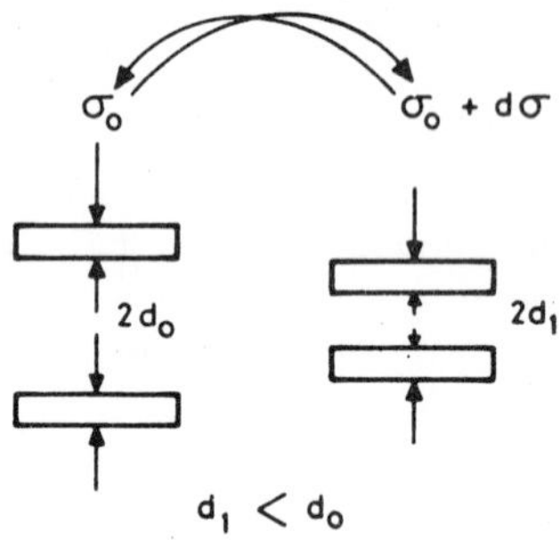

Fig. 3.17 Loading and unloading processes at microlevel for ideal condition (after Srinivasa Murthy et al., 1991).

have been purely elastic. But if in the process of loading, there is a grouping of particles as shown in Fig. 3.18, in addition to reduction in 'd' spacing from d_1 to d_2, the volumetric compression due to such grouping may not be recovered upon unloading back to p_1. This is because, once the particles have come to the close range of attractive force field, they can not be separated by mere unloading but

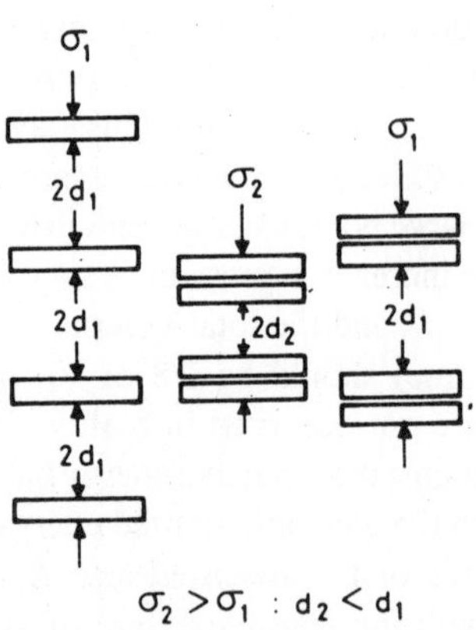

Fig. 3.18 Loading and unloading processes at microlevel, $(d_1 - d_2)$ is elastic compression; grouping of particle is plastic compression (Srinivasa Murthy et al., 1991).

require considerable input of energy to force them apart. The same is true even for the real microstructure with large enclosed water pores. The reduction in the size of the water pores from R_1 to R_2 is not reversible because the pores have no compulsion to dilate as they can be in equilibrium under the smaller size itself. The equation $p = 2T/R$ only specifies the maximum possible size of the pore under a given pressure but pores smaller than this are equally possible. However, if the particle groups are broken into smaller units by any process, the soil can dilate back and this is what happens during shearing. It can be stated that the volumetric compression due to changes in the '*d*' spacing between particles is elastic and the compression due to changes in the size of the large pores associated with grouping of particle units is plastic in nature.

Under the application of a deviatoric stress, the large water pores have to distort to an elliptical shape (Fig. 3.19) in order to sustain the deviatoric stress, the pore pressure still being hydrostatic and equal at all points at equilibrium. i.e., the pore water is held by a suction equal to the mean principle stress at every point in the soil mass.

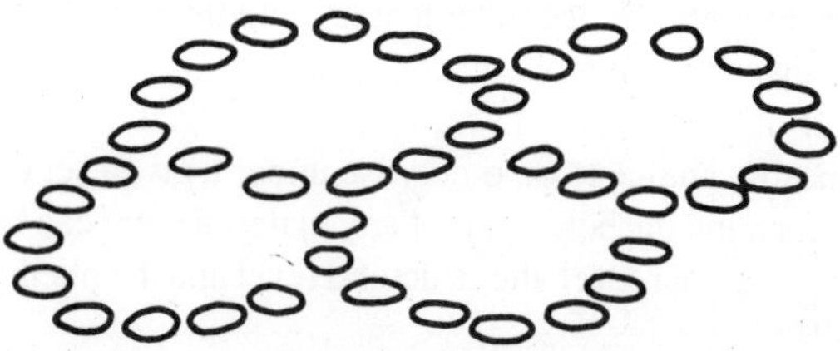

Fig. 3.19 Microstructural changes due to deviatoric stress.

There appears to be a desirable level of aggregation of particles for any soil which is defined for different stress levels. The volume of voids that the soil can cover with this aggregation corresponds to the critical state void ratio under that stress. If the current state of aggregation is more than this specific value, the soil would tend to have a break down of aggregates into smaller units whenever there is the required input of energy, as for example during shearing. This would result in a dilation of the soil. If on the other hand, the current state of aggregation is less than this specific value at the given pressure, any input of energy, would tend to bring about a grouping of particles resulting in compression of the soil.

With this knowledge that the type of interactions and mode of stress transfer is similar for all clay minerals, we shall examine in the next chapter, if there is any possibility of generalizing the macro behaviour of fine grained soils.

4

DEVELOPMENT OF A GENERALIZED APPROACH

4.1 Introduction

Of all the civil engineering construction materials, soils exhibit the greatest degree of variability and uncertainty. Hence judgment and predictions play a pivotal role in geotechnical engineering. Hypothetically speaking, it should be possible to predict the behaviour of any material because every material in reality respond in a defined way and has no choice to do otherwise under a given set of conditions.

Generally, to understand the behaviour of any material there can be three levels of investigation, the molecular level, the structural level and the phenomenological level (Klausner, 1991).

The phenomenological level considers the material as an entity and deals with the phenomenon as it appears, and not with the properties, actions and interactions of the large number of smaller units or molecules of the matter. The mathematical formulations at this level of investigation would be considerably simple and the boundary conditions and the required material parameters can be obtained from experiments at a grosser level with simpler and less sensitive instruments. The models based on continuum mechanics, principles of elasticity or plasticity are some examples. Often, for a complex material, it may not be possible to explain the observed behaviour in entirety and consistently with this level of examination. For eg. whether a partly saturated soil upon inundation undergoes heave or collapse and by what magnitude, cannot be answered by models considering soil mass as an entity or, the strain softening behaviour of cemented soils cannot be explained with the established principles of elasto-plastic theory. We may have to adopt subtler levels of investigation.

The next level of investigation is the structural level in which the various components of the material such as air, water, mineral particles, are studied for their individual behaviour and mutual interaction to finally get the overall response. This approach is a via-media between other two approaches.

The subtlest level is the molecular level in which the behaviour of each individual molecule is predicted from the known physical laws and their net effect is computed to get the total response of the material. In principle, this is the most exact level but the difficulty is in comprehending, exploring and defining all the intricacies and interactions occurring at this level. The mathematical formulations to describe the behaviour of numerous number of units would be extremely complex. Measurements and observations at this level of investigation will have to be highly sensitive and refined and sometimes beyond our technical means. The Gouy-Chapman diffuse double layer theory, for colloids and its application to predict the compressibility of clays is an example in soil mechanics field. The rate process theory (Mitchell et al., 1968) wherein the activation energy required to displace each flow unit at molecular level is related to the shear and creep response of fine grained soils at macrolevel is another example. Recently the behaviour of coarse grained soils is being studied by considering the forces and deformations at each grain for the known initial packing of the grains.

Apart from the above scientific approaches, as is true in any field where practice or application precedes scientific understanding, in geotechnical engineering field also, a few empirical relations correlating mechanical properties with simple inferential parameters have been in vogue, and have almost become indispensable tools because of their simplicity. These empirical relations are essentially based on experimental observations without probing into why or how of these correlations. The fact that these methods have no scientific basis restricts their usage to limited type or location or condition of soils and further generalisations could be sometimes dangerous. To reiterate, some of the widely used relations are:

(i) Skempton's (1944) compression index equation and its modified forms to describe the compressibility of soils,

(ii) c_u/p vs I_P relationship (Skempton, 1953) to get the undrained strength of natural soils and

(iii) remoulded strength vs liquidity index relationship (Houston and Mitchell, 1969).

Though there have been attempts by both of the above approaches, i.e., molecular and phenomenological approaches independently, to understand and model the soil behaviour, these attempts have not resulted in completely satisfactory models. What is desirable and practicable is perhaps a combination of the two approaches, i.e., having a qualitative understanding of the mechanism involved at micro level, and identifying the reflection of these processes at macro level, and finally arriving at the mathematical formulation at macro level itself involving easily measurable macro parameters.

A large number of publications from the Indian Institute of Science in the recent past reflect the research work in this direction of developing phenomenological models to generalize and predict the behaviour of fine grained soils based on micro-mechanistic considerations. This chapter discusses the development of this generalised approach.

As discussed in Chapter 3, it is not possible to develop a unified approach for all types of soils like clays, silts and sands and gravels. It has been recognized that mechanistically the soil can be grouped into coarse and fine grained soils or noninteracting and interacting soils. This classification is essentially based on the non-dominance or otherwise of the physico-chemical forces of interaction.The coarse grained soils have physically very simple micro-mechanism in the sense that the particles sliding over each other can be simulated easily with a friction model, but have a very complex micro structure and hence a complex macro-behaviour, the description of which is still not satisfactory. Further, the description of this micro-structure can only be done with a large number of randomly varying parameters. Though the medium can be treated as continuum at macro level it is necessarily discontinuous at micro level. These difficulties and discrepancies in the nature between the two levels of the medium make the development of the generalized approaches very difficult if not impossible for the coarse grained soils.

In this chapter, a detailed discussion on the phenomenological approaches for generalisation and prediction of behaviour of 'fine grained soils' will be presented. A discussion on the behaviour of coarse grained soils will be presented in a later chapter. Since the variability within the class of fine grained soils is also quite large, it is obvious that a single model may not be feasible. Hence, as a first step, the different complex states of fine grained soils are distinguished as:

(1) Normally consolidated, uncemented saturated soils.
(2) Over consolidated uncemented saturated soils (stress dependent)
(3) Naturally cemented saturated soils
   (i) soft and sensitive soils
   (ii) stiff cemented soils
(4) Partly saturated uncemented soils.
(5) Cemented partly saturated residual soils
(6) Layered/ fissured/ varved - insitu soils.

In the above classification, the order of the state of the soil reflects the increasing complexity. The differences in the above states are due to inherent variations and divergent factors induced by stress, time and environment and hence the parameters required to delineate their behaviour are also different with increased number of parameters for soil states of higher complexity. For example, for the overconsolidated saturated uncemented state, the preconsolidation stress, σ_p' may be an additional parameter over and above those required for a normally consolidated state. For cemented soils, the cementation bond strength and for partly saturated states the degree of saturation may be the required additional parameters.

4.2 Effective Stress in Fine Grained Soils

Since, the mechanical behaviour of soils are always described in terms of 'effective stress', we will examine what this effective stress is in fine grained soils before discussing the generalized approach.

Historically, since the advent of effective stress principle by Karl Terzaghi in 1936, the solutions to the geotechnical engineering problems have been visualized in terms of the effective stress. The principle, as cited earlier relates the effective stress σ' to the total stress, and to the pore water pressure, u by a simple equation:

$$\sigma' = \sigma - u \tag{4.1}$$

and further states that all the measurable effects of a change of stress such as compression, distortion and a change of shearing resistance are exclusively due to changes of effective stress which has its seat in the solid phase. Though this principle could explain the behaviour of coarse grained soils under most conditions, researchers had some difficulty in explaining the behaviour of fine grained soils under certain conditions. The main reason attributed for this discrepancy is that the basic physico-chemical forces in terms of repulsive and attractive forces at particle level have not been accounted for in this equation. This led to the formulation of several other forms of effective stress equation incorporating these physico-chemical forces. Lambe (1960) wrote the basic equilibrium equation of all the forces operating between the two particles in the form:

$$\sigma = \bar{\sigma}\, a_m + p_a\, a_a + u\, a_w + (R - A) \qquad 4.2$$

where σ is the total external force applied on unit area,
$\bar{\sigma}$ is the mineral to mineral contact stress,
p_a is the pore air pressure,
u is the pore water pressure,
$(R - A)$ is the net.interacting force on unit area,
a_m, a_a and a_w are the fractions of the total unit area in contact with each phase of mineral, air and water respectively.

For a saturated system, $a_a = 0$ and hence the equilibrium equation (4.2) can be written as

$$\sigma = \bar{\sigma}\, a_m + u\, a_w + (R - A)$$

then the conventional effective stress can be expressed as

$$\sigma' = (\sigma - u\, a_w) = \bar{\sigma}\, a_m + (R - A) \qquad 4.3$$

Sridharan (1968) thought that the effective stress increase with increase in the net repulsive force in a system is inconsistant and hence modified this equation in the form:

$$\bar{C} = \bar{\sigma}\, a_m = \sigma - u\, a_w - (R - A) \qquad 4.4$$

where $\bar{C}$ is the effective mineral to mineral contact stress.

These forms of effective stress equations were based on the superposition of both the stress transferred through solid to solid contacts and that through physico-chemical interactions with the adsorbed fluid. It is unlikely to have both the zones of adsorbed fluid complex and that devoid of the same elsewhere on the surface. It is known that very high stresses, which are not encountered in engineering practice are required to remove the last few layers of adsorbed water from the clay surface.

From the discussions presented on the fundamentals of soil behaviour it is very clear that the internal force field mobilized between the structural units to balance the external force system is due to either short range or long range forces. In the above equations $\bar{\sigma}$ or $\bar{C}$ reflect the short range force components while $(R - A)$ is that due to long range forces. It has been brought out that at particle spacings at which long range forces are operative, the short range forces are insignificant and vice versa. This leads to writing the effective stress equation in two forms as applicable to respective class (Nagaraj, 1981).

for $\bar{C}$ being zero, $\qquad \sigma' = \sigma - u_w = (R - A) \qquad 4.5$

and for $(R - A)$ being zero, $\sigma' = \overline{C} = \sigma - u_w$ 4.6

In fine grained soils, where only long range forces are operative the former equation is the effective stress equation.

Sridharan and Rao (1971, 1973 and 1979) have further shown with an extensive experimental programme that $\overline{C}$ controls the mechanical behaviour of kaolinite type of clays while $(R - A)$ controls the behaviour of montmorillonitic clays. But these conclusions were based on the opposite trends in the volume change response of the soil under changed physico-chemical environments, i.e., different pore fluids such as acetone, benzene, methanol. The kaolinite soil showed an increase in volume while montmorillonite soil showed a decrease in volume with decrease in the dielectric constant of the pore fluid. A decrease in volume is in agreement with the diffuse double layer theory as reiterated above in the case of soils with montmorillonite clay mineral. But subsequently for the same data, it was shown (Nagaraj, 1992) that for water as the pore medium, both soils show (Fig. 4.1) identical behaviour in shear mobilisation. With changes in the pore fluid although

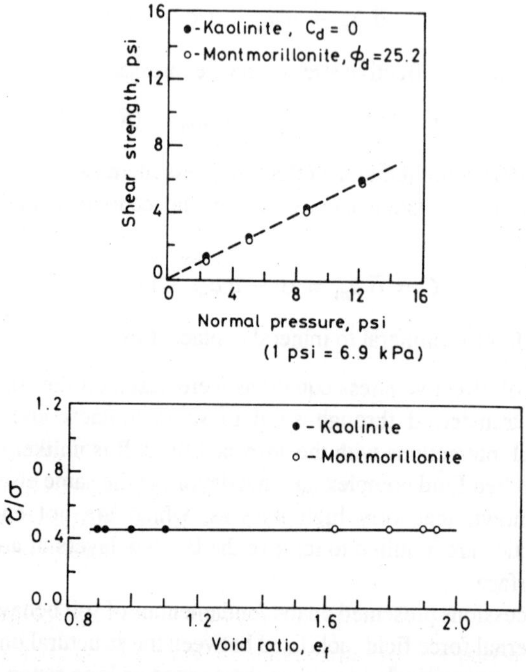

Fig. 4.1 Reanalysis of experimental strength data of Sridharan and Rao, 1979 (Nagaraj, 1992).

the nature of change in the initial state is opposite for the two soils, further changes in volume with changes in stress is found to be similar and proportional to the initial volume for both the clays.

Just as clay minerals and their proportions can vary over a wide range with the fluid phase remaining the same, it is interesting to examine the flow lines of clays with different pore fluids. Figure 4.2 illustrates the marked changes in the relative

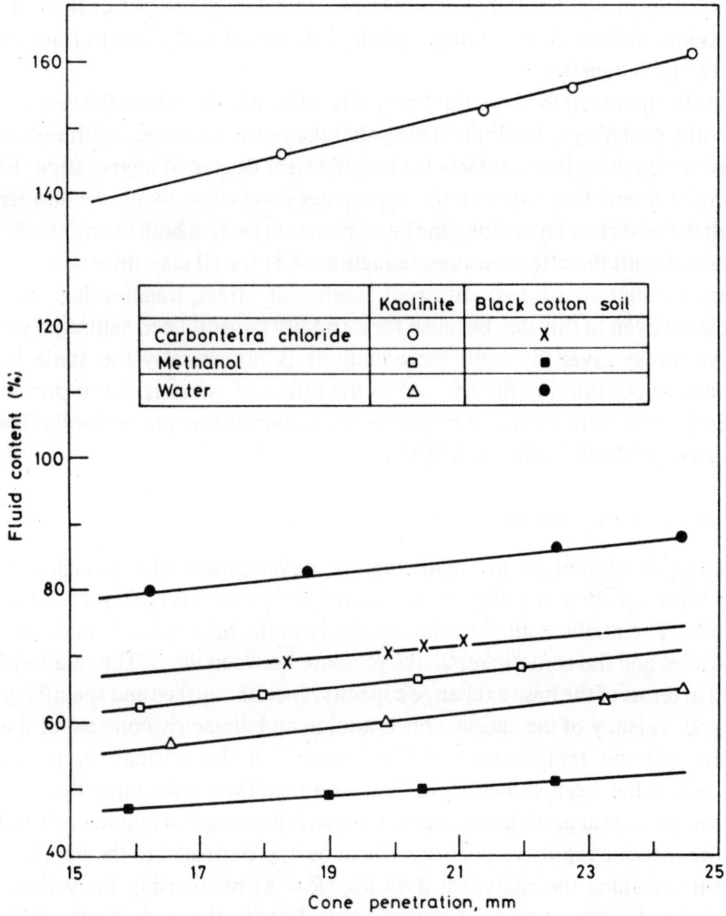

Fig. 4.2a Fluid content versus cone penetration plots with different pore fluids.

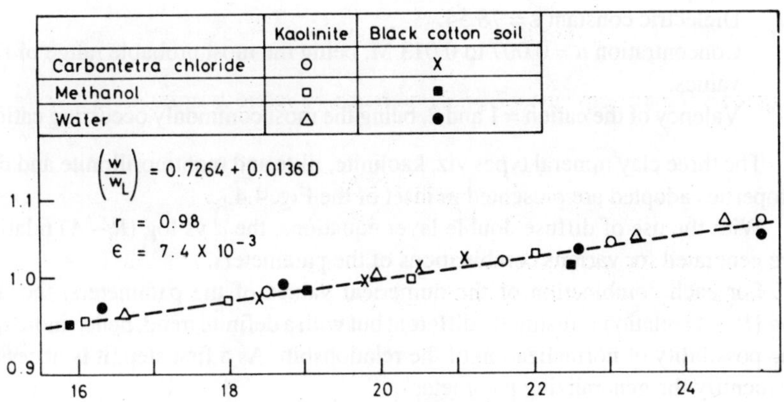

Fig. 4.2b Generalisation of fluid content versus cone penetration plots.

disposition of flow lines as affected by pore fluids, even trends getting reversed when the clay minerals are different. Despite this fact, the slopes of flow lines are proportional to their fluid holding capacity i.e., the ratio of slope to fluid holding capacity, w_F is constant.

Thus the nature of the pore fluid may affect the initial fabric of the two minerals differently, probably in kaolinite it may alter the positive charges on the edges, and thus reduce edge-to-face contacts leading to lesser degree of aggregation. But the subsequent interaction between the aggregates or clusters would be similar, i.e., through the adsorbed layer along the basal plane surface, in both the minerals. Thus we proceed with the effective stress equation (4.5) for all clay minerals.

This distinction of two different modes of stress transfer has not been appreciated even to this day because for normally consolidated saturated soils the effective stress given by both the equations is numerically the same but the advantage in considering the $(R - A)$ as the effective stress is that it provides an independent predictive capability for the behaviour of fine grained soils (Nagaraj and Srinivasa Murthy, 1983 and 1986).

4.3 Micro-Analytical Model

A detailed discussion on the mathematical development of interacting diffuse double layer equation in a clay water electrolyte system has been presented in the Appendix. The mathematical formulation relates the half space distance between the particles and the osmotic repulsive pressure between them. The relationship is defined in terms of the base exchange capacity, specific surface and specific gravity of the soil, valency of the cation, concentration and dielectric constant of the pore medium and the temperature of the system. In the normal environmental conditions, it has been shown (Sridharan and Jayadeva, 1982) that the attractive force is negligible at particle distances at which R forces are of engineering interest. Thus, the osmotic repulsive pressure, R can be approximated to the net $(R - A)$.

Let us examine the analytical d vs log $(R - A)$ relationship for various field conditions and different types of clay minerals. For this the method proposed in the appendix has been adopted. The physical properties considered in the analysis are:

Temperature $T = 298°$ K.

Dielectric constant $\varepsilon = 78.54$.

Concentration $n = 0.007$ to 0.013 M, being the most probable range of field values.

Valency of the cation = 1 and 2, being the most commonly occurring cations.

The three clay mineral types viz. kaolinite, illite and montmorillonite and their properties adopted are presented in inset of the Fig. 4.4.

With the use of diffuse double layer equations, the d vs log $(R - A)$ relations are generated for various combinations of the parameters.

For each combination of the numerical values of the parameters, the d vs log $(R - A)$ relation is distinctly different but with a definite trend, hence indicating the possibility of normalization of the relationship. As a first step, it is necessary to identify the generalizing parameter.

It is a general principle in any field that generalizations of a particular behaviour of similar materials can be achieved through normalization by a standard or reference state common to all the materials. One such example is the attempt to generalize the deformation behaviour of homologous temperatures of ice and other materials of similar grain structure by normalization with their respective melting points (Sinha, 1981).

In soil engineering, for saturated soils without any pre- stress effect and cementation bonds, it is possible that equilibrium state under any unique stress condition can be considered as a normalizing parameter. We shall consider liquid limit state to represent this, since it has been found that under an isotropic effective consolidation pressure of about 6 kPa, all soils equilibrate at their liquid limit water contents (Russel and Mickle, 1970; Wroth and Wood, 1978 and Whyte, 1982). Further, the shearing resistance, c_{uL} of soils with liquid limit varying between 36 and 159 have been measured using a laboratory vane (Federico, 1983). The measured values of c_{uL} fall within the limits of 1.7 to 2.8 kPa with a concentration of values around 2.3 kPa. These values agree closely with data published in the literature even though different techniques were used. Thus water content at liquid limit state of soils can be regarded as a generalizable state parameter since that equilibrium condition is due to the same order of consolidation/ suction pressure and shearing resistance. Recently, Burland (1990) has tried the equilibrium void ratio, e_{100} at 100 kPa stress level as the normalising parameter. But this requires a compression test to determine e_{100} for each soil, where as the method of determining liquid limit water content is much simpler.

First of all, let us consider the d vs log $(R-A)$ relations for the three clay minerals for the same physico-chemical environment. It can be seen (Fig. 4.3) that this relation is practically unique for all the clays. However, the d vs log $(R-A)$ curves are different for any given clay under different pore fluid environment (Fig. 4.4).

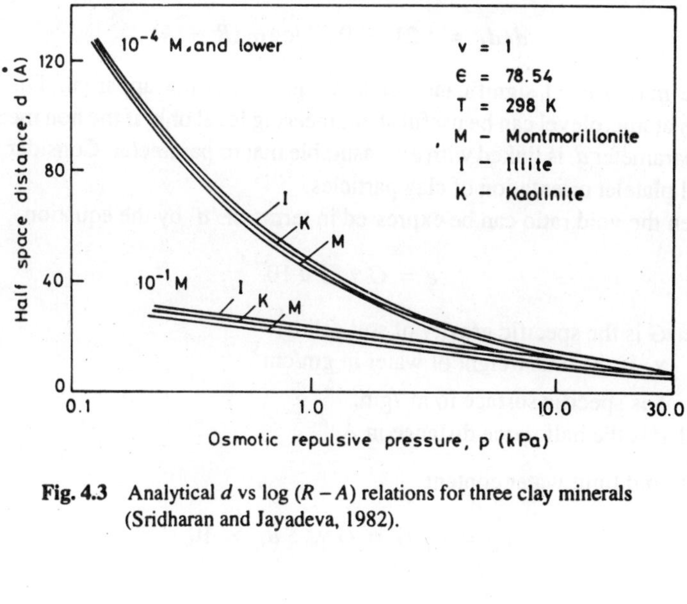

Fig. 4.3 Analytical d vs log $(R-A)$ relations for three clay minerals (Sridharan and Jayadeva, 1982).

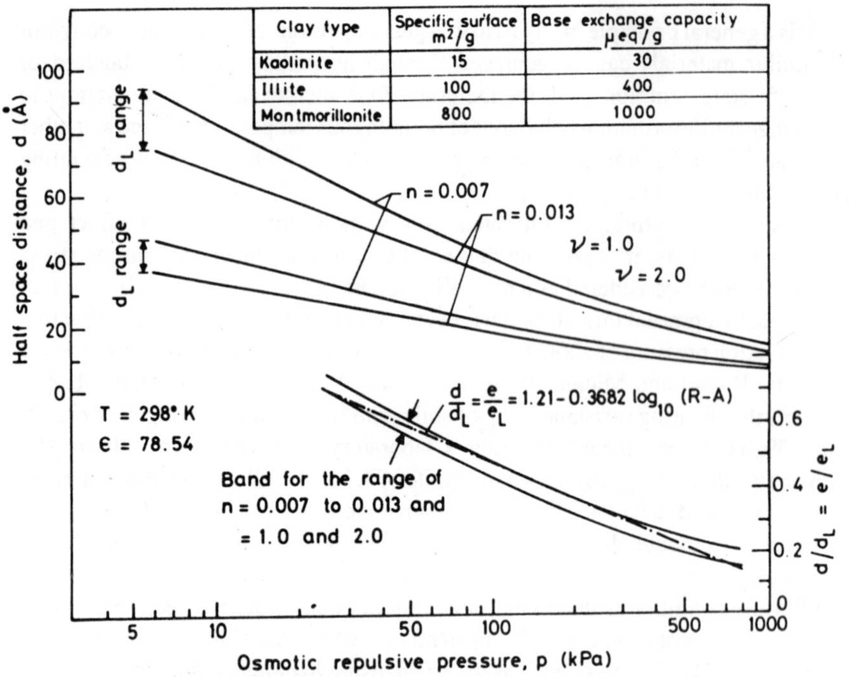

Fig. 4.4 Normalisation of analytical d vs log p' relations (Nagaraj and Srinivasa Murthy, 1986).

Now, in d vs log $(R - A)$ relations, the value of d at $(R - A) = 6$ kPa, which corresponds to the liquid limit state, is considered as d_L. Then normalizing the d vs log $(R - A)$ curves using their respective d_L values, results in a unique d/d_L vs log $(R - A)$ relation (Fig. 4.4) of the form :

$$d/d_L = 1.21 - 0.37 \log_{10} (R - A) \qquad 4.7$$

The macro level significance of this equation is not apparent. This unique relation at microlevel can be useful at engineering level only if the non measurable microparameter d, is linked with a measurable macro parameter. Consider an ideal parallel platelet orientation of clay particles.

Then the void ratio can be expressed in terms of 'd' by the equation

$$e = G \gamma_w S d \, 10^{-4} \qquad 4.8$$

where G is the specific gravity of soil solids,

γ_w is the unit weight of water in gm/cm^3.

S is specific surface in m^2/gm,

and d is the half space distance in Å.

At liquid limit water content

$$e_L = w_L G = G \gamma_w S d_L \times 10^{-4}$$

Then

$$e/e_L = G\gamma_w Sd/G\gamma_w S d_L = d/d_L = 1.21 - 0.37\log_{10}(R - A) \qquad 4.9$$

At equilibrium ($R - A$) being equal to σ' or p',the equation reduces to

$$d/d_L = e/e_L = 1.21 - 0.37\log_{10}p' \qquad 4.10$$

which could be the generalized compressibility equation.

In fine grained soils, as already brought out in chapter (3), an ideal parallel plate orientation, assumed above is not a reality. In such a situation does this micro-level equation really represent a macro-behaviour, needs further examination. This is because the mircro-structure presented earlier clearly indicates that '*d*' is not directly related to the void ratio, *e* as per the equation (4.8). For all practical purposes *d* can be regarded as inter cluster separation distance. For the large enclosed pores, the diameter of the pore, *D* is dependent on the applied pressure controlled by the surface tension equation

$$p = 4T/D$$

Since, both *d*, the half space distance between the basic structural units and the diameter of the large pores are related to the same pressure, *p* it is reasonable to assume a defined relationship between *d* and *D*. Now, since the void volume covered by the large pores constitutes a major fraction of the total void volume, the void ratio can be related with '*d*', though not in the same form of equation 4.8.

4.4 Liquid Limit as Generalization Parameter

There are a few other facts to reinforce that liquid limit is a proper parameter for generalization and that the volume change relation with increase in pressure for different soils is generalisable. An examination of the pore size distribution curves presented by Griffiths and Joshi (1989), indicates the following:

(1) At liquid limit void ratio, the maximum diameter of the large pore is unique and also the pore size distribution pattern is nearly same (Fig. 3.14, Ch. 3) for all the soils.

(2) At void ratios for a soil corresponding to different maximum consolidation pressures, the maximum diameter of the large pores decreases with increase in maximum pressure.

(3) The maximum diameter of the large pores is nearly same for all the soils at a given maximum consolidation pressure.

Another interesting data worth examining at this stage is the permeability test data which gives an indirect reflection of the micro structure. Figure 4.5 shows the permeability test data for different soils. The coefficient of permeability *k* is found to be nearly same for all the four soils at their liquid limit states in spite of a six fold variation in the liquid limit water contents.

It is not apparent as to how the permeability can be of the same order for states with such large variation in void ratios. Unfortunately, we geotechnical engineers are tuned to infer in terms of void ratio or per unit weight of solids rather than in

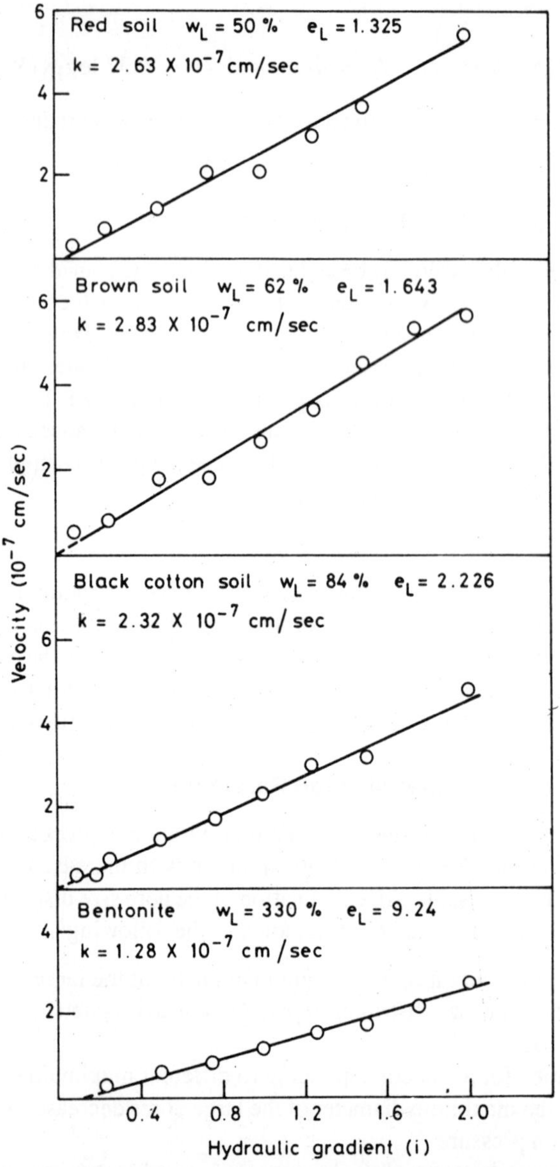

Fig. 4.5 Rate of flow vs hydraulic gradient relations

terms of porosity or unit volume of the soil. If the state of soil is considered in terms of volume basis instead of weight basis, the weight of solid particles present in a unit volume, would be inversely proportional to the surface area or in turn the liquid limit water contents of the soils i.e., the weight of solid particles in unit volume would be such as to provide the same order of surface area and hence the same order of physico- chemical potential for all soils. Hence the volume of large pores enclosed, (which essentially controls the volume changes under further loading or permeability) will also be nearly same. The constancy of physico-chemical potential per unit volume at liquid limit for all soils logically explains the macrolevel constant magnitudes of soil suction or consolidation pressure, constant shear strength and same order of permeability. These unique conditions at liquid limit reaffirms that it represents a datum state in relation to which all other state and stress conditions can be normalized.

4.5 Examination of other Index Parameters

A few other inferential parameters such as plastic limit or plasticity index, I_P liquidity index, I_L, colloidal activity, A are sometimes being used in geotechnical engineering to generalize or to correlate mechanical properties of soils. We shall examine the potential of these parameters for generalisation.

The methods of determination of plastic limit by shaping of wet clay to balls and threads and subjecting them to further manipulations are prone to personal impressions. Further, it is found that the plastic limit water contents for different soils cannot be related to a unique state of stress or shear strength to the same narrow range of acceptability as the stress state at liquid limit water contents. If the recent trend of (Wroth and Wood, 1978) redefining and determining plastic limit as the water content at a defined stress is adopted (100 times the stress at w_L) it may be possible to use plastic limit also as a parameter for generalisation. But with this new definition of plastic limit, (and for the known, linear relation between void ratio and logarithm of pressure for all soils. which will be discussed later in Chapter 5) the plastic limit happens to be a function of liquid limit of the soil, and not an independent parameter in which case liquid limit alone would be sufficient. The existence of a functional relationship between I_P and w_L of the form

$$I_P = A\,(w_L - B) \qquad 4.11$$

has been experimentally verified for a large data set of 520 soils (Nagaraj and Jayadeva, 1983) and subsequently by Paul (1984) for a larger data of 1100 soils, although the constants A and B varied a little from the two studies. The variability is probably due to the inherent variability in the plastic limit values determined by thread rolling method. It is likely that using plastic limit values determined from constant stress state definition, would give a better defined relation.

The same disadvantage that the determination of plastic limit which in its present day practice, is susceptible for personal errors, and is a superfluous task, while liquid limit alone can serve the same purpose, applies to all the three derived parameters involving plastic limit i.e., plasticity index, liquidity index and colloidal activity.

Plasticity index represents the range of water contents over which a defined level of change in shear strength takes place. Liquidity index which involves the state parameter w in addition to w_L and w_P, can be again expressed in terms of only the two independent parameters, since w_P is a function of w_L. For the data of natural water content and consistency limits of Komornik and David (1969) the relation between I_L and w/w_L is shown (Fig. 4.6) to be (Srinivasa Murthy et al., 1986) of the form;

$$I_L = 1.55\, w/w_L - 0.56 \qquad 4.12$$

Hence, liquid limit itself appears to be adequate for normalization of the states.

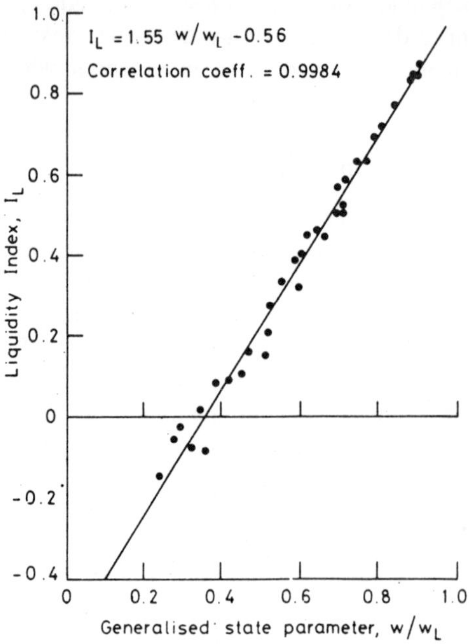

Fig. 4.6 Relation between liquidity index and generalised state parameter (Srinivasa Murthy et al., 1986).

Regarding colloidal activity (I_P /Per cent clay size), as before the plastic limit in the numerator is a function of liquid limit and it is well known that the physico-chemical interactions due to the clay fraction is completely reflected in the liquid limit. Activity, being the % clay fraction required to result in a given plasticity, would give a reflection of the type of clay mineral. But what is required to quantify the net physico chemical potential is both the type and amount of clay minerals for which we will have to use A along with amount of clay fraction which will reduce it to again plasticity index only (Pandian and Nagaraj, 1990).

Plastic limit may serve a secondary purpose i.e., to identify the nature of clay minerals in some cases. For the usual inorganic clay minerals, the plasticity index vs liquid limit data plot along a line as shown earlier (Eq. 4.11) which runs nearly along Casagrande's A line. But for some soils with non-sheet type clay minerals like Hallosite the plastic limits would be too high (usual values are like $w_L = 140\%$, $w_P = 80\%$) and the points would plot very much below the A line. The generalisation proposed in this book may not be applicable to such soils. Apart from this, in silty clays plastic limit may indicate the limiting water content above which only this mode of stress transfer through double layer interactions prevail.

4.6 Cohesive Matrix Concept

So far, our discussions, i.e., double layer interactions and the resulting water holding capacity, with large enclosed pores etc. were all pertaining to clay minerals.

But most natural soils are not made up of clay fraction only but also contain coarser particles of silt and sand size. Then, it needs to be examined whether the generalisations proposed earlier will be applicable to such natural soils.

It has been recognized since long that the clay fraction controls the behaviour of a soil to an extent much more than its percentage content in the total solids. The reason for this is that the clay particles, because of their very high specific surface, form a coating around the coarser particles, thus preventing a direct contact between them, or, the coarse particles float in a matrix provided by the clay particles, so that finally the interaction is always between two clay surfaces: With an extensive liquid limit and consolidation experiments on different soils with different percentages of coarse particles, it has been shown (Srinivasa Murthy et al., 1987) that the presence of coarse particles only dilutes the physico-chemical potential of the soil proportionately and does not alter the basic mode of stress transfer, i.e., if for a clayey soil of liquid limit 100%, and compression index of 0.5, 20% by weight of coarser particles are mixed, then for the mixed soil, liquid limit would be 80% and C_c would be 0.4. Figures 4.7 and 4.8 indicate some of the test results of flow curves and compression paths of soils with different percentages

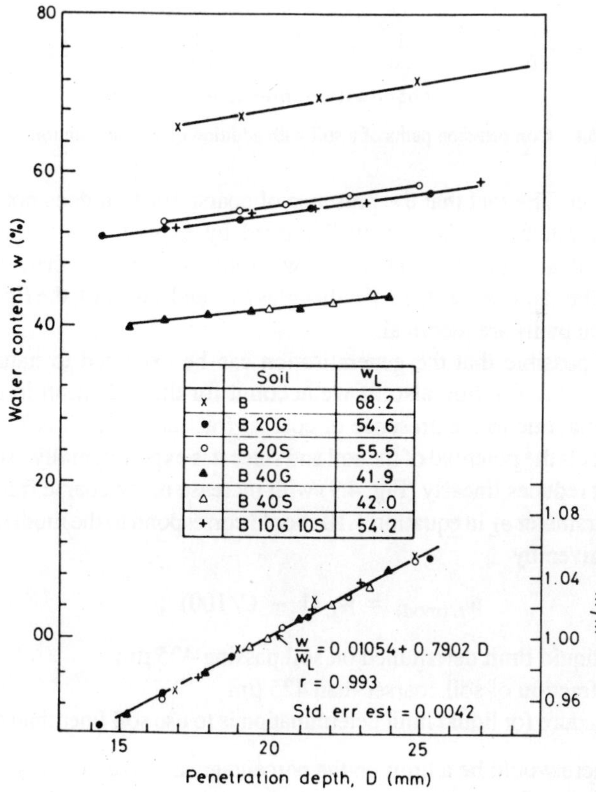

Fig. 4.7 Flow lines of a soil with presence of coarser fraction (Srinivasa Murthy et al., 1987).

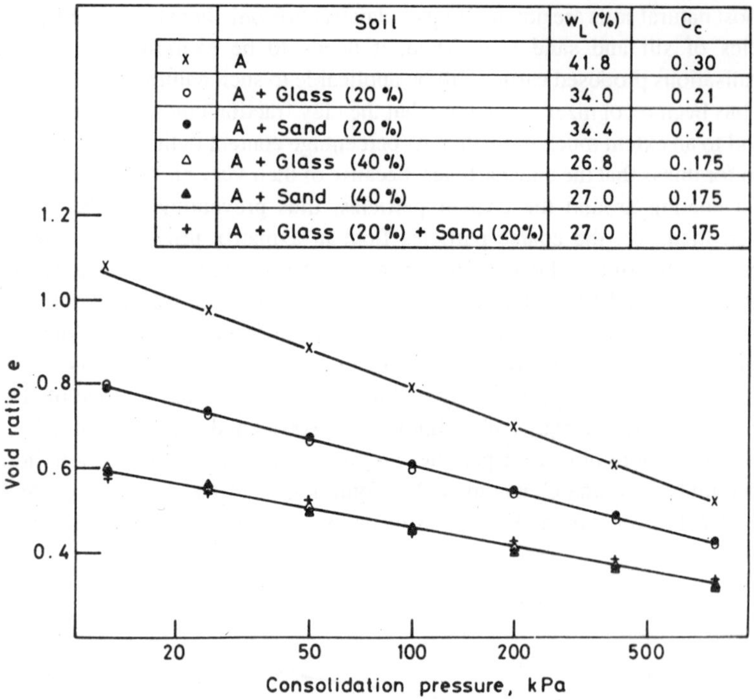

Fig. 4.8 Compression paths of a soil with addition of coarse fraction.

of coarse fraction. The fact that the presence of coarse fraction does not alter the basic nature of interaction is very well proved by experiments using coarser particles of sand and glass which have two fold variation in their frictional properties. For the same percentage of either glass or sand particles, the flow curves and compression paths are identical.

Hence it is possible that the generalisation can be extended to natural soils containing non-clay fraction also, if we account for the reduction in physico-chemcial potential due to the presence of coarser fraction. Since it is known that liquid limit reflects the potential of the soil and since it is experimentally established that liquid limit reduces linearly (Fig. 4.9) with increase in the coarse fraction, the generalizing parameter e_L in equation 4.10 should correspond to the modified value of liquid limit given by

$$w_{L\,(\text{mod})} = w_L\,(1 - C/100). \qquad 4.13$$

where w_L = liquid limit determined on soil passing 425 μm.
C = fraction of soil coarser than 425 μm

(The usual procedure for liquid limit determination is to use soil finer than 425 μm).

However there would be a limit on the percentage and size of coarse particles for the cohesive matrix condition to prevail. It may not be valid for gravel size particles. It may be tentatively suggested that the double layer theory can be applied if the modified w_L is atleast 30%.

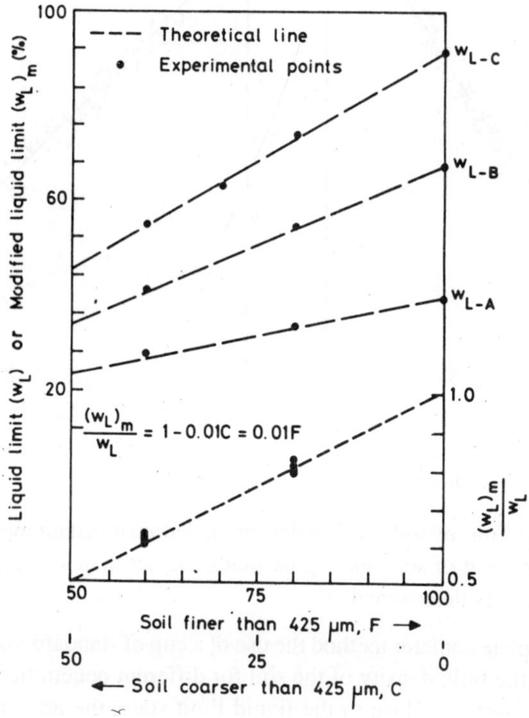

Fig. 4.9 Modified liquid limit with percentage of coarse fraction (Srinivasa Murthy et al., 1987).

It has been observed by many early investigators (Skempton, 1953; Seed et al., 1964; Nagaraj and Jayadeva, 1981 and several others) that the composite effects of surface activity of fine grained soils and the physico-chemcial properties of the interstitial fluids are reflected to a large extent in their liquid limit water contents. Terzaghi (1926) intuitively had expressed that two soils of similar origin will have similar behaviour if their liquid limits are identical.

The preceeding discussions clearly point towards the possibility of adopting the liquid limit water content as a generalizing parameter. This is consistent with the generalisation of micro-behaviour from micromechanistic considerations.

4.7 Liquid Limit Determination - Further Simplified

Having realized that water content at liquid limit would enable to take into account the physico-chemical potential of soils, it is now examined about the methods of determination of the same. In recent years the cone penetration method is gaining popularity and acceptance. Wroth (1979) provides adequate supporting evidence in favour of the cone method, as a simple strength test, instead of the more elaborate percussion device. Recent investigations (Nagaraj et al., 1987) have enabled to advance two more specific advantages in this method.

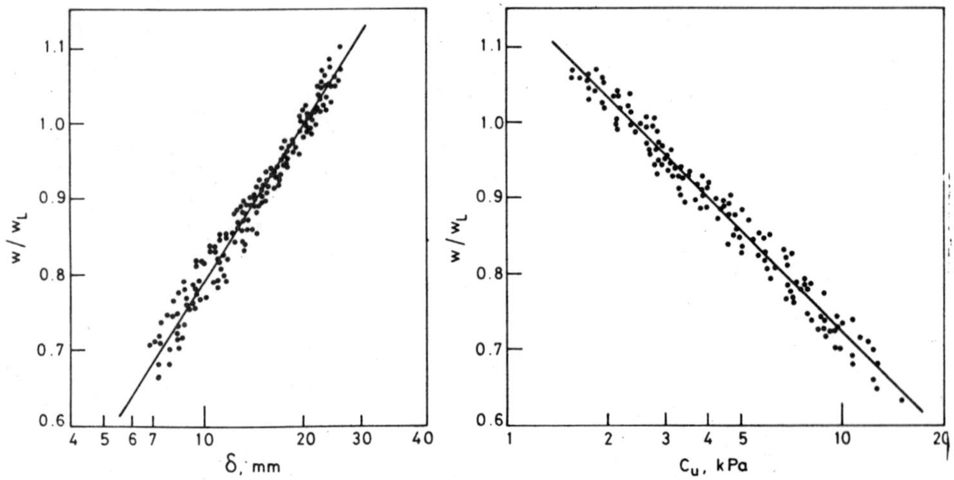

Fig. 4.10 Generalized state parameter, w/w_L versus cone penetration or undrained strength (Federico, 1983).

(i) The liquid limit of soils can be determined almost instantaneously.

(ii) The liquid limit of soils having particles coarser than 425 μm (0.6 – 2 mm) can be directly determined.

In the cone penetrometer method the use of a cup of standard volume is helpful in determining the bulk density of the soil for different penetration values. In the range of water contents close to the liquid limit state, the amount of soil paste transferred to the cup is a function of both water content and the liquid limit of that soil.

In the range of water contents around liquid limit it is further shown that the degree of saturation of soil in the cup is around 98 percent. From the above observations the water contents at different penetrations can be estimated. The errors associated with the assumption of a value for the specific gravity of the soil in the range of 2.65 to 2.8 are well within the acceptable levels of accuracy. This method of assessment of water content by bulk density determinations permits one to independently check the values obtained by the conventional method or by the use of microwave oven.

In order to minimize the number of operations in the estimation of water content corresponding to 20 mm penetration, one point method can be adopted. A scientific basis to determine liquid limit by one point method has been provided by Nagaraj and Jayadeva (1981).

The following two relationships have been developed to determine water content at liquid limit from cone penetrometer method in the above investigation.

$$w_L/w = 1/(0.77 \log D) \tag{4.14}$$

$$w_L/w = 1/(0.65 + 0.017\,D) \tag{4.15}$$

where D is the depth of cone penetration in mm.

The work of Federico (1983) has shown that for many soils of Southern Italy covering a wide range of plasticity, the regression equation of the straight line of w/w_L against log δ is (Fig. 4.10)

$$w/w_L = 0.102 + 0.688 \log \delta \tag{4.16}$$

with $r^2 = 0.959$, where δ is in mm.

This equation is very much similar to the monomial relationship

$$w/w_L = 0.77 \log D$$

where D is the penetration written as δ in the equation (4.16). Both the above relations satisfy the condition $w/w_L = 1$ corresponding to a penetration of 20 mm. Further works reported by Srinivasa Murthy et al., (1987) and Nagaraj et al., (1987) reaffirm that one point method can be used for determination of water content at liquid limit by cone penetrometer method. Hence the one point method can be extended even to bulk density method of determination of liquid limit. By two trials it would be possible to check the value of liquid limit of soils by cone penetration and inturn forms an instantaneous independent method of liquid limit determination.

5

SATURATED UNCEMENTED SOILS

5.1 Introduction

The application of the load at or below the surface of the soil induces stresses in the soil. Most of the studies in geomechanics are involved in understanding, analysing and predicting the responses of the soil to these stresses. Two types of calculations are regularly performed in geotechnical engineering; stability analysis and settlement computations. Settlement calculations are concerned with the stiffness or compressibility of the soil mass under applied loads. Stability calculations are concerned with the ultimate failure of soil mass, with large deformation occurring on rupture planes, accompanied by collapse of geotechnical structures. For a long time, the settlement calculations have been based on linear elasticity whereas the stability calculations have been based on rigid plasticity. Partly for simplicity in practice and partly because of historical development, these problems in soil mechanics are treated separately and in unrelated ways. But now it is well known that soils exhibit large inelastic deformations over a wide range of working stresses before they reach a state of total failure. This behaviour is described by the recent work hardening elasto-plastic constitutive relations which define unique relationships between general states of stress and strain, encompassing both the settlement and stability problems (Chen and Mizuno, 1990).

The engineering property predictions of soils required in the analysis of geotechnical engineering problems are identified as:

(1) compressibility and consolidation behaviour,
(2) shear strength and stress-strain behaviour and
(3) permeability

Whenever saturated fine grained soils are loaded, they deform. The rate of deformation is time dependent and depends on the permeability of the soil. Under a given increment of loading, due to the relative incompressibility of the bulk water, excess pore water pressure mobilizes, which sets in a hydraulic gradient for the flow of water out of the stressed soil. This state of instantaneous equilibrium of the soil written in terms of total stress, effective stress and excess pore water pressure is termed as undrained state. As the water flows out, the excess pore water pressure reduces and after a certain time, the soil reaches an equilibrium state at which the excess pore water pressure is completely dissipated. Then the total stress is the effective stress. This process of time dependent dissipation of excess pore water pressure and consequent change in pore volume is known as consolidation. Between two equilibrium states, the process of change in volume brought about by the change in effective stress is referred as compression. The equilibrium state after complete dissipation is referred as the drained state.

In geotechnical engineering, the changes in the states of stress and their corresponding strains are conveniently represented by the stress invariants, the effective mean principal stress, p' and the octahedral shear stress, q and the corresponding volumetric and deviatoric strains, ε_v and ε_s (vide sec. 2.7).

We discuss for simplicity, the contents of this chapter, in terms of the axisymmetric stress conditions though it is possible to extend the thoughts for general three dimensional stress conditions. The loading path described by $(q/p') = 0$, refers to isotropic consolidation while (q/p') equal to any other constant refers to an anisotropic consolidation, of which the one dimensional consolidation is a particular case with $q/p' = (1 - K_o)/(1 + 2K_o)$. A path with variable (q/p') refers to shearing.

The simplest state of soil in nature is the saturated uncemented state. It could be either normally consolidated or over consolidated. We present in this chapter, the generalisation of compressibility, shear strength and permeability behaviour of saturated uncemented soils, based on the generalized approach developed in the previous chapter. We first introduce, the one dimensional compressibility behaviour of normally consolidated and over consolidated soils, as it is the simplest and the most common loading path.

5.2 Generalised Compressibility Behaviour

5.2.1. Normally consolidated state:

In the previous chapter, it has been brought out that the microlevel relationship between the half space distance, d and the internally mobilised net electrical force per unit area, $(R - A)$ between two parallel platelets can be generalised for all the soils and for all environmental conditions, to result in a unique relationship. Assuming, similar initial mircostructure and its change with loading for all soils, this unique micro level relation has been transformed into a macro level relationship of the form:

$$e/e_L = 1.21 - 0.368 \log \sigma_v' \quad 5.1$$

In this equation σ_v' is the normal stress applied on the parallel plates, which is equal to the internally mobilised $(R - A)$. The term (e/e_L) represents equilibrium states of all soils under a stress of σ_v', and hence it can be termed as the generalised state parameter of the soil. Let us examine experimentally the possibility of obtaining such a unique relationship for soils. We have considered linearised $e - \log \sigma_v'$ plots (Fig. 5.1a) of eleven, normally consolidated uncemented soils (Nagaraj and Srinivasa Murthy, 1983 and 1986). To avoid any prestress effect, only such test results of the soils where the consolidation has been started from water contents greater than their respective liquid limit water contents, have been considered. The soils have a wide range of liquid limit water contents from 36% to 160%, being, the normal range of variation encountered with natural soils. The curves are spatially spread out, with the position of each curve being in the order of the liquid limit water content of the soil. Similar plots, for soils having different liquid limit water contents was indicated by Lambe and Whitman (1969). Figure 5.1b, represents the plots of the same curves of Fig. 5.1a, normalised with their

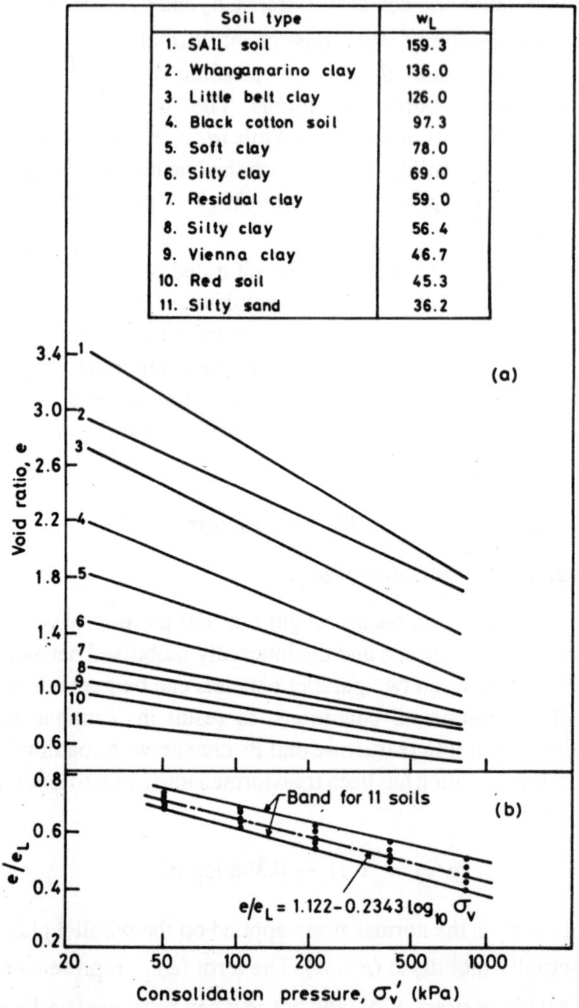

Fig. 5.1 Experimental compression paths and their generalisation. (after Nagaraj and Srinivasa Murthy, 1986).

respective void ratios at liquid limit, e_L. All the normalised points fall with in a narrow band and can be fitted with a linear equation, with in the working stress range of 25 to 800 kPa, in the form:

$$e/e_L = 1.122 - 0.234 \log_{10} \sigma_v' \qquad 5.2$$

with a correlation coefficient of 0.962 and standard error of estimate of 0.025.

The generalised equation (Eq. 5.2) is of the same form as equation 5.1, but with considerable difference in the values of constants. This implies that, the generalisations proposed in Chapter 4, may be applicable, but may not be in its entirety. It is necessary to examine the reasons for the deviations, which could be due to the simplifying assumptions made while generating the equation 5.1. From a more detailed investigation on this aspect (Nagaraj and Srinivasa Murthy, 1986), it has been logically shown that the deviation from the assumptions viz.,

(a) orientation of particles being different than the parallel orientation,
(b) presence of considerable percentage of coarse particles in the soil and
(c) the existence of particle to particle contact stress,

really do not cause such variations in the values of the constants. The possible reason has been shown to be a gradual reduction in the operating specific surface, resulting from grouping of particles into larger clusters with increase in stress. While deriving the analytical micro-mechanistic equation, using Gouy-Chapman diffuse double layer theory, it has been implicitly assumed that the particles are single particles. But during consolidation, when some of the particles cross the energy barrier as they come to closer proximity, they form a relatively stable entity called aggregates or clusters. Now the operating specific surface for the physico-chemical interaction between such clusters, is defined as the external surface area of the clusters per unit weight, which is less than the initial specific surface. This phenomenon of reduction in specific surface has been further substantiated (Nagaraj and Srinivasa Murthy, 1986a), by a comparison of the computed and experimental plots of Na-montmorillonite (Klausner and Shainberg, 1967 and 1971). The computation of e vs log σ_v' relation has been carried out for the physico-chemical environment of the experiment, for different number of particles in a cluster, assuming the physico-chemical interaction between the clusters. The computational method followed is essentially, same as those proposed by Klausner and Shainberg (1967 and 1971) and Sridharan and Jayadeva (1982). Figure. 5.2 represents these plots and the superposed experimental plot. It is clear from Fig. 5.2 that the experimental curve cuts across the computed curves for different number of particles in a cluster. This indicates that the number of particles in a cluster increases with increase in consolidation pressure. There is ample evidence in published literature, which has been obtained from mircostructural observations, that there is aggregation of soil particles forming tactoids or clusters and the size of the tactoid or cluster increases with increase in effective stress (Aylmore and Quirk, 1960 and 1962; Scott, 1963; Sloane and Kell, 1966; Smart, 1967; Dickson and Smart, 1978; Stepkowska, 1982; Delage and Lefebrve, 1984; and others). It has also been indicated that these clusters break down during shearing, a phenomenon which we will discuss later to explain the shear behaviour of soils. Further, Nagaraj and Srinivasa Murthy (1983), have logically argued that, for any soil the reduction in specific surface at any stress level is a function of its original specific surface itself, i.e., for different soils with different specific surfaces, the changes in operating specific surface with stress, occur proportionately.

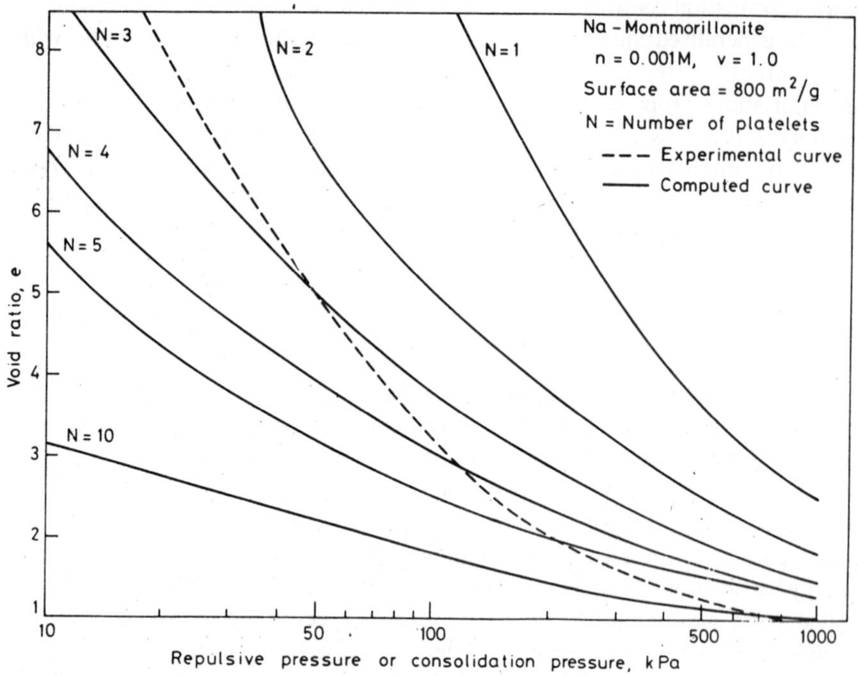

Fig. 5.2 Theoretical compression paths of Na-Montmorollonite for different assumed number of particles in a cluster.

In the above analysis, the specific surface computations were done for an ideal parallel plate configuration of particles or clusters, we have already seen in Chapter 3 that the microstructure of soils is more complex than that described by the parallel plate model. But we have also seen that the process of grouping of particles is a feature necessary to bring about irrecoverable volumetric compression whether it is a parallel plate microstructure, or the structure with large enclosed pores. This process of grouping of particles, being a stress induced phenomenon, and the initial microstructure per unit volume being similar for different soils, it is likely that the changes in specific surface due to grouping or the changes in the microstructure as a whole, will be similar for all soils, for a given change in stress. Thus the specific surface at any stress level is a definite fraction of the original specific surface for all soils. Hence the void ratio at liquid limit e_L which reflects the initial specific surface of the soil itself can be used for normalising the states under subsequent loading although there is a change in the specific surface during loading.

Since, equation 5.2, is generated from experimental results and all effects of the deviations from the assumptions made while deriving equation 5.1 are in built, the equation 5.2 can be treated as the generalised compressibility equation for uncemented normally consolidated saturated fine grained soils.

This generalised compression path in Fig. 5.1b has a practical significance. It can be used to identify the nature or complexity of the state of existence of the field

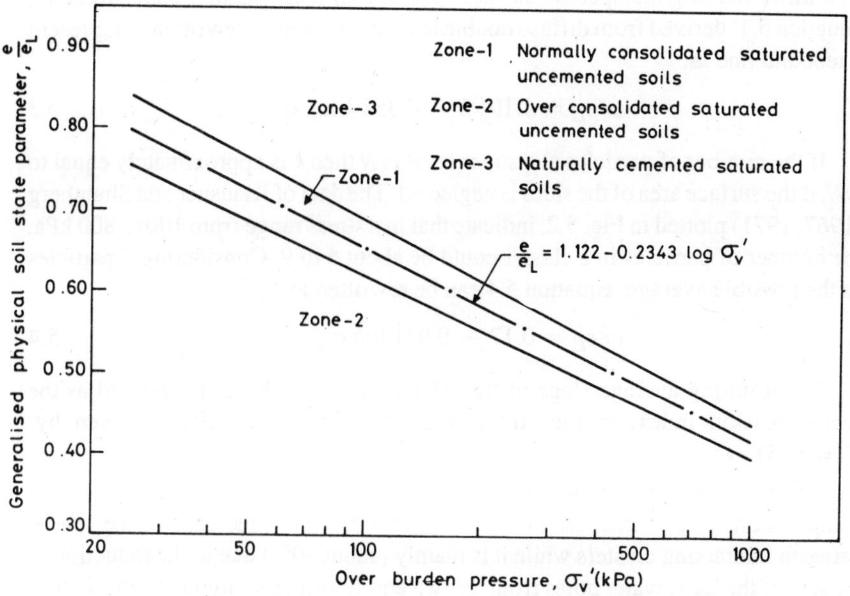

Fig. 5.3 Identification of soil states.

soil. For a field situation of saturated soil, with known effective overburden pressure, natural water content and liquid limit water content, a point can be located in Fig. 5.3. If this point falls within the prescribed band width of data points, the state of the soil can be inferred as normally consolidated. However this will not indicate whether it is uncemented or cemented. Since most of the cemented soils exhibit sensitivity, undisturbed and disturbed field vane tests can be used to compute the sensitivity. If the sensitivity is near unity, the soil state is normally consolidated uncemented state, then the generalised equation 5.2 can very well be applied for predicting the compressibility behaviour.

If the field state point plots in the region above the band width of data points, the soil state can be definitely identified as soft sensitive/ cemented soil. If the point is plotted in the region below the band width, it reflects an overconsolidated state, may be cemented or uncemented, the delineation of which is relatively difficult and requires strength data. We will discuss the compressibility model for the cemented states in a later chapter.

5.2.2 Over consolidated state:

The diffuse double layer theory again does not distinguish between the normally consolidated and over consolidated states. From mircostructural considerations, we have presented that there is a gradual grouping of particles during consolidation and this process reduces the operating specific surface. It we assume that at a particular consolidation stress level the operating specific surface is k times the original specific surface such that $0 < k < 1$, and during unloading from this stress level the clusters will not breakdown and thus the specific surface remains constant as k times the original specific surface, then for an ideal parallel plate structure, equation 5.1, derived from diffuse double layer theory can be rewritten to represent a rebound line as:

$$e/e_L = 1.191\,k - 0.358\,k \log \sigma'_v \qquad 5.3$$

If the number of particles at a stress level is N then k is approximately equal to $1/N$, if the surface area of the sides is neglected. The data of Klausner and Shainberg (1967, 1971) plotted in Fig. 5.2, indicate that in a stress range from 100 to 800 kPa, the number of particles in a cluster could be about 5 to 9. Considering 7 particles as the possible average, equation 5.4 can be rewritten as:

$$e/e_L = 0.17 - 0.051 \log \sigma'_v \qquad 5.4$$

The resulting average slope of the rebound line which can be defined as the recompression index, in the stress range of 100 to 800 kPa, is given by: $C_r = 0.051\ e_L$.

But we have seen that for the realistic microstructure, the volumetric compressions due to increase in load is not so much by decrease in the spaces between interacting clusters while it is mainly (about 90%) due to the reduction in the size of the large water pores from its own equilibrium requirement. This in turn forces the grouping of particles and hence reduction in specific surface. Theoretically the soil need not undergo further volume changes under further unloading or reloading, (but for the small elastic volume changes due to changes in spacing between particles) since the water pores do not have a compulsion to increase in size upon stress release, and hence to decrease again upon reloading. However the particles may have a tendency to group or to breakdown in order to reach the critical state of aggregation under the given stress, which depends on the fartherness of the current state from the critical state. Thus the overconsolidated soils may show some plastic volume changes although much less than the normally consolidated state, for the same change in stress state.

To examine the possibility of obtaining the unique recompression index, the linearised rebound lines of soils of Fig. 5.1a in e vs $\log \sigma'_v$ plot from different maximum stress levels have been considered (Fig. 5.4a). In some cases the linearisation is only approximate. Since, only the slope of the generalised line is of interest, the normalised lines from different stress levels have been plotted in Fig. 5.4b with a shift of origin to $e/e_L = 1$ at $\sigma'_v = 1$ kPa. All the data points fall in a very narrow band and the average linearised slope has a value of 0.046 with a correlation coefficient of 0.98. The position of the rebound line in $e - \log \sigma'_v$ plot depends on the maximum past pressure and hence it will be a point on the normally consolidated line which can be defined using equation 5.2. Let σ'_p be the maximum past pressure, then the rebound line through this point will have an equation of the form (Nagaraj and Srinivasa Murthy, 1985 and 1986a)

$$e/e_L = 1.122 - 0.188 \log \sigma'_p - 0.046 \log \sigma'_v \qquad 5.5$$

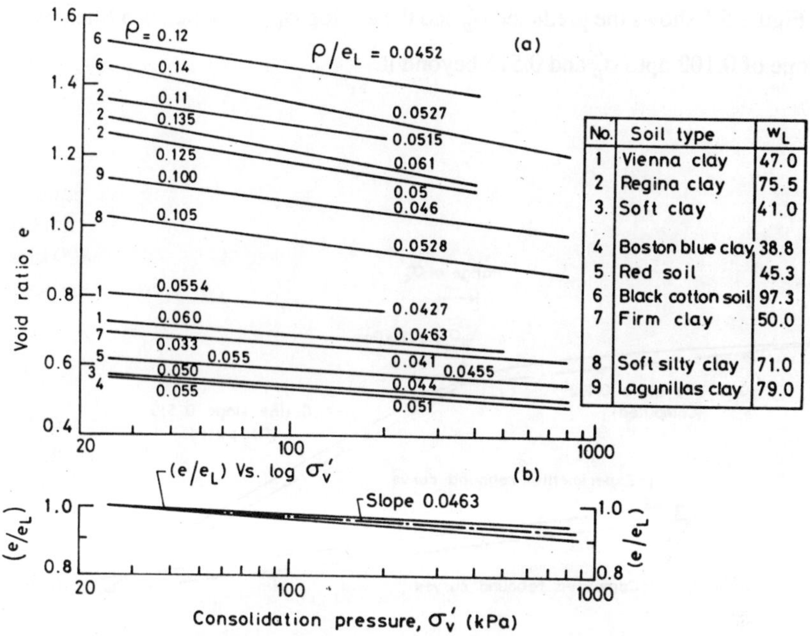

Fig. 5.4 Experimental average rebound - recompression paths of soils in Fig. 5.1 and their generalisation.

The practical advantage of equation 5.5 is that the preconsolidation pressure can be easily estimated without having to conduct a consolidation test and the entire curve can be predicted.

Example:

For a field situation given below, estimate the preconsolidation pressure and plot the possible compression curve.

* Ground water table at ground surface	
* Effective over burden pressure	23 kPa
* Field water content	46.8%
* Liquid limit of the soil	80%
* Specific gravity	2.75

Solution:

Now $e/e_L = w/w_L = 46.8/80 = 0.585$

From equation 5.5, $\sigma_p' = 332$ kPa.

To generate the $e - \log_{10} \sigma_v'$ curve, we know that upto the preconsolidation pressure equation 5.5 can be used to obtain the recompression path and beyond that since, it is normally consolidated, equation 5.2 can be used.

For the above data,

At $\sigma_p' = 332$ kPa, $e/e_L = 0.532$ and $e = 1.17$

Figure 5.5 shows the predicted σ_p' and the $e - \log \sigma_v'$ plot which is bilinear with a slope of 0.102 upto σ_p' and 0.515 beyond it.

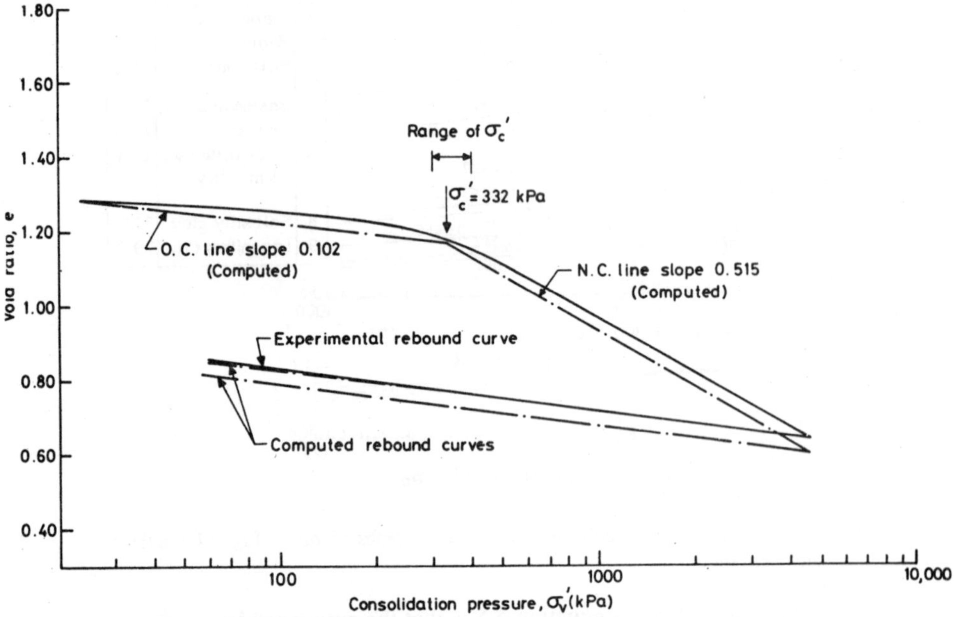

Fig. 5.5 Comparison of predicted and experimental compression paths of Spring field varved clay. (Data - Casagrande and Wilson, 1953).

5.2.3 Generalised compression index C_c

The compressibility of a soil is usually defined in terms of the negative slope of the $e - \log \sigma_v'$ plot, and this slope is defined as: $C_c = - de/d\,(\log \sigma_v')$

Now from the generalised equation 5.2, the compression index can be expressed as:

$$C_c = - de/d\,(\log_{10} \sigma_v') = 0.234\, e_L = 0.234\, G_s\, w_L \qquad 5.6$$

This equation predicts a constant slope for a given soil being a function of its liquid limit water content. Skempton (1953) has empirically shown that for remoulded soils, the compression index of the soil can be related to its liquid limit water content in the form:

$$C_c = 0.007\,(w_L - 10) \qquad 5.7$$

Terzaghi and Peck (1967) modified this equation to include some low to medium sensitive soils, in the form:

$$C_c = 0.009\,(w_L - 10) \qquad 5.8$$

A comparison of these equations indicates that the compression index calculated by equation 5.6 is in between the values obtained from equations 5.7 and 5.8, upto a liquid limit of 80%. It may be noted here that the specific gravity of the soil linearly affects the compression index as per equation 5.6.

More recently, Burland (1990) has tried to characterise the intrinsic compressibility behaviour of normally consolidated soils in terms of a new soil parameter, the void index $(e - e_{100})/(e_{100} - e_{1000})$ where e_{100} and e_{1000} are the void ratios at consolidation pressures of 100 and 1000 kPa respectively. From an analysis of an extensive set of data of 33 soils (Table 5.1) it has been shown that

Table 5.1 Intrinsic constants of compressibility for reconstituted natural clays (Burland 1990)

Soil	w_L	w_P	G_s	e_L	e_{100}	C_c	Reference
Lower Cromer Till	25	13.0	2.650	0.663	0.503	0.154	Gene (1982)
Boulder clay	28	14	2.69	0.753	0.520	0.120	Skemption (1944)
Silty clay	28	20	2.72	0.762	0.603	0.136	Ramiah (1959)
Magnus clay	35	17.2	2.73	0.956	0.760	0.270	Jardine (1985)
Grangemouth	35	21	2.78	0.973	0.659	0.229	Burland (1990)
Ton V	36	18	2.71	0.976	0.740	0.250	Skempton (1944)
Weald clay	39	19	2.73	1.065	0.770	0.240	Skempton (1944)
Boston blue clay	39	23	2.78	1.084	0.800	0.210	Skempton (1944)
Red soil	45.3	22	2.667	1.208	0.785	0.270	Nagaraj et al., (1986)
River Severn alluvium	46	25	2.59	1.191	0.800	0.210	Skempton (1944)
Wiener Tegel	46.7	22	2.76	1.288	0.859	0.297	Hvorslev (1937)
Oxford clay	53	27	2.57	1.362	0.960	0.300	Skempton (1944)
Ton IV	58	26	2.85	1.653	0.970	0.320	Skempton (1944)
Residual clay	58	27	2.74	1.589	1.024	0.337	Ramiah (1959)
London clay	62.3	24.3	2.73	1.707	1.200	0.446	Jardine (1985)
Belfast estuarine clay	67	30	2.66	1.782	1.000	0.320	Skempton (1944)
London clay	67.5	26.5	2.71	1.829	1.227	0.494	Som (1968)
Ganges delta clay	69	28	2.77	1.911	1.220	0.420	Skempton (1944)
Gosport clay	76	29	2.67	2.029	1.200	0.480	Skempton (1944)
London clay	77	28	2.71	2.087	1.280	0.490	Skempton (1944)
Brown London clay	88	32	2.65	2.332	1.320	0.560	Skempton (1944)
Black cotton clay	97.3	32	2.73	2.656	1.744	0.690	Nagaraj et al., (1986)
Kleinbelt Ton	127	36	2.77	3.518	2.180	0.910	Hvorslev (1937)
Argile plastique	128	31	2.58	3.302	1.820	0.810	Skempton (1944)
Whangamarino clay	136	61	2.78	3.740	2.440	0.797	Newland and Allely (1956)
SAIL soil	159.3	46	2.826	4.443	2.769	1.050	Nagaraj et al., (1986)

the required soil parameters e_{100} and C_c can be related to the liquid limit void ratio e_L of the soil in the form:

$$e_{100} = 0.109 + 0.679\, e_L - 0.089\, e_L^2 + 0.01\, e_L^3 \qquad 5.9$$

$$C_c = 0.254\, e_L - 0.04 \qquad 5.10$$

The resulting C_c and e_{100} values are shown to agree very well (Fig. 5.6) with those predicted from the generalised equations 5.2 and 5.5 over a wide range of liquid limit values.

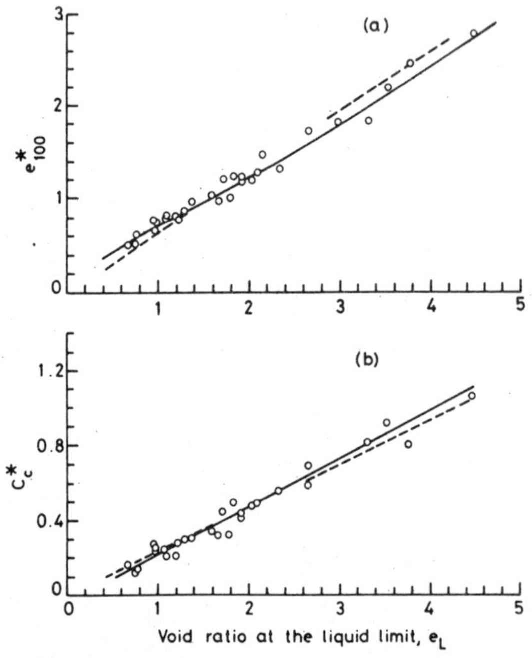

Fig. 5.6 Relationships of e_{100} vs w_L and C_c vs w_L (after Burland, 1990).

A feature to be noted in the above data is that all the soils presented by Burland and those presented earlier to arrive at the generalised compressibility relation, typically plot along the Casagrande A - line in the plasticity chart (Fig. 5.7). Most inorganic soils with sheet type clay minerals do follow the A - line. Clay minerals with tubular structure e.g., Hallosite or Allophane and also soils containing considerable amount of organic matter or silty soils may deviate much from the A - line. It has been found that for such soils, the above generalisations do not hold good. In fact, it can be theoretically shown that the generalisations will be valid only if w_P is a defined function of w_L. But, it appears that for the group of soils which follow a defined relation in the plasticity chart among themselves such as the three points circled on Fig. 5.7, the generalisation with respect to liquid limit

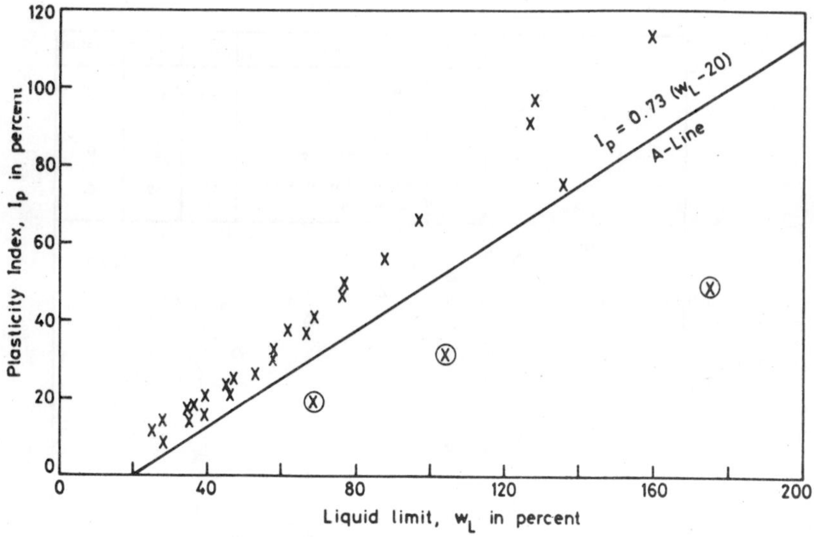

Fig. 5.7 Plot of Plasticity chart of soils in Fig. 5.6.

would still be possible, although the constants appearing in the relations may be different. This is very well indicated by the compression paths of the same three soils in Fig. 5.8a and their generalisation on e/e_L vs log σ' plot in Fig. 5.8b. It would be interesting and worthwhile to examine the possibility of generalisations of different soils following other defined plasticity relations.

Quite often, the changes in environmental conditions bring about a permanent change in the soil potential. The marine clay from Cochin area in India is a typical example of this category. The liquid limit of this soil reduces drastically when subjected to drying. The liquid limit reduced from 106 to 91 upon partial and to 85% upon complete air drying at 25°C. It is interesting to note that the compression paths of the natural and air dried samples can still be generalised with respect to their respective liquid limit values (Fig. 5.9).

The equation 5.2 has been obtained from test results of one dimensional consolidation. The diffuse double layer theory does not distinguish between one-dimensional, isotropic or any other loading. It only relates the distance between particles to the net repulsive pressure between them. From microstructural considerations we have seen that double layer interactions take place between the clusters placed on the periphery of large pores, perhaps which is similar in both one dimensional and isotropic stress states, except for the shape of the large pore. In every case, the net repulsive pressure between particles and hence the suction in the water phase will be equal to the applied mean principal stress $p' = (\sigma'_1 + \sigma'_2 + \sigma'_3)/3$. Hence we expect similar generalised behaviour for isotropic stress states also. Fig. 5.10a shows the e – log p' plots of 4 isotropically consolidated soils and Fig. 5.10b their collapse into a narrow band upon normalization with their respective void ratio at liquid limit water content. The data

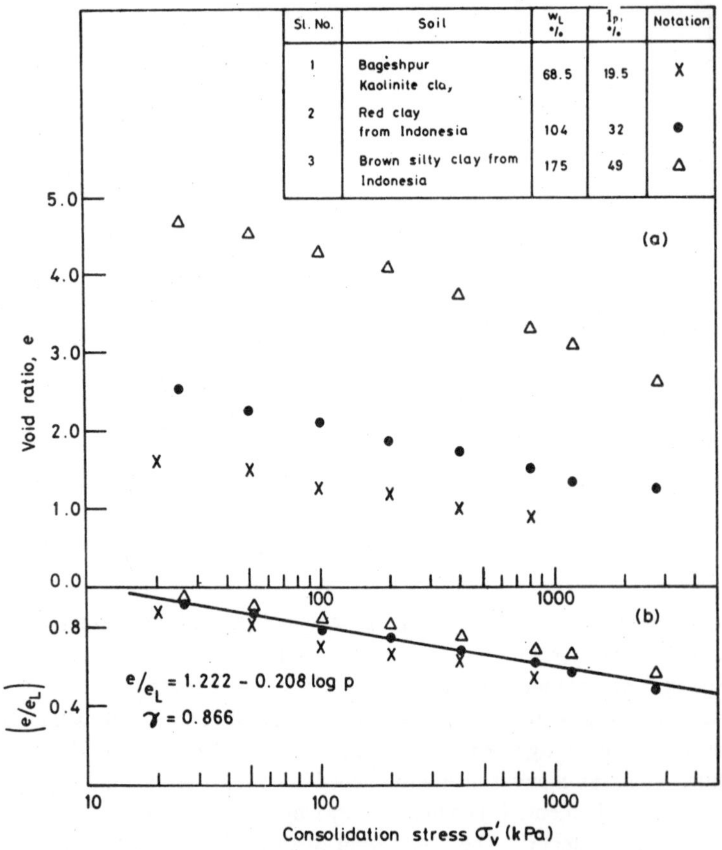

Fig. 5.8 Compression paths and their generalisation of soils deviating from A – line.

points of the generalised plot has been linearlised with in the stress range of 100 to 800 kPa in the form;

$$e/e_L = 0.99 - 0.21 \log p' \qquad 5.11$$

With a correlation coefficient of 0.98 and a standard error of estimate of 0.015. Since the consolidation paths with different constant q/p' values are all known to be parallel (as per critical state concepts) we should have expected the same slope for both isotropic and one dimensional compression lines. But the data show a small difference in the two slopes. The reason could be some overconsolidation effect in the samples isotropically consolidated to 100 kPa and lesser, since these samples were obtained by an initial one-dimensional consolidation sufficiently high to enable handling of triaxial samples. Hence we have assumed the slopes of the two paths to be equal for all practical purposes.

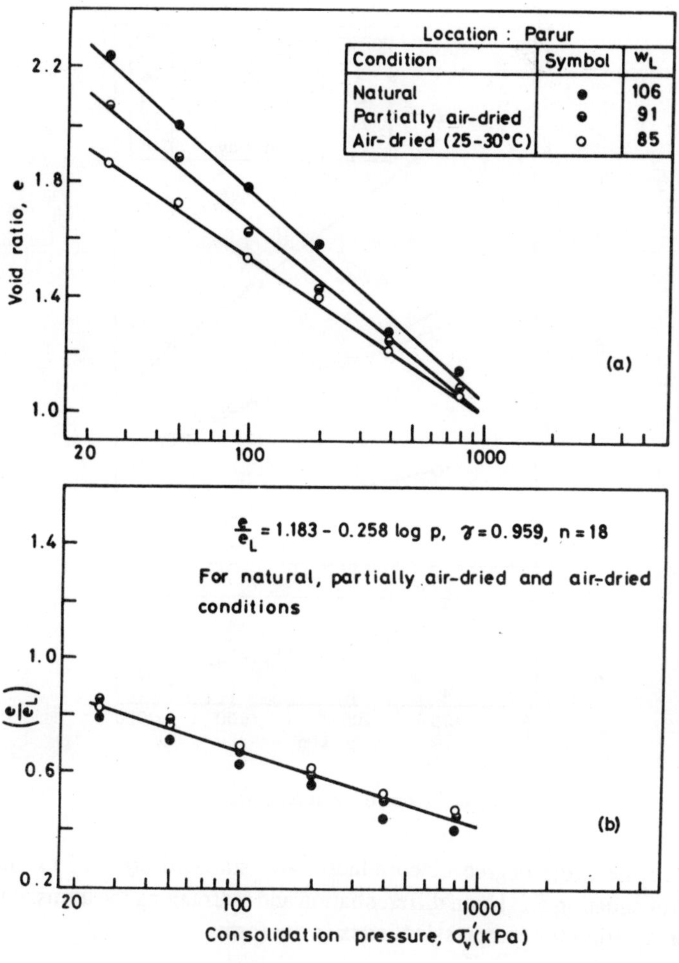

Fig. 5.9 Generalisation of compression paths of Cochin Marine clay under differnt environmental conditions

This equation 5.11 represents the generalised compressibility equation under isotropic stress conditions. Similar to one dimensional case, the equation for over consolidated state under isotropic condition can be written as;

$$\left\{\frac{e}{e_L}\right\} = 0.99 - 0.164 \log p_p' - 0.046 \log p' \qquad 5.12$$

where p_p' is the past maximum mean principal stress.

The plots of Fig. 5a and 5b upon plotting in double logarithm scale (Fig. 5.11), results in yet another linear equation of the form:

$$\log (e/e_L) = 0.1433 - 0.168 \log \sigma_v' \qquad 5.13$$

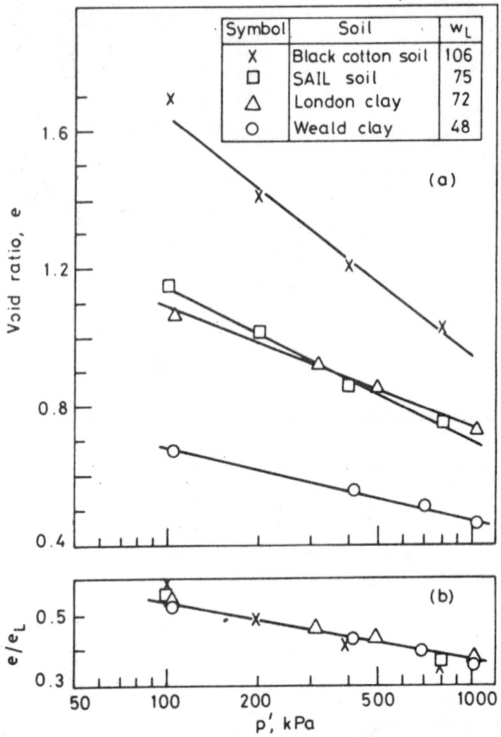

Fig. 5.10 Generalisation of isotropic compression paths.

with nearly the same degree of correlation and standard error of estimate as obtained for equation 5.2. Upon differentiation and rearranging the terms, it results in another equation for compression index in the form:

$$C_c = -de/d(\log_{10}\sigma_v') = 0.39\,e \qquad 5.14$$

This equation is quite different from equation 5.7. Sufficient evidences are available in literature (Koppula, 1981; Bowles, 1979 and Nishida, 1956) that empirical correlations are possible linking the compression index and the natural water content or void ratio for normally consolidated soils. For a general value of specific gravity of soils of 2.65, equation 5.14 can be written as:

$$C_c = 0.0103\,w_n \qquad 5.15$$

where w_n is the natural water content in percent. The practical significance of this equation is that the natural water content expressed as a decimal, itself represents the compression index, if the soil is normally consolidated and uncemented. i.e., for $w = 40\%$, $C_c = 0.4$ or for $w = 60\%$, $C_c = 0.6$ etc.

There are several other empirical equations to predict the compressibility of natural soils. They have been obtained based on specific soil data from different

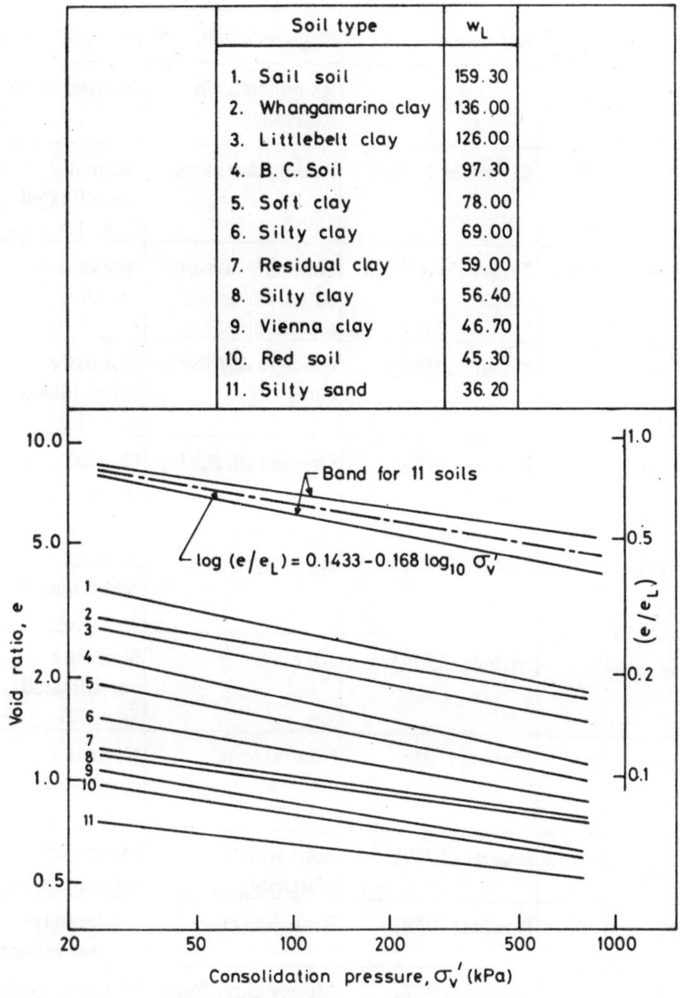

Fig. 5.11 Compression paths of soils in Fig. 5.1 in log e vs log σ'_v plot.

regions. Nagaraj and Srinivasa Murthy (1986), have made a thorough comparison of these equations and have identified the general state to which these equations are valid. Table 5.2 presents the details of this work.

5.3 Generalised shear strength behaviour

If the load or stress on a foundation or an earth slope is increased until the deformations become large and unacceptable, we say that the soil in that situation has failed. Here we are referring to the strength of the soil which is the maximum or ultimate stress the soil can sustain. As geotechnical engineers, we are concerned about the shear strength of the soil because in most of our engineering problems, failure results from excessive applied shear stresses. In the following section, we

Table 5.2 Compression Index equations (Nagaraj and Srinivasa Murthy, 1986)

Equation	Reference	Regions/Conditions of Applicability	
		As indicated in Reference	Inferred by authors
$C_c = 0.007\,(w_L - 10)$	Skempton (1944)	Remoulded clays	Normally consolidated $S_t < 1.5$
$C_c = 0.009\,(w_L - 10)$	Terazaghi & Peck (1967)	Normally consolidated, Moderately sensitive	Moderately sensitive $S_t < 5$
$C_c = 0.01\,w_n$	Kopulla (1981)	Chicago & Alberta clays	Normally consolidated $S_t < 1.5$
$C_c = 0.0115\,w_n$	Bowles (1979)	Organic silt & clays	Normally consolidated $S_t < 1.5$
$C_c = 1.15\,(e - e_o)$	Nishida (1956)	All clays	Normally consolidated $S_t < 1.5$
$C_c = 1.15\,(e - 0.35)$	Nishida (1956)	All clays	Normally consolidated $S_t < 1.5$
$C_c = 0.54\,(e_n - 0.35)$	Nishida (1956)	Natural soils	Normally consolidated $S_t < 1.5$
$C_c = 0.75\,(e_o - 0.50)$	Bowles (1979)	Soils with low plasticity	Moderately sensitive $S_t < 5$
$C_c = 0.0046\,(w_L - 9)$	Bowles (1979)	Brazillian clays	Moderately overconsolidated
$C_c = 1.21 + 1.055\,(e_o - 1.87)$	Bowles (1979)	Motley clays from Saopaulo city	Highly sensitive $S_t > 5$
$C_c = 0.30\,(e_o - 0.27)$	Hough (1957)	Inorganic silty sandy-silty clay	Overconsolidated
$C_c = 0.208\,(e_o - 0.0083)$	Bowles (1979)	Chicago clays	Moderately Overconsolidated
$C_c = 0.156\,e_o + 0.0107$	Bowles (1979)	All clays	Moderately Overconsolidated
$C_c = 0.5\,(\gamma_w/\gamma_d^2)^{1.2}$	Oswald (1980)	Soil systems of all complexities and types	Not applicable to any condition

Note: C_c – compression index; w_L – liquid limit water content; γ_d – dry density of soil at which C_c is required; e_o – initial or insitu void ratio; w_n – natural water content; γ_w – unit strength of water; S_t – sensitivity of the clay

will discuss, how the soil responds under applied shear stresses, and examine the possibility of generalising the shear behaviour of different soils.

During the application of shear stress it is possible to have either undrained or drained condition. Before discussing the details of generalisation, let us observe the stress-strain behaviour of both normally and over consolidated states under both undrained and drained conditions.

Consider an isotropic compression and rebound stress path as shown in Fig. 5.12. Let us have three specimens on this path. Specimen (1) normally consolidated

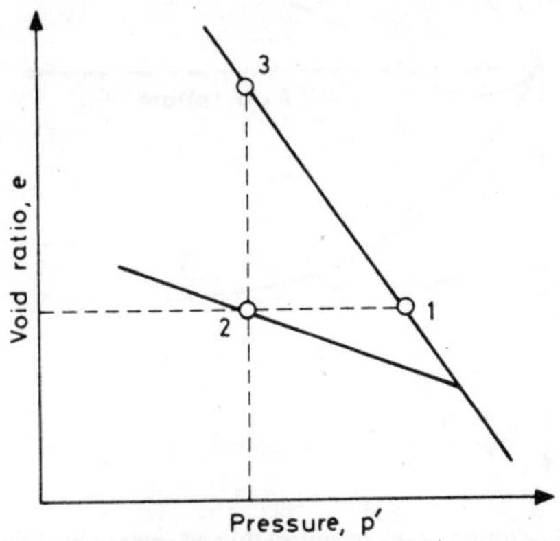

Fig. 5.12 Typical normally consolidated and over consolidated states under identical pressure and volume condition.

to a pressure p'_1 and void ratio e_1. Specimen (2) on the rebound line from a maximum past pressure of p'_p having the same void ratio of e_1 i.e., $e_1 = e_2$, and the corresponding confining pressure is p'_2. The specimen (3) is on the normally consolidated line with the consolidation pressure same as p_2 and the corresponding void ratio e_3 [$e_3 > (e_1 = e_2)$] and [p'_1 ($p'_2 = p'_3$)]. Let us apply deviatoric stress, incrementally, under undrained condition on samples (1) and (2). Figure 5.13, shows schematically the responses of the two specimens during shearing, in the usual q versus ε_a (axial strain) plot and the pore pressure, u versus ε_a plot. For specimen 1 both shear stress and the excess pore water pressure (positive) increase with increase in axial strain. For specimen (2), the shear stress increases in the same way as for specimen (1) but may be with a little slower rate due to lower confining pressure. However the final strength value is nearly same. But the pore pressure in specimen (2) is positive and increases initially but after a particular strain level decreases and even reaches negative with increase in axial strain.

Now let us apply incrementally deviatoric stress on two new specimens, specimen (2) and specimen (3) allowing complete drainage of water so that the excess pore water pressure in the sample at any stage is zero. The samples undergo volume changes. Figure 5.14a represents the q vs ε_a and ε_v vs ε_a plots for both the specimens. The shear stress for specimen (2) rapidly increases with a higher

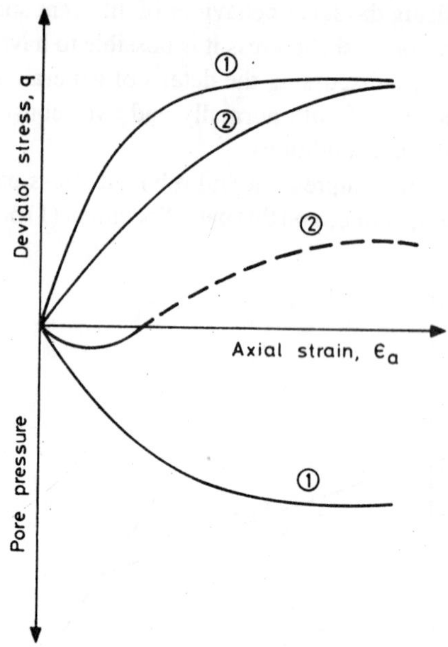

Fig. 5.13 Typical stress-strain curves of normally and over consolidated states at the same void ratio in undrained test.

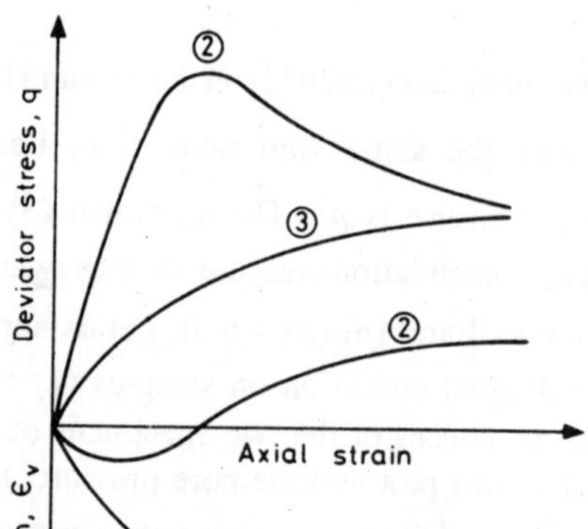

Fig. 5.14 Typical stress-strain curves of normally and over consolidated states at the same confining pressure in drained test.

modulus, reaches a peak and softens beyond, with increase in axial strain. The specimen (3) will have shear stress non linearly increasing to reach an asymptotic value at large strains. It is interesting to note that, both the specimens have the same ultimate shear strength. Regarding volumetric strain - axial strain plot, there is a continuous compression for specimen (3) while, specimen (2) undergoes initial compression and then after a particular strain level expansion or dilation. It is equally interesting to note that the final void ratio of both the specimens are nearly same though they had very different initial void ratios. This final state is normally referred to as critical void ratio.

It can be noted that an observable peak strength exists only in the case of drained shearing on overconsolidated sample. For all other cases the ultimate strength is the maximum strength which corresponds to the critical state.

It is an observed fact that the behaviour of a given soil can be generalised from one stress level to the other. Figure 5.15a schematically shows the undrained behaviour of a normally consolidated soil at three confining pressures. Figure 5.15b

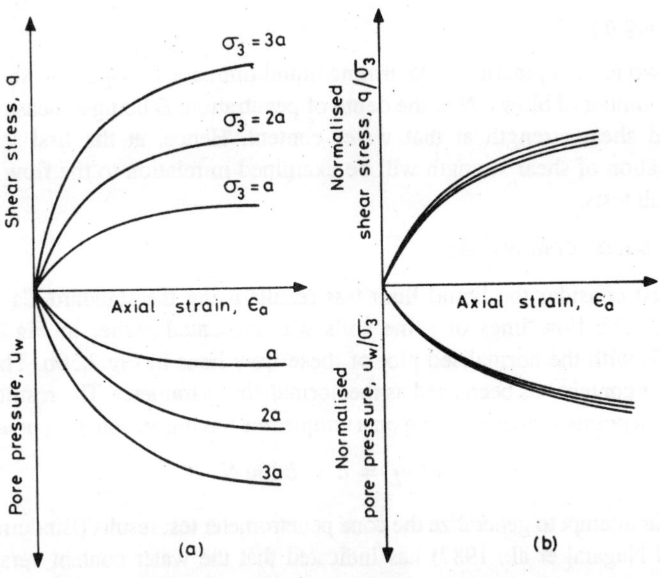

Fig. 5.15 Stress-strain curves under different confining pressures and their normalisation.

shows their generalisation upon normalising the deviator stress with their respective confining pressure. Similar generalisations are possible even for drained shearing. More details about these aspects are covered later.

The differences in behaviour observed earlier, are so wide in their variety but, the recent attempts of generalisation using elasto-plastic models have indicated the possibility of reducing them into a single frame work. Further we attempt to bring a broader generalisation encompassing the different soils also. We first consider the generalisation of the ultimate strength or strength at critical state.

5.3.1 Normally consolidated state

It is well known that the shear strength of a particulate system is a function of the normal stress. The mobilisation of shear strength is strain dependent. At large strains the strength is referred to as ultimate strength. The term 'residual strength' is often used which is supposedly different (much lower) from the ultimate strength, and is thought to be realised after very large failure deformations. But we are of the opinion that the two are not different. Once the ultimate or the critical state is reached, there can be no further changes in stress states. The values reducing beyond ultimate to finally reach lower values reported in literature may be due to the formation of large shear planes, with the soil on either side sliding as rigid blocks, and hence the entire soil mass not taking part in the shearing resistance.

The critical state concept has indicated that for any soil the consolidation lines in e vs log p' plot with different constant (q/p') values will all be parallel and also they are parallel to the void ratio vs shear strength line. Since, (e/e_L) versus log p' relation is a generalised relation, we can expect similar generalisation in (e/e_L) vs log q plot.

The two tests in practice to determine liquid limit are analogous to shear tests, with the number of blows, N or the depth of penetration, D being a measure of the undrained shear atrength at that water content. Hence, at the first level, the generalisation of shear strength will be examined in relation to the flow lines of liquid limit tests.

Generalisation of flow lines

Let us first consider the liquid limit test results using the standard Casagrande apparatus. The flow lines of some soils was presented earlier in Fig.2.50a of Chapter 2, with the normalised plot of these flow lines in Fig.2.50b. The liquid limit water content has been used as the normalising parameter. The resulting plot has the data points virtually falling on a unique line having an equation of the form:

$$w/w_L = a - b \log N \quad 5.16$$

Similar attempt to generalize the cone penetrometer test results (Bindumadhava, 1985 and Nagaraj et al., 1987) has indicated that the water content versus cone penetration plots can also be normalised (Fig. 5.16) using their respective water contents at 20mm penetration and the resulting equation is of the form:

$$w/w_L = 0.77 + 0.012\, D \quad 5.17$$

where D refers to the cone penetration in mm.

The additional benefit of equations 5.16 or 5.17 is that it enables to determine the liquid limit of a soil with just one trial, which is referred as the one point method of liquid limit determination. In the above equation, for measured values of w and N or D, obtained from a single trial, liquid limit water content can be obtained.

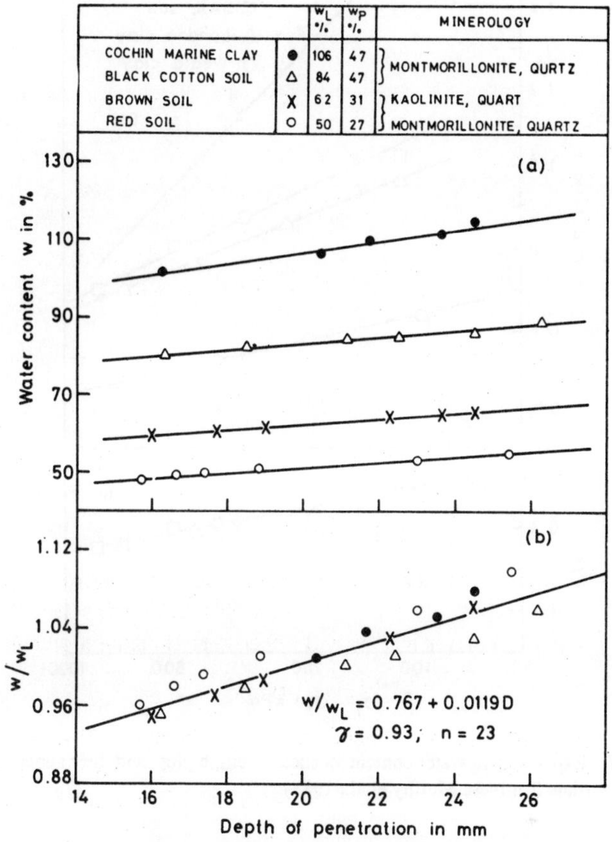

Fig. 5.16 Generalisation of flow curves.

The possibility of generalisation of ultimate shear strength of soils at lower water contents/higher consolidation pressures has been examined (Srinivasa Murthy et al., 1988) by considering the data from Henkel (1960) on Weald clay and London clay together with the data of two other soils. The isotropic consolidation data for these soils have been presented in Fig. 5.17a and their generalised plots in Fig. 5.17b. Figure 5.17a shows the plot of the void ratio versus shear strength, plotted in semi log scale for the same 4 soils. Figure 5.17b represents the normalised plots of (e/e_L) versus log q relation. All the data points fall within a narrow band and the best fit linear equation is of the form:

$$e/e_L = 0.932 - 0.21 \log q_f \qquad 5.18$$

Similar excercise on the mean principal stress at failure has yielded an equation of the form:

$$e/e_L = 0.952 - 0.21 \log p'_f \qquad 5.19$$

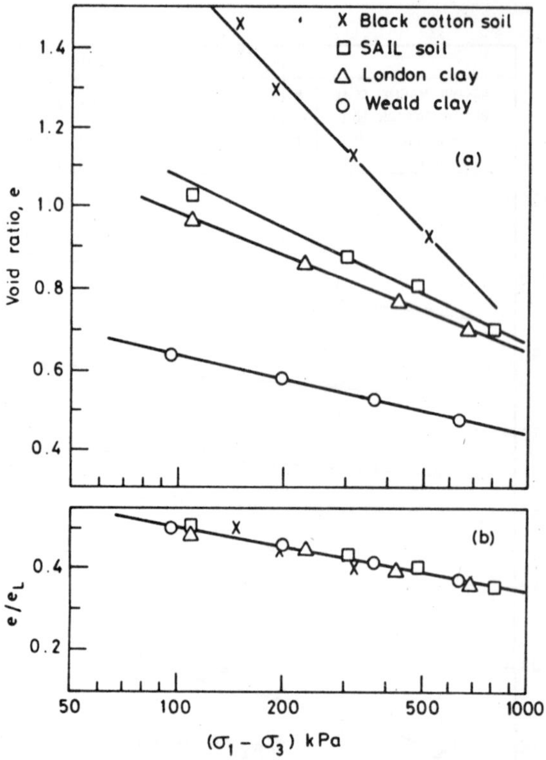

Fig. 5.17 Experimental water content vs shear strength plots and their generalisation (data Srinivasa Murthy et al., 1988).

Both these equations have very high correlation coefficient (0.98) and low standard error of estimate (0.015). This clearly brings out that the ultimate shear strength behaviour of normally consolidated fine grained soils can be generalised.

While attempting (Srinivasa Murthy et al., 1988) for this mode of generalisation for Henkel's data on Weald clay and London clay, the e/e_L vs log p' (or the e/e_L vs log q) plot for the two soils, which was expected to be a unique curve, all though had the same slope for both the soils, showed a small spatial shift in the e/e_L axis. An examination of water content contours of these soils obtained from both drained and undrained shear test paths (Henkel, 1960) had indicated, for the same level of stress or mobilisation of deviatoric stress, the ratio of water contents for the two soils to be nearly a constant value of 0.67 at all stress levels. If this logic is extended back to the level of liquid limit state, which also reflects an identical state of stress for all soils, the ratio of liquid limits for these two soils should have been 0.67. The reported liquid limit values for these two soils are 43 and 78 whose ratio is 0.55. But, Norman (1958) has reported elsewhere liquid limit values of 48 and 72 for the same two soils, whose ratio is exactly 0.67. Based on all these, for all further generalisations, these slightly changed values of liquid limits were used (Srinivasa

Murthy et al., 1988). Even without these changes, the generalisation would have been still possible but with a higher dispersion.

In essence, the above generalisations mean that in fine grained soils the equilibrium pressure and consequently the mobilised shear strength are the functions of the half space distance between the clusters or particle units. With the complex microstructure having three levels of pores, it can still be argued that the above generalisations are acceptable. It has been brought out (Srinivasa Murthy et al., 1991), that during shearing the spherical pores will be distorted to an ellipsoid to maintain equilibrium under deviatoric stress conditions. This together with the change in the mean principal stress dictate the shape and size of the elloposid. The interparticle spacing between the clusters, placed on the periphery of the ellipsoidal large pore depends on the effective mean principal stress, which inturn induces the same p' in the large pore. The changes in the perimeter and the equilibrium mean principal stress dictate the grouping or breakdown of clusters. Since all these manifestations are due to the changes in the applied stress which is reflected in the stress conditions of the continuous fluid phase, the generalisations have been possible.

The shear strength of a soil is conventionally defined in terms of two components viz., a cohesion, c which is normal stress independent and a frictional component, which is normal stress dependent. The latter component is expressed interms of angle of internal friction, ϕ'. But now it is fairly accepted that there is truly nothing like cohesion unless the soil is chemically cemented. The cohesion component which is commonly observed in overconsolidated states, or partly saturated soils and in most of the undisturbed soils is only when peak stress values are considered. Also it is known that these components are influenced by several factors including the drainage condition and method of testing.

The consequence of the two plots, (e/e_L) vs log q and (e/e_L) vs log p'_f being parallel is far reaching. In the development of critical state concepts, the ratio of (q/p') at failure, which is found to be a constant for a soil, is referred as the friction factor. This factor $(q/p')_f = M$, is related to the angle of internal friction in the form:

$$M = 6 \sin \phi'/(3 - \sin \phi')$$

Now from equations 5.18 and 5.19, (e/e_L) can be eliminated to obtain $(q_f/p'_f) = M = 0.803$. Since this has resulted from the generalised equations, it implies that it must be unique and same for all soils, i.e., the effective angle of internal friction is same and is about 21°. Figure 5.18, shows the Mohr's envelope for the same soils obtained from the same tests. It is clear that an approximately constant value of ϕ' may be assigned for all the soils. To support this further, Table (5.3) presents the data from Schofield and Wroth (1968), for five proven normally consolidated soils at their critical states. The maximum variation in the value of M is only from 0.845 to 1.02 which corresponds to a variation of ϕ' from 22 to 26°. Considering the range of variation in the soil type (with liquid limit water contents varying from 43 to 127%), this variation in M value is not considerable and hence constant M value can be accepted. However in literature values of ϕ' varying from 12° to 35° have been reported. The deviations from the constant value could be due to:

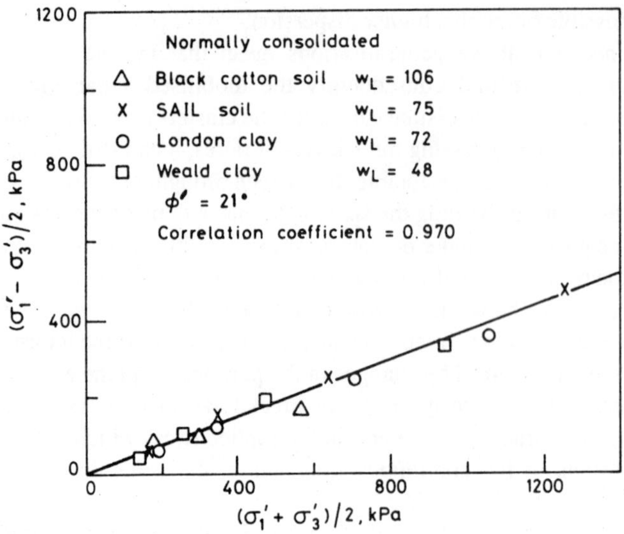

Fig. 5.18 Mohr's envelope for the data in Fig. 5.17.

(1) the soil may be sensitive with cementation,
(2) the state of soil may not reflect the true effective stress as in the case of partly saturated states,
(3) critical state might not have reached as in overconsolidated states and
(4) strain rate may be significantly different.

Table 5.3 Liquid limit, M and ϕ' values of Normally consolidated clays (Data from Schofield and Wroth, 1968).

Soil	w_L	M	ϕ'
Klein Belt Ton	127	0.845	21.74
Weiner Tagel	47	1.010	25.61
London clay	78	0.888	22.75
Weald clay	43	0.950	24.21
Kaolin	74	1.020	25.84

The existence of a unique friction factor is also reflected from the assumptions made in deriving the generalised equations. For the case of continuous water phase and absence of direct particle - particle contact, the shear failure plane ought to pass through the interacting fluid phase and then the shearing resistance could be independent of the mineral type. The mobilization of shearing resistance in such a physico-chemical force field which is electrical in nature, has been conceptually explained by Trollope (1960) and has been termed as colloidal friction. This is analogous to the case of two magnets of like poles placed at a distance apart with the help of an external force which needs certain amount of force for lateral displacement.

Two factors decide the magnitude of shearing resistance, viz, the resistance to displacement between two interacting particle units and the number of such units available over a unit cross sectional area of shearing. The theoretical d vs log $(R - A)$ relation at a given physico- chemical environment, being unique for all clays indicates that the surface charge activity is perhaps of the same order for all clay minerals. Also, the failure plane being essentially through the fluid phase, it is perhaps the magnitude of mid plane potential that controls the resistance to shearing between two particles, which is the same for all clay minerals for a given pore fluid environment. Thus the resistance to shearing between two particles is likely to be of the same magnitude for different clay minerals. Secondly as already seen in Chapter 3 the micro-structure i.e., distribution of particles and of pores being nearly same for all clay types, the overall frictional resistance would also be of the same order for all soils. However it is quite possible that the frictional resistance can be different for different microstructural configurations for the same soil, say for e.g., the compacted soil with its greater extent of particle clustering, may have a higher friction angle than a remoulded clay.

Of course, it may be noted here that the angle of internal friction, ϕ' or the friction factor, M even from colloidal friction point of view, may not be a pin pointed value. It is generally observed that, the angle of internal friction is slightly higher for soils with massive clay minerals like kaolinite. But this difference would be small and definitly cannot vary over 12 to 35 for the usual sheet type clay minerals.. In any case the constant value of ϕ' obtained for different soils and depicted in Fig. 5.18 appears to be the minimum that any soil can mobilise. Hence, for most of the soils a constant value of $\phi' = 21°$ i.e., $M = 0.8$ may be adapted in analysis and designs.

The basic structural constituets of all clay minerals belong to the same family of alumina and silica and hence it is likely that all clay minerals have the same surface frictional characteristics. This being the fact, it is not possible to explain the following two observed behaviors with the conventional contact model.

(a) Two soils can mobilize different strengths at the same void ratio and
(b) Two soils can have the same strength at widely different void ratios.

The consideration of the mode of stress transfer in fine grained soils (discussed earlier), through physico-chemical interaction can explain the above facts.

The strength mobilization is essentially a function of the distance between the interacting particle units rather than the void ratio. Figures 5.19 and 5.20, show the explanation based on the generalised behaviour.

5.3.2 Empirical Relations - Critical Examination

Some of the empirical relations for predicting the shear strength of soils are popular with the professional engineers. In this section, we will examine the validity of, or possible scientific basis for these relations.

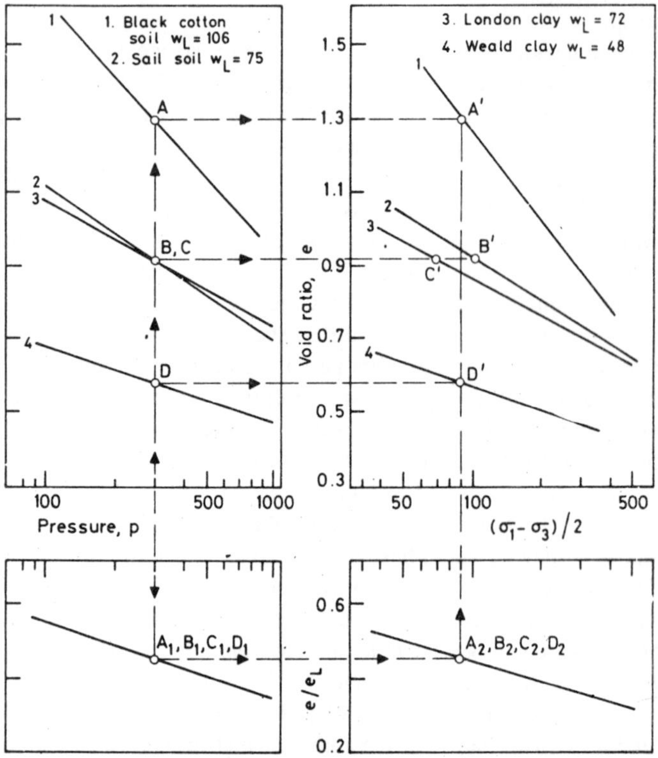

Fig. 5.19 Shear strength of different soils at the same consolidation pressure.

(1) For normally consolidated undisturbed clays (Skempton, 1953 and Skempton and Henkel, 1953)

$$c_u/\sigma_v' = 0.11 + 0.37\, I_P \qquad 5.20$$

where c_u is the undrained shear strength
σ_v' is the effective overburden pressure
I_P is the plasticity index represented as a decimal

(2) Liquidity index versus remoulded shear strength relation (Houston and Mitchell, 1969; Schofield and Wroth, 1968)

c_u/σ_v' vs I_P relationship

There have been several studies (Bjerrum, 1954; Grace and Henry, 1957; Wu, 1958; Osterman, 1960; Metcalf and Townsend 1960; Leonards 1962, Bishop and Henkel, 1962; Lumb and Holt, 1968; Cox, 1970 and others), in the past to examine the validity of Skempton's relation. A large number of data agree with this relation and there is an equally large volume of data that contradicts it. Most of the above examinations have been only experimental. Only a few (Kenney, 1959 and 1960)

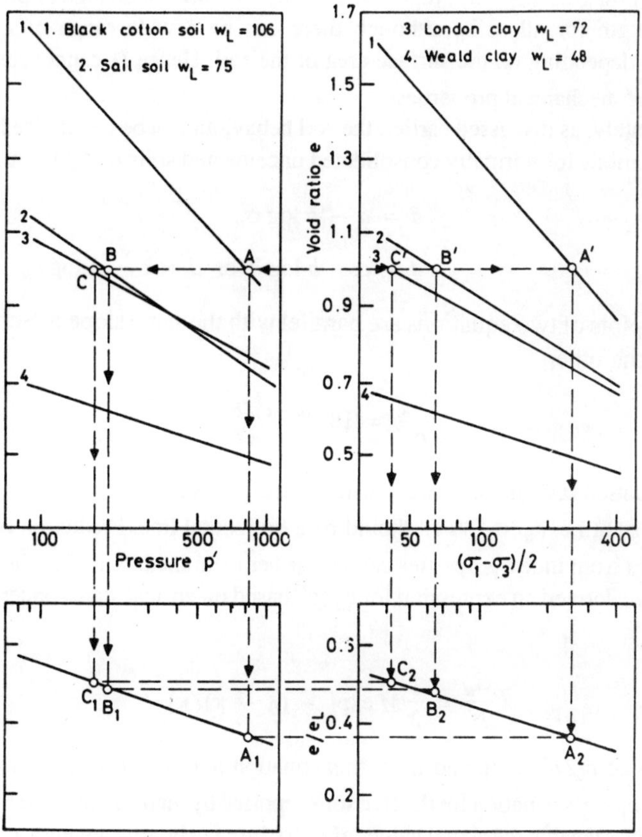

Fig. 5.20 Shear strength of different soils at the same void ratio.

have concluded that the c_u/σ_v' ratio for undisturbed soils is strongly dependent on geological and time history of the soil and cannot be reflected by two inferential parameters ($I_P = w_L - w_P$) determined on completely remoulded soil. Sridharan and Narasimha Rao (1973) have concluded that no linear increase in c_u/σ_v' occurs with increase in I_P and instead both theoretical and experimental results tend to show a decrease in c_u/σ_v' with increase in I_P. But no attempt has been made to understand the basic considerations.

Plasticity index, being the numerical difference between liquid and plastic limits can be expressed in terms of the d spacings corresponding to these states as

$$I_P = kd_L - kd_P \tag{5.21}$$

$$= k\,(d_L - d_P)$$

where k is the functional relationship between d spacing and the water content.

Since d_L and d_P are uniquely related to specific values of equilibrium pressures and shear strengths, $(d_L - d_P)$ represents a constant range of change of pressure or shear strength for all soils, although there can be a wide range of I_P values $(w_L - w_P)$ depending on the surface area of the soil. Hence I_P cannot reflect any variation of mechanical properties.

Alternately, as discussed earlier, the soil behaviour can be generalised to yield unique relations for normally consolidated uncemented soils in the form

$$d = a - b \log \sigma_v' \quad 5.22$$

$$d = c - b \log q \text{ or } d = c - b \log c_u \quad 5.23$$

where the plots of two equations are parallel with the same slope b. Subtracting one from the other

$$\frac{c_u}{\sigma_v'} = 10^{-(a-c)/b} \quad 5.24$$

In this equation a, b and c being constants irrespective of the soil type, $c_u/\sigma_v'$is also a constant i.e. c_u/σ_v' vs I_P should be a horizontal line. Prediction of c_u/σ_v' magnitudes from index properties can be further examined. Schofield and Wroth (1968) have derived an expression for c_u/p', based on critical state concepts in the form:

$$\frac{C_u}{P'} = \frac{1}{2} M \exp[-(\lambda - \kappa)/\kappa] \quad 5.25$$

Values of c_u/p' computed from this equation and the corresponding values using Skempton's equation for the five soils reported by them are indicated in Table 5.4. It can be seen that the actual values of c_u/σ_v' are fairly constant with an average of 0.24 as against a variation of 0.2025 to 0.4467 given by Skempton's equation, for the reported values of λ, κ, M, w_L and w_P. In eqation 5.25, p is the isotropic consolidation pressure. For K_o consolidation pressures, values of c_u/σ_v' would be slightly higher but still would be of the same range for all soils. Further, Skempton's relation predicts higher strengths for highly plastic soils than for low plastic soils at the same confining pressure. which is contrary to the generally observed behaviour.

From the above discussions it can be concluded that c_u/σ_v' vs I_P relation is not tenable for uncemented normally consolidated soils. From a thorough examination of Skempton's work, Srinivasa Murthy et al. (1986) have shown that the data used to arrive at this relation were of sensitive soils, and have also shown that the possibility of such a relation is much less reasonable for sensitive soils. This is because in sensitive soils, c_u is nearly constant for all confining pressures upto the yield stress, σ_y and hence c_u/σ_v' varies with the level of σ_v', even for a single soil. Despite the above limitations this relation is being used extensively by geotechnical engineers , this being the only available simple tool.

Table 5.4 Comparison of c_u/σ'_v values

Soil	Klein Belt ton	Weiner Tegel.V	London clay	Weald clay	Kaolinite
w_L	27	47	78	43	74
w_P	36	22	26	18	42
λ	0.356	0.122	0.161	0.093	0.260
κ	0.184	0.026	0.062	0.035	0.050
M	0.850	1.010	0.890	0.950	1.020
c_u/p' (eq. 5.25)	0.261	0.230	0.240	0.245	0.227
c_u/σ'_v (eq. 5.20)	0.447	0.203	0.271	0.203	0.228

I_L vs remoulded strength

Based on the observation of a definite trend in the relation between Liquidity index and remoulded strength of several clays (Skempton and Northey, 1953; Schofield and Wroth, 1968; Houston and Mitchell, 1969), Wroth and Wood (1978) have linearly idealised this relation. However, no scientific basis for such a relation was identified.

Liquidity index being a ratio of the difference in natural water content and plastic limit to the plasticity index, can be expressed in terms of the 'd' spacing between particles as

$$I_L = \frac{w_n - w_P}{w_L - w_P} = \frac{kd - kd_P}{kd_L - kd_P} = \frac{d - d_P}{d_L - d_P} \tag{5.26}$$

Since the functional form 'k' which reflects the soil type vanishes, and d_L and d_P are constants for all soils, I_L reduces to be a function of 'd' alone.

As seen earlier, the generalised state parameter, e/e_L is also a function of 'd' and hence there must be a relation between I_L and e/e_L. From the data of consistency limits as reported by various sources and at different assumed water content values, Srinivasa Murthy et al., (1986) have shown that the relation between the two can be expressed by the equation (Fig. 4.6 in Ch. 4)

$$I_L = 1.55\ (w/w_L) - 0.56$$

with a correlation coefficient of 0.992. As it has been proved based on Gouy-chapman theory, that there is a unique relation between e/e_L and q, I_L versus remoulded strength relation should also be tenable and will be applicable for normally consolidated states. The resulting equation will be of the form;

$$I_L = 0.89 - 0.33 \log q \tag{5.27}$$

But then e/e_L is a simpler parameter to correlate the mechanical properties than I_L, because this does not involve the additional parameter plastic limit.

5.3.3 Overconsolidated state

The basic considerations presented earlier clearly indicate that the shear failure plane passes through the interacting fluid phase and hence there cannot be any difference in the shearing resistance of the normally and overconsolidated states. However we know that there is a difference in the level of interaction because of reduced operating specific surface due to increased aggregation in the overconsolidated states, which also reduces the magnitude of plastic deformation modulus. In addition, it is likely that the clusters formed during consolidation may undergo breakdown gradually with shearing. At every stage of shearing, the d spacing should be the same for both normally and overconsolidated states. However at large strains, i.e., at critical states, when the dismembering of clusters is complete, the angle of shearing resistance will be same for both normally and overconsolidated states. To verify this, shear test results of Weald clay and London clay (Henkel, 1960), overconsolidated to a maximum pressure of 120 psi (846 kPa), and those of SAIL soil and Black cotton soil (Srinivasa Murthy et al., 1988), over consolidated to a maximum pressure of 800 kPa, have been examined. Figure 5.21 shows the Mohr's envelope at large strains. It is clear that the angle of internal friction is nearly same as that for normally consolidated soils (about 22°). The shear strength at critical state can be represented by the same equation (equation 5.18), provided we know the (e/e_L) value at that state. For undrained tests, the initial value of (e/e_L) itself is the value at critical state. For drained tests, however we can calculate the shear strength directly using the known stress path and the generalised Mohr's strength envelope (Fig. 5.21).

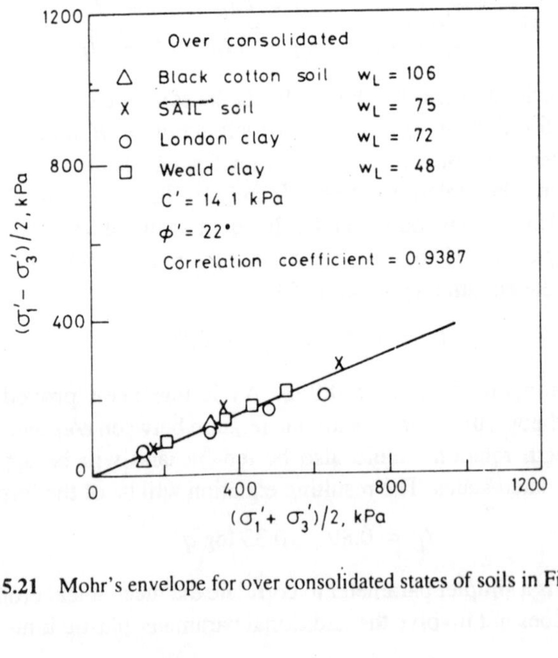

Fig. 5.21 Mohr's envelope for over consolidated states of soils in Fig. 5.17.

Often in engineering designs, involving overconsolidated soils we need to know the peak strength values. This value depends on the over consolidation ratio and the maximum past pressure. Since, the consolidation rebound lines of different soils could be generalised interms of an unique slope, we believe that the peak shear strength can also be generalised. To examine this possibility, we considered the test data of two soils (SAIL and Black cotton soil) consolidated to a maximum past pressure of 800 kPa. The $e - \log q$ plots of these two soils indicated a constant slope of 0.076 e_L. More work is required with samples of higher OCRs and rebounded from different maximum pressures before proposing a generalised equation.

5.4 Stress - Strain Relations

The above discussed one-dimensional or isotropic compression paths are only the description of soil response under particular stress paths. The shear failure conditions correspond only to particular states. We must be able to know the response of soil under any increment of general stress state, and at any stage of shearing i.e., even at prefailure stages. Although this book is mainly intended to bring out the generalisations in the behaviour of differnt soils, obtaining the response of a soil under any required stress path, knowing its response under some other stress path is an equally important aspect of prediction. This has been made possible by the recent developements in the constitutive relations of soils based on sound principles of continuum mechanics. It is now well established that the behaviour of soils is well described by elasto-plastic theories. The entire work in this direction was signalled by the pioneering work of Roscoe and his coworkers of Cambridge University in late fifties. The original elasto-platic model (Roscoe et al., 1958 and Schofield and Wroth, 1968) was called the Cam-clay model. Today there are hundreds of versions of elasto-plastic models, which are basically modifications or improvements over the original model. But the fundamental theory behind all these models is the same. In the following section, we will present a brief discussion on only the Cam-clay model, since our aim is mainly to bring out the possiblity of generalisation of the constitutive relations for different soils.

The development in the field of constitutive modelling has followed a rather unhealthy trend. Any observed feature of soil behaviour at macrolevel is being modelled by introducing additional parameters, with the result that the present day refined models (for fairly general purpose) are ending up to be too very complicated involving too many input parameters to be determined. It may be noted that the original Cam-clay model needed only 5 soil parameters while the present day models require in the range of 25-35 parameters, most of which do not have any physical meaning. Scott (1989) has rightly warned against such a trend and has emphasized the need for probing into the fundamental behaviour and understand the mechanisms involved, which alone can result in a simple, and comprehensive modelling. The 'Revised Cam-clay' model (Srinivasa Murthy et al., 1991) was perhaps a first step in this direction, in which an attempt was made to consider the mechanisms involved at microlevel, which is in continuation with what has been discussed in earlier chapters. Hence a special discussion on only this Revised

Cam-clay model will be made although there are several other models known to be predicting the soil behaviour reasonably well.

5.4.1 Basic plasticity theory

A perfectly plastic material deforms continuously at constant stress with no change in volume, once the yield stress has been reached. Soil and other particulate materials are rarely perfectly plastic but are either work hardening or work softening. They undergo plastic volume changes and the material with the changed volume may behave all together as a different material with its own yield properties. Most of the current day elasto-plastic theories are extensions of perfect plasticity by incorporating the effects of work hardening.

The basic features of an elasto - plastic model for a work hardening material are (Chen and Mizuno, 1990):

(1) The existence of an initial yield surface and subsequent loading surfaces.
(2) Formulation of an appropriate hardening rule that describes the evolution of the subsequent loading surfaces.
(3) A flow rule that specifies the general form of stress- strain relationship.
(4) A failure criterion which defines the limiting or ultimate stresses.

The first requirement specifies that a stress function which defines the limit of the elasticity of the material and the onset of plastic deformations, exists. It is called the 'initial yield function', before the plastic deformations occur. The yield function is called the 'loading function' beyond the initial yielding.

The second requirement is about the hardening rule. Three types of hardening rules, viz., isotropic hardening, kinematic hardening and mixed hardening have been applied to soils. In the isotropic hardening the yield surface expands uniformly without distortion (Fig. 5.22a) as plastic flow occurs. In the kinematic hardening the loading surface translates as a rigid body in the stress space during plastic deformation, keeping the size, shape and orientation of the initial yield surface same (Fig. 5.22b). A combination of the two previous modes of hardening is

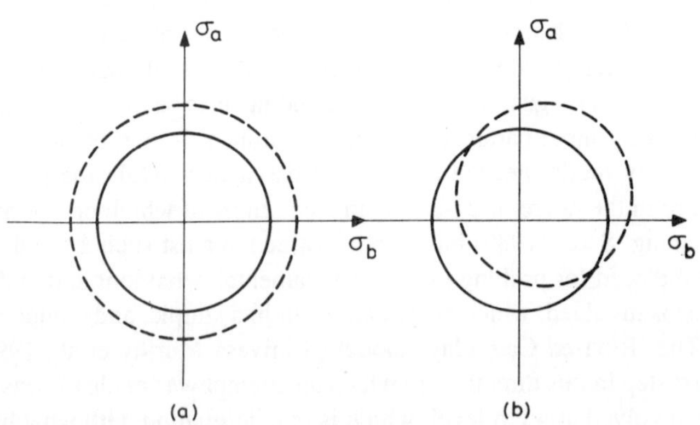

Fig. 5.22 Hardening of yield locus. (a) Isotropic, (b) Kinematic.

referred to as mixed hardening. Most of the later versions of the elasto-plastic models like nested yield surface models, or the multisurface models, the bounding surface models are based on this hardening rule. The choice of a specific hardening rule depends on the ease with which it can be applied and its ability to represent the hardening behaviour of the material being considered.

The third requirement defines the direction of the plastic strain increment vector, $d\varepsilon_{ij}^{p}$ and it will be normal to the plastic potential surface, g. (Fig.5.23) i.e.,

$$d\varepsilon_{ij}^{p} = d\lambda \; \{\partial g/\partial \sigma_{ij}\} \tag{5.28}$$

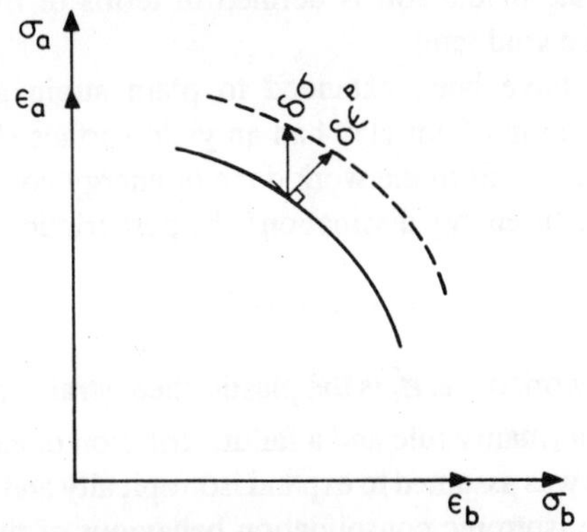

Fig. 5.23 Normality and Flow rule.

where $d\lambda$ is the proportionality factor determining the magnitude of plastic strain increments. If the potential function, g takes the same form of the yield function, then the above flow rule is called associated flow rule and is given by

$$d\varepsilon_{ij}^{p} = d\lambda \; \{\partial f/\partial \sigma_{ij}\} \tag{5.29}$$

The term $d\lambda$ can be written in its general form:

$$d\lambda = 1/H' \,(\partial f/\partial \sigma_{mn} \cdot d\sigma_{mn}) \tag{5.30}$$

where H' is the hardening modulus. As can be seen, the magnitude of $d\lambda$ depends on the magnitude of H' and the normal component of the stress increment to the yield surface. For a stable work hardening material satisfying the Drucker's stability criteria, it can be shown that the associated flow rule or the normality condition is a must.

An accepable plasticity model should meet the following theoretical requirements.

(1) The mathematical formulation should result in a stable stress-strain relation
(2) The constitutive relation should reflect the key characteristics of experimental data.
(3) The relationship should involve well defined fewer parameters which can be determined by standard tests.

5.4.2 Cam-Clay Models

The basic Cam-clay model was developed for axisymmetric coordinates, i.e., $\sigma_2' = \sigma_3'$, with spherical and deviatoric components given by the invariants of the form:

$$p' = (\sigma_1' + 2\sigma_3')/3 \text{ and } q = (\sigma_1 - \sigma_3)$$

and corresponding components of strain

$$\varepsilon_v = (\varepsilon_1 + 2\varepsilon_3) \quad \text{and} \quad \varepsilon_s = (2/3)\,(\varepsilon_1 - \varepsilon_3)$$

and the volumetric state of the soil is defined in terms of the specific volume $v = 1 + e$, where e is the void ratio.

The formulations have been extended to plain strain and general stress conditions also. The original Cam-clay had an yield surface (Roscoe surface) in the shape of bullet, derived from the work done or energy consideration, with an assumption that the plastic energy dissipation is by pure friction following the law:

$$W = Mp' \mid \varepsilon_s^p \mid \qquad 5.31$$

where W is the plastic work done, ε_s^p is the plastic shear strain and M is the friction factor. They used the normality rule and a failure criterion of extended von Mises type. The yield surface was assumed to expand isotropically and the hardening rule was obtained from the isotropic consolidation behaviour of the soil, noting that each point on the v vs ln p' curve corresponds to states with $q = 0$, on successively hardening yield surfaces. Linearisation of consolidation and rebound behaviour has been assumed in v – ln p' space (Fig. 5.24).

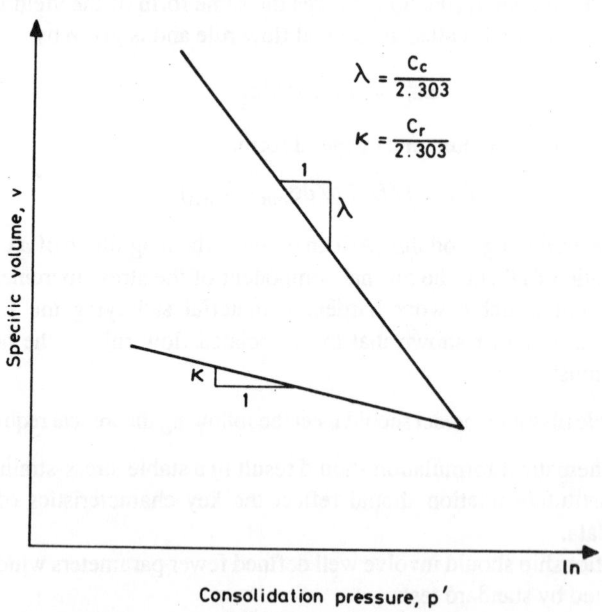

Fig. 5.24 Representation of compression and swelling paths.

The failure states defined by the critical state line was shown to be a unique curve in $p' - q - v$ space. The projection of the critical state line on $p' - v$ plane and $p' - q$ plane indicated the following relationship respectively.

$$v = \Gamma - \lambda \ln p'_f \qquad 5.32$$

$$q_f = Mp'_f \qquad 5.33$$

where Γ and λ are the intercept at unit pressure and slope of the critical state line respectively, M is the slope of this line in $q - p'$ space. This implies that the projection of critical state line on $p' - v$ plane is a line parallel to normal consolidation line (i.e., $v = N - \lambda \ln p'$) and in $p' - q$ plane it is a line passing through origin. The practical implications of these equations are that, if for a soil the initial state $(p', q, v)_i$ are known, the failure conditions $(p', q, v)_f$ are uniquely determined for any stress path. The compacting or dilating state was identified based on:

for $q/p' < M$, compression (wet state and weak at yielding)

for $q/p' > M$, dilating (dry state and strong at yielding).

The model assumed pure elastic behaviour for stress states within the yield surface, and also assumed that elastic shear strain is zero.

The elastic bulk modulus was obtained from the rebound part of the consolidation curve as $K = v\, p'/\kappa$ where κ is the slope of the rebound line in v vs $\ln p'$ plot.

The derived yield curve was of the form:

$$q/Mp' + \ln (p'/p'_x) = 1 \qquad 5.34$$

where p'_x is the value of p' at the intersection of the yield curve with the critical state line. A family of all such yield curves at different specific volumes forms a yield surface in $p' - q - v$ space.The final expression for the yield surface is of the form:

$$q = M\, p'[\Gamma + (\lambda - \kappa) - v - \lambda \ln p']/(\lambda - \kappa) \qquad 5.35$$

On the dry side of critical state, ie. for overconsolidated states,the yield surface was slightly modified later (Atkinson and Bransby, 1978) to match better the experimental results, by a planar surface of the form of Hvorslev's failure envelope. The complete yield surface is sometimes termed the 'state boundary surface' since it encloses all possible states of existance of a soil. Figure 5.25 schematically shows the yield surface in $p' - q - v$ space, the Roscoe and Hvorslev surfaces on either side of the critical state line.

A differentiation of equation 5.35 with respect to v and separation of elastic strain, would result in an expression for plastic volumetric strain increment in the form:

$$\delta\varepsilon_v^p = (\lambda - \kappa)((M - q/p')\, \delta p' + \delta v)/Mp'v \qquad 5.36$$

Applying normality rule, the shear strain component would be

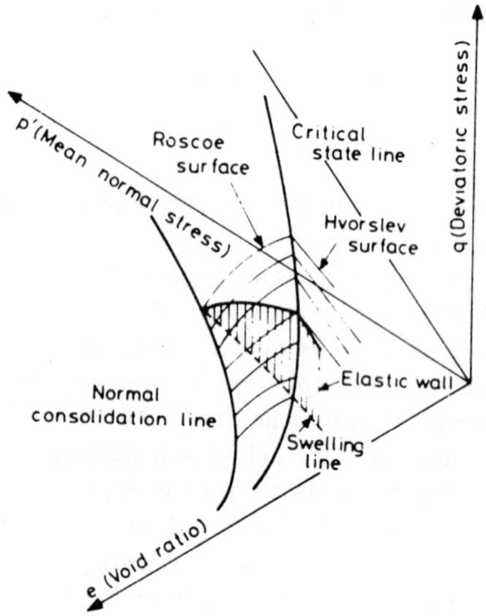

Fig. 5.25 Three dimensional representation of cam clay yield surface (Roscoe and Hvorslev surfaces).

$$\delta\varepsilon_s^p = \delta\varepsilon_v^p/(M - q/p') \tag{5.37}$$

Adapting an incremental working procedure, using equations 5.35, 5.36 and 5.37, together with the elastic volumetric strain component, $\varepsilon_v^e = \kappa\delta p'/p'v$, the entire stress strain behaviour for any loading path can be obtained. However the model predicts larger shear strains than the observed values at small shear stress level. To overcome this limitation, a modified version of the Cam-clay model, with a different energy dissipating mechanism was suggested. This is referred to as 'Modified Cam-clay model' (Roscoe and Burland, 1968).

Modified Cam-clay model

To account for the energy dissipated in plastic volume changes also, a slightly different mode of plastic energy dissipation was assumed as:

$$W/v = p'\left[(\delta v^p/v)^2 + (M\delta\varepsilon_s)^2\right]^{1/2} \tag{5.38}$$

This form, on application of normality rule, resulted in the yield curve:

$$(p' - p_x')^2 + q^2/M^2 = (p_x')^2 \tag{5.39}$$

where $p_x' = p_o'/2$.

The hardening rule was obtained from the consolidation behaviour as follows:

$$\delta\varepsilon_v^p = (\lambda - \kappa)\, \delta p'/vp'$$

Then $$\delta p_x' = \delta p_o' = vp'\delta\varepsilon_v^p/(\lambda - \kappa) \quad 5.40$$

As can be seen, from the above expression (eqns 5.35 to 5.37 or 5.39 and 5.40) the only soil parameters appearing are λ, κ, Γ and *M*. Fig. 26, represents

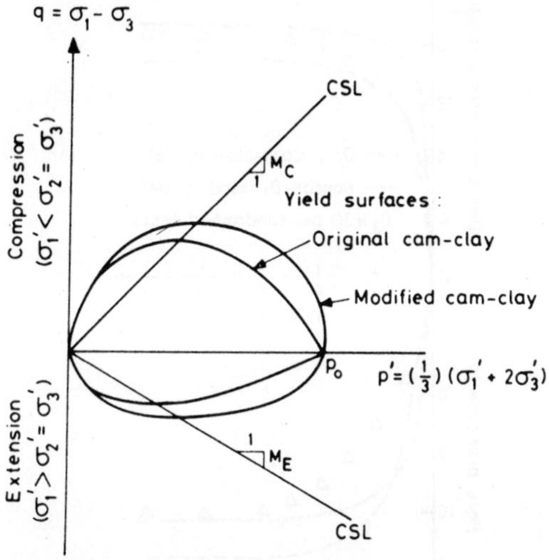

Fig. 5.26 Cam clay yield surfaces in $p' - q$ plane.

schematically the original and modified Cam-clay yield surfaces in triaxial plane.

Let us examine the predictive capability of these models using the test data of Weald clay (Henkel, 1959). The material constant are : λ = 0.095, κ = 0.035, Γ= 2.128 at one kPa and *M* = 0.92. With these model parameters, for a confining pressure of 206.75 kPa (30 psi) with a specific volume of 1.621, being in its normally consolidated state, the behaviour is predicted by both original and modified Cam-caly models. The predicted stress-paths in $q - p$ space are compared with the experimental path in Fig. 5.27. The comparison indicates that the

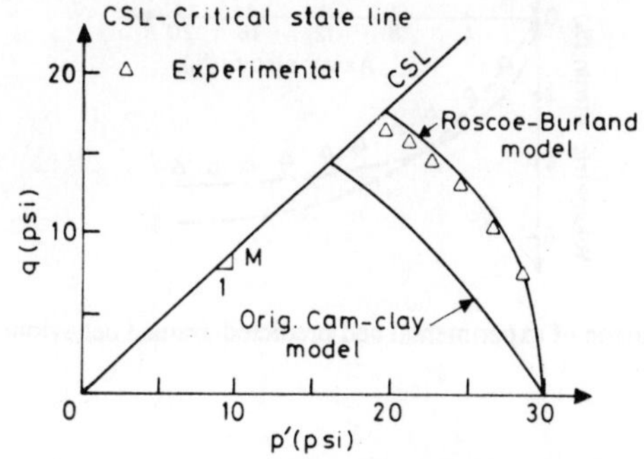

Fig. 5.27 Comparison of experimental and predicted stress paths of Weald clay.

prediction of stress path is quite good by the modified model while considerable deviation is observed in the predicted path by original Cam-clay model. Figures 5.28 and 5.29 show the results of stress-strain behaviour from undrained and drained tests respectively on the same soil state. It is seen that the original Cam-clay model over predictes the strains while the modified model under predicts the same.

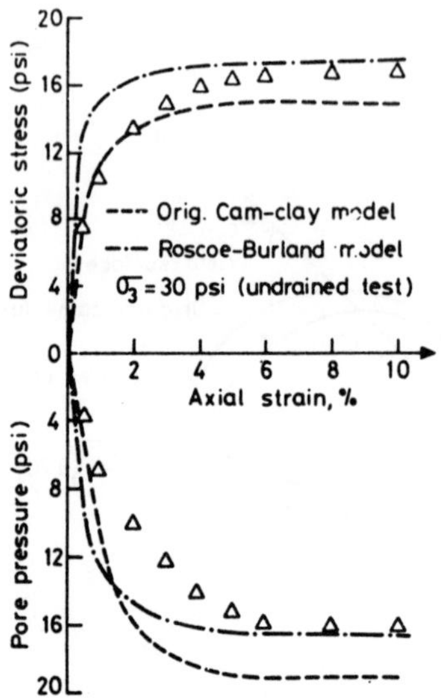

Fig. 5.28 Comparison of experimental and predicted undrained behaviour of Weald clay.

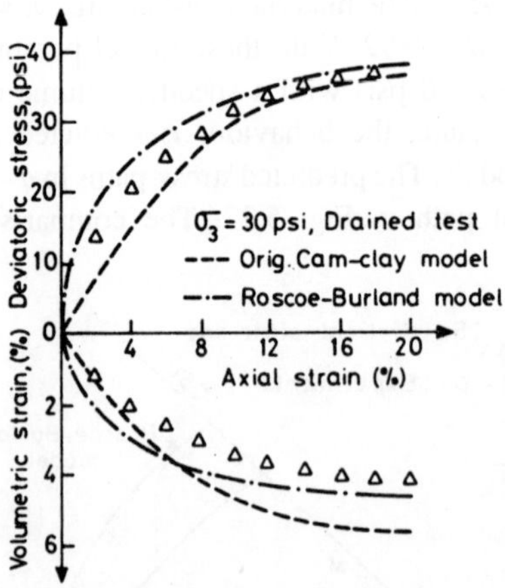

Fig. 5.29 Comparison of experimental and predicted drained behaviour of Weald clay.

Since the modified model predicted well the yield paths in $q - p'$ plane, but showed deviations only in strains, researchers have attempted to further modify the strains by admitting non-associated flow rules. It has been shown (Jain, 1981), that the maximum possible plastic energy dissipation as required by the laws of thermodynamics, occurs in a standard process obeying normality and that in any non-associated flow, the stresses required to cause a given strain increment should be greater than those in a standard process. This imposes a restriction that the strain increments from a non associated flow should always be less than what they would be in an associated flow under the same stress increments.

But we have seen earlier that the modified Cam-clay model under predicts the shear strain for Weald clay and hence the strains need to be increased. Then, is the soil material behaving against the established laws of thermodynamics? We can see that, if we adapt the original Cam-clay yield surface, which overpredicts the strains, as the yield surface and if we can find a logical means to increase the stresses to the required level, the laws of thermodynamics will not be violated, since in this case we have to reduce the strains. The 'Revised Cam-clay model' (Srinivasa Murthy et al., 1991) was developed in this direction.

5.4.3 Revised Cam-clay model

By observing the soil behaviour of fine grained soils at microlevel, from the considerations of physico-chemical interactions, we find that the stress transfer is essentially through the interacting diffuse double layers. The effective stress is dictated by the half space distance between the particles or interacting clusters. We have seen that during consolidation, increased level of particle grouping, resulting in clusters is a possible mechanism to explain the plastic compression of soils. If the aggregates or clusters breakdown due to some mechanism, under a constant volume condition, the half space distance between the broken clusters decreases, thus increasing the effective stress. And in a drained condition, to maintain the stresses at the prescribed level, the broken units would move apart thus resulting in the dilation of the soil. Now, had there been complete rigidity of these clusters during shearing perhaps the original Cam-clay model would have correctly predicted the behaviour. Also we have seen that evidences are available to indicate that during shearing the clusters break into smaller units. This introduces additional effective stress into the system which has not been accounted in any of the earlier models. We quantify the magnitude of this stress incrementally and incorporate its effect in the original Cam-clay model. The original Cam-clay model did not consider the energy dissipated due to plastic volume changes. The modified Cam-clay model did consider this aspect but treated this component of energy also similar to the frictional component. It now appears that we have to consider the two modes of energy differently. The plastic volumetric changes are due to the micro-structural changes of grouping or breakdown of clusters. Both the processes of particle grouping or cluster breakdown require an input of energy (probably of the same magnitude), but in the process of breakdown of clusters, there is a release of energy locked with in the clusters during grouping. Thus the process may not be a dissipative one in the strict sense. It is neither elastic because the energy is not released immediately upon mere unloading. In the Revised Cam-clay model the normality rule is considered only for the frictional component of power dissipation while the effect of work done on grouping and dismembering of particles is considered separately.

Normally Consolidated Soils

Let us start with the assumption that the original Cam clay surface $f = 0$ (equation 5.35) is the surface defining the yielding of the soil and that normality is applicable to soils. Let us now also assume that the difference in the actual experimental undrained path and the predicted path at any stage of shearing is the additional stresses induced due to microstructural changes. The exact magnitude of such stresses at any strain level can be known if we know the experimental undrained path. Infact this does not require any additional testing. The same shear test used for determining the friction factor M can be used to obtain the complete undrained path also. The results of Weald clay and London clay indicate that the undrained paths in $q - p'$ space can very well be fitted by the equation

$$(p' - p_o'/2)^2 + q^2/K^2 = (p_o'/2)^2 \qquad 5.41$$

with the value of the constant K being about 1.15 (This equation although is very much similar to the Roscoe-Burland yield surface, it has got nothing to do with the plasticity model or energy dissipation. It is only a curve fitting equation defining the undrained paths in $q - p'$ space). Once the experimental undrained path is known, the method to incorporate the additional internal stresses induced is quite simple. Let us consider the undrained and drained cases separately for the sake of clarity, although in principle there is actually no difference between the two cases.

Undrained Shearing: Consider the soil at A isotropically consolidated to a pressure p' with a specific volume v. The soil parameters λ, κ, Γ and M to define the yield surface are known. The experimental undrained path $u = 0$ (Eqn. 5.41) is also known. We give a stress increment AB' along the yield curve $f = 0$ (Fig. 5.30) and compute the strain increments associated with this yield curve. During this

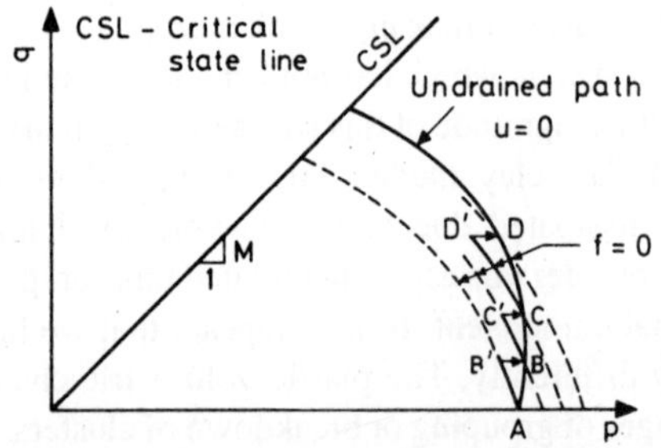

Fig. 5.30 Incremental working scheme for prediction of undrained behaviour.

increment of shear strain, the micro structural changes induce additional stresses $B'B$. It is likely that any internally mobilised stress is always isotropic. Hence $B'B$ will be horizontal. The magnitude of $B'B$ is known since B should reflect the true stress on the soil and hence should lie on the actual experimental path $u = 0$. The yield curve will have hardened at this stage from $(f = 0)_A$ to $(f = 0)_B$ as an effect of the same microstructural changes, although there are no overall plastic strains externally.We shall note this hardening in terms of a pseudo reduced specific volume v_B. (Of course it would be better if the hardening of the yield surface can be defined in terms of the microstructure but since at this stage we have no means to measure it, we will retain specific volume for the purpose). We now give the next increment *BC'* along this hardened yield surface and compute the associated strains. Microstructural changes during this strain increment induce additional stress *C'C* being an isotropic stress, with *C* lying on the experimental path $u = 0$. The yield surface will harden further to $(f = 0)_C$.The process is continued till the critical state is reached, each time giving stress and strain increments along the yield surface $f = 0$ but shifting on to the actual experimental path $u = 0$.

The process may seem to be nothing but a non associated flow (with $u = 0$ being the yield surface and $f = 0$ being the potential surface) but there is a subtle difference. Stress increments are not given along $u = 0$ but instead,an associated yielding is considered along $f = 0$, giving stress increments also along this surface and then shifting the stress state on to $u = 0$ due to the increased internal stresses. This in effect is that plastic flow is considered only for frictional component and the effect of dismembering of clusters is taken care of seperately.

Drained Shearing: Let us consider a standard triaxial drained test so that the applied path has a slope of 3 in 1. For the initial isotropic state at *A* (Fig. 5.31), we give a stress increment *AB* such that $dq/dp' = 3/1$ and compute the volumetric and

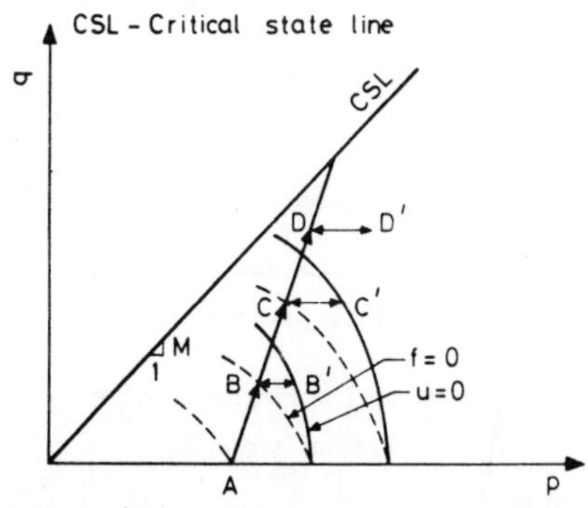

Fig. 5.31 Incremental working scheme for prediction of drained behaviour.

shear strain increments associated with the yield surface $(f = 0)_A$. At this stage the dismembering of clusters will cause an increase in effective stress at constant volume, from *B* to *B'*, such that *B'* lies on an undrained test path $u = 0$ corresponding to the current void ratio or volume. Again it is considered that the internally mobilized stress *BB'* is isotropic which instantaneously has to be in the form of negative pore pressure. But since drainage is permitted, the soil dilates so that these mobilized stresses are dissipated. This being a case of unloading from *B'* to *B* magnitude of dilation can be computed from usual elastic equations. The current yield curve through B may still be specified by a pseudo specific volume v_{oB} which will be less than the actual volume at that instant, because this additional volume change is a result of micro structural changes without external plastic strains. Now the current stress state is B and the next increment BC is applied along the path with slope 3 in 1 and the corresponding strains associated with the yield surface $(f = 0)_B$ are computed. There will be an isotropic increase in stress (corresponding negative pore pressure) due to dismembering equal to CC', C' lying on undrained path $u = 0$ corresponding to the reduced specific volume at that instant. The dissipation of this pore pressure which is an unloading from C' to C causes an elastic dilation. The current stress state is C and the corresponding yield curve is specified by a pseudo specific volume v_{oc} which is different from the actual specific volume at that instant. Thus the experimental undrained path $u = 0$ expands only with changes in pseudo volume since the dilation component is realised due to changes in micro structure. The process is continued till the critical state is reached.

To verify the above scheme, drained and undrained test results of Weald clay at 30 psi confining pressure have been predicted and compared with the actual test results in Figs. 5.32 and 5.33. For this analysis the chosen values of the model parameters are same as those discussed earlier. The predictions using this scheme are quite good for both drained and undrained test results. The observed small deviations are within the choice of the parameters λ, κ and M. Test results of Kaoline clay presented by Jain (1981) have also been predicted in Fig. 5.34.

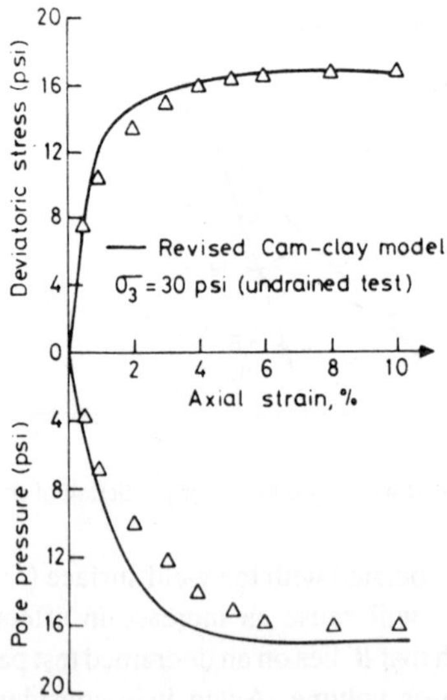

Fig. 5.32 Prediction of undrained behaviour of Weald clay.

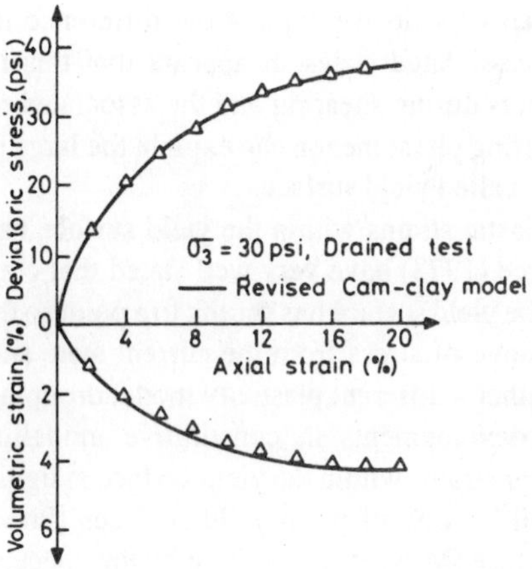

Fig. 5.33 Prediction of drained behaviour of Weald clay.

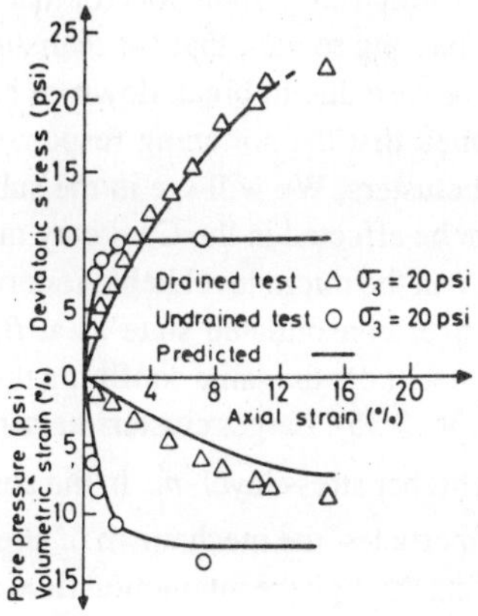

Fig. 5.34 Prediction of drained and undrained behaviours of Kaolin clay (Data - Jain, 1981).

Overconsolidated Soils

Even after a considerable development in the plasticity models for soils, over the past few years, the behaviour of over consolidated states (especially the heavily over consolidated ones which are on the dry side of critical) has not yet been modelled satisfactorily. We do not have a proper yield function defined on the dry side of critical. The Cam-clay model assumes a planar yield surface derived from Hvorslev's failure envelopes at constant water contents, and assumes purely elastic behaviour for stress states within this yield surface. But when we examine the possible behaviour at micro level, we find that the mechanism of stress transfer is essentially same in both normally and over consolidated states. Hence we should expect yielding behaviour to be similar in both the cases, i.e., we need not have a different yield surface for the dry side. Every state of soil should yield only along a Roscoe surface. Then how do we explain the difference in the behaviour of normally and over consolidated states. It appears that this is again due to the dismembering of clusters during shearing and the associated release of locked-in energy. The dismembering phenomenon can explain the large plastic shear strains observed within the so called yield surface.

The presence of plastic strains within the yield surface has been noted since long. Palmer and Pearce (1973) have very well stated that clays can be idealised as materials in which the yield surface has shrunk to a point so that all deformations are plastic for any change of stress from the current state and even suggest the development of altogether a different plasticity model dropping the concept of an yield surface. Later developments in constitutive modelling did attempt to incorporate plastic shear strains within the yield surface assigning varying degrees of plasticity to the soil by way of multi-yield surfaces (Prevost, 1980) or by a moving inner yield surface (Mroz et al., 1979) or by the concept of a mapping rule with the image point on the outer bounding surface (Dafalias et al., 1980). But in all these approaches, defining the growth and movement of the inner yield surfaces or the mapping rule was quite a formidable task because they did not look into the basic processes at micro level which are responsible for macro level plastic strains. Dafalias (1986) has made a passing remark that for realistic predictions, a model should consider the energy release due to break down of bonds between particle groups. He has also mentioned that the softening response can be explained by such break down of particle clusters. We will see in the subsequent sections how the above considerations can be effected in the Cam clay model.

Recalling the discussions on the micro level behaviour of fine grained soils we have seen that a soil in an over consolidated state as at *B*, is different from the normally consolidated state at *A* under the same confining pressure, only in its level of aggregation of particles (Fig. 5.35). Larger clusters have been formed in the soil at *B* during loading upto the higher stress level, p'_p. In the absence of direct contact between particles/groups of particles, the mechanism of stress transfer is the same in both the states *A* and *B*, being through the interacting fluid phase. Dafalias (1986)

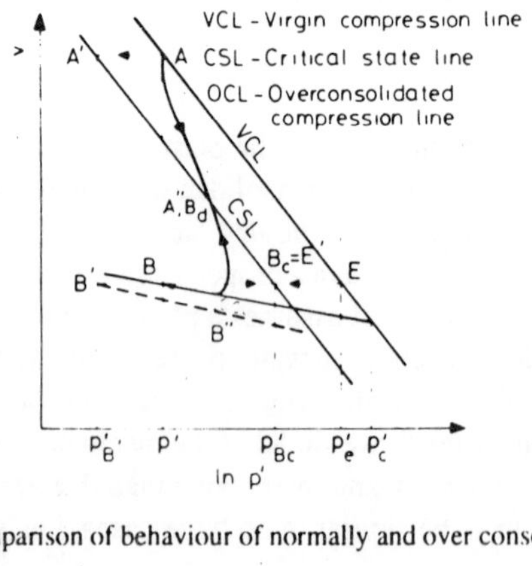

Fig. 5.35 Comparison of behaviour of normally and over consolidated soil states.

indicates a similar thought that the same framework of elasto-plasticity/visco-plasticity can be used for soils with or without clusters. We have also seen that the effective stress is a function of the distance between particles/clusters. In that case shearing resistance being only a function of the normal stresses, should also be the same for both A and B.

Thus, we would expect the soil at B to reach a critical state at B' in an undrained test, following a path similar to AA' of the normally consolidated state. But in actual tests, the specimen shows an increase in p and moves towards the critical state at B_c. Obviously, this can also be explained in terms of increase in effective stress due to dismembering of clusters during shearing. Similarly, in the drained test, we would expect the soil at B to trace a path BB'' similar to AA'', but in actual tests the soil dilates and moves towards the critical state $B_d = A''$, which the normally consolidated soil at A would reach. Again the dilation of the soil can be explained as an effect of the same phenomenon of break down of clusters.

The above logic leads to an interesting deduction. If the soil is thought to be yielding at every stage only along Roscoe surfaces, it can never reach a state with q/p' ratio > M. But experimental results do show soil states with $q/p' > M$. For the particular case of Weald clay, values of q/p' as high as 1.02 in undrained tests and 1.06 in drained tests can be observed for samples with OCR = 24, whereas the average M for the normally consolidated states is about 0.92 or 0.93. May be this simply indicates that the overconsolidated soil does have a higher M itself. This may be quite possible or reasonable because higher energy may be required in overconsolidated soils to break down the clusters without which they cannot attain the micro-level configuration required to sustain the deviatoric stress. Hence in this model, states dry of critical are also considered to have q/p' less than M, M itself being higher, say M'.

It now appears that we can get the true response of over consolidated soils by superposing the usual frictional yielding response and the effects of dismembering of clusters, i.e., an additional component of effective stress.

For analysis we need to know the magnitude and the rate of additional stress induced due to dismembering. Considering the point B (Fig. 5.35), if we can by some means cause a dismembering of clusters, without shearing and keeping the volume constant, the soil would reach the point E on the normally consolidated line with an effective pressure p', hence, the difference $(p' - p_e')$ is the additional pressure that would be induced due to dismembering.

However, the break down of clusters does not occur in one step during shear but happens gradually with strain. Instead, hypothetically if we can maintain the clusters stable during shearing, the soil at B would reach a critical state at B', and the normally consolidated state (E) at the same void ratio would reach $E' = B_c$. If we dismember the clusters after reaching the critical state at B' to the same level as at B, the induced stresses would be, $(p_{BC}' - p_B')$ i.e., at the ultimate critical state $(p_{BC}' - p_B')$ is the increase in pressure due to dismembering, $(p_{BC}' - p')$ is the reduction in pressure due to usual frictional yielding along the Roscoe surface (similar to normally consolidated state), and their difference $(p_{BC}' - p_B')$ is the net change in pressure or net pore pressure observed in actual tests. Thus we definitely know the end value of the additional stresses induced but we do not know its distribution with strain. We can only say that this distribution is a function of (p'_e) to start with and will be a function of $p'_{BC} - p'_B$ towards the critical state. We can obtain a suitable distribution from the results of the typical shear tests.

Effect of Dismembering on Material Properties

The next question is how to obtain the frictional yielding response of the soil at state *B*. Obviously it is not the same as that of the soil at state *A*. The soil at *B* has a reduced operating specific surface than at *A* due to grouping of particles. This means a reduced plasticity reflected by a lower value of λ. Infact the state at *B* would be non plastic or purely elastic ($\lambda = \kappa$) to start with, if the clusters formed under the pressure p_p' remained perfectly stable upon unloading to *B* as assumed in the Cam-clay model. But a rebound line with gradual increase in steepness (Parry and Amerasinghe, 1973), indicates that some of the clusters are dismembered during unloading itself, i.e., the soil at *B* will have regained certain degree of plasticity ($\lambda_B > \kappa$ but $< \lambda$). With further dismembering of clusters during shearing, specific surface increases and thus the soil gradually regains its original plasticity. Similarly,the. friction factor *M* which may be higher initially due to the higher energy required for breakdown of clusters, may gradually decrease with shearing as the rate of breakdown decreases to finally reach the same value as that of the normally consolidated state at very large strains.

Overconsolidation ratio defines the degree of deviation of a soil state from the normally consolidated state. Hence the values of λ, κ and *M* of an overconsolidated soil depend on OCR. The OCR of the soil reduces with continued shearing and can be computed at any stage. Analysis of a few test data has shown that it is possible to relate and to obtain generalised patterns for the variation of λ, κ and *M* with changes in OCR. However this requires extensive data to establish the actual functional forms.

The method of introducing the additional stresses for overconsolidated case is very much similar to that for the normally consolidated soil except that the dismembering component is greater (Fig. 5.36). We do not know the experimental effective stress path in $q - p'$ space, as we fortunately had in the normally consolidated case (i.e., $u = 0$), to help us in estimating the dismembering

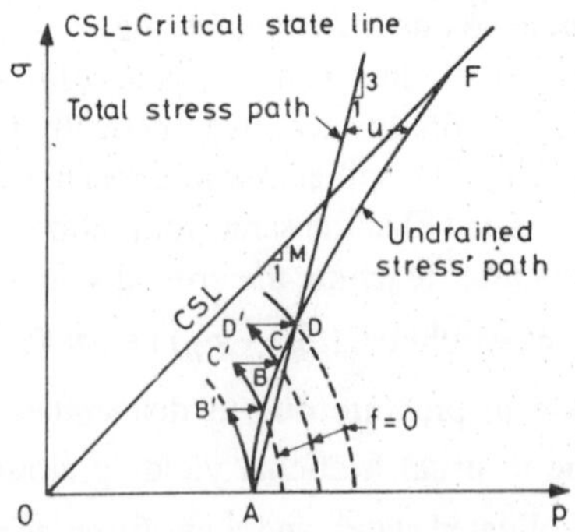

Fig. 5.36 Incremental working scheme for prediction of undrained behaviour for overconsolidated states.

component. Hence, we will have to assume a suitable distribution of this component with shear strain before hand, may be using a typical experimental data. Further a suitable variation for λ, κ and M with shear strain has to be assumed.

To verify the validity of the above approach, we have tried to predict the experimental stress-strain curves of Weald clay. For undrained test with 5 psi confining pressure, the stress strain relations have been predicted using the known experimental stress path. From this we have obtained the distribution of extra component of stress with shear strain which has been used for predicting the results of undrained test with $\sigma_c' = 10$ psi, and drained tests at 5 and 10 psi confining pressures. For these computations we have used an initial λ value of 0.045, and a κ value of 0.0225 to start with. The values of both λ and κ have been varied with shearing as:

$$\lambda_i = \lambda_{(i-1)} + (0.095 - \lambda_{(i-1)})\, C_1$$

$$\kappa_i = \kappa_{(i-1)} + (0.035 - \kappa_{(i-1)})\, C_1$$

$$M_i = M_{(i-1)} + (M_{(i-1)} - 0.92)\, C_2$$

effected at the end of shear strain increments of 0.25%. Here C_1 and C_2 are constants chosen to suite the test results. The values .095, .035 and .92 are the λ, κ, and M values for normally consolidated state. Figures 5.37 and 5.38 indicate the thus analysed test data of Weald clay. The predicted values indicate a close agreement with the test results. The fact that the assumed distribution of additional stresses could result in good agreement in both q vs ε_s and u vs ε_s plots simultaneously indicates that the assumed distribution is not orbitrary but is a unique solution.

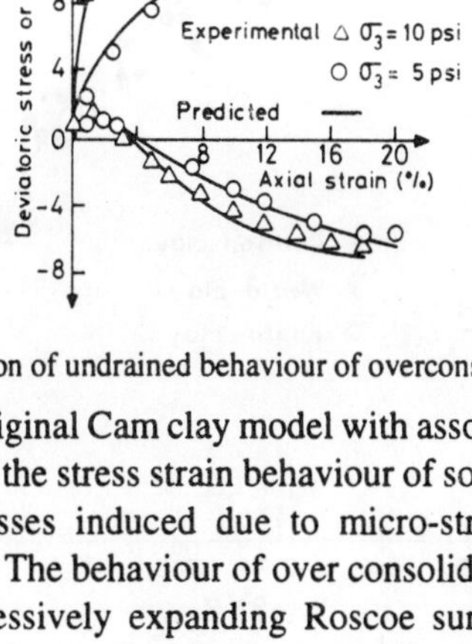

Fig. 5.37 Prediction of undrained behaviour of overconsolidated Weald clay.

In conclusion, the original Cam clay model with associated flow rule itself can be used to predict better the stress strain behaviour of soils by incorporating some additional internal stresses induced due to micro-structural changes without causing external strains. The behaviour of over consolidated soil can be described as yielding along successively expanding Roscoe surfaces (and not Hvorslev surfaces) with gradually increasing plastic properties and with an increased value of friction factor M.

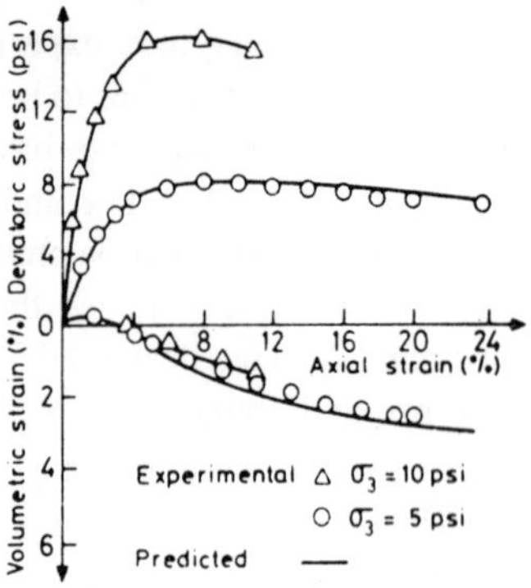

Fig. 5.38 Prediction of drained behaviour of overconsolidated Weald clay.

Generalisation of Cam-clay model

We have seen that, the compression paths and the paths of void ratio versus ultimate failure stresses (both q and p') can be generalised within the level of engineering accuracy. Since, the isotropic compression path and the ultimate failure stress constitute extreme limits of stress-strain behaviour during shearing, and since both of them can be generalised, it must be possible to generalise the entire stress path and stress-strain behaviours. This has been examined in detail and described by Srinivasa Murthy et al., (1988).

Figure 5.39 shows the normalised stress paths (q/p_e' versus p/p_e') of several undrained and drained tests on Weald clay and London clay (Henkel, 1959) and the average path of several tests on kaolinite clay (Atkinson and Bransby, 1978).

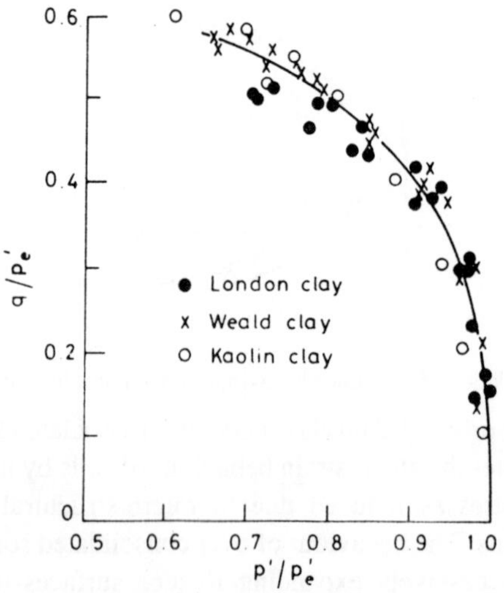

Fig. 5.39 Normalised stress paths of normally consolidated soils.

In this normalisation, which has been valid for both drained and undrained conditions p'_e refers to the equivalent consolidation pressure on the normally consolidated state at the same void ratio as that under consideration. It can be observed that the path is nearly unique for all the three soils. This indicates that normalisation with equivalent consolidation pressure, p'_e not only accounts for the variation in stress level, but also encompasses different soils. This is so because in the stress plots all the stress parameters are functions of (e/e_L) and the points $[(q/p'_e), (p/p'_e)]$ will be independent of soil type, the friction factor M being nearly same for all soils.

To confirm this mode of generalisation, we can examine the shearing test paths in $v - p'$ space. Figure 5.40a shows the shear test paths in $v - p'$ space of London clay and Weald clay, for both drained and undrained conditions, at two stress levels. (The vertical axis has been plotted with water content instead of specific volume). Figure 5.40b shows the generalised path, having been normalised with their liquid

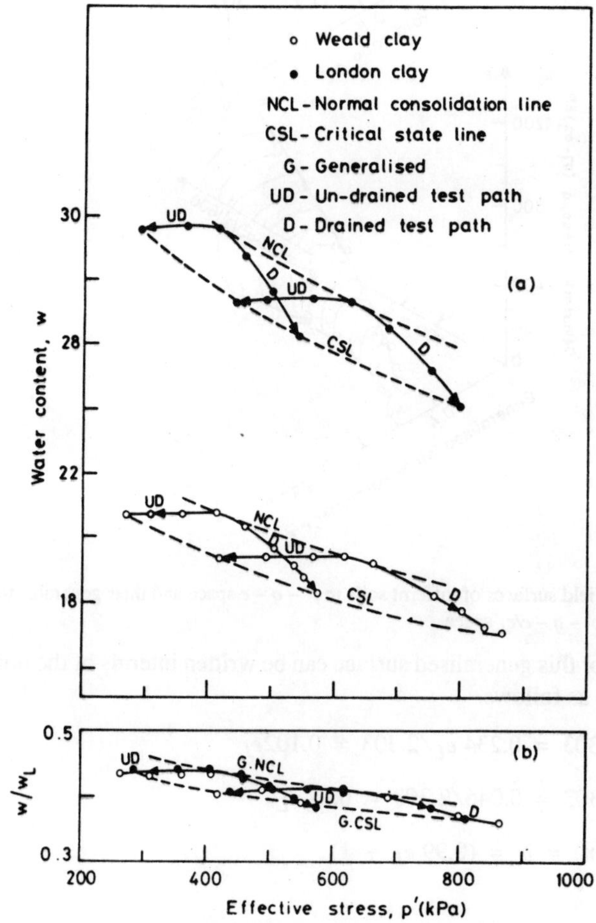

Fig. 5.40 Generalisation of stress paths in v vs p' plane.

limit water contents. It is interesting to note that the stress paths can be generalised in both $q - p'$ and $v - p'$ planes, thus indicating the possibility of generalising the yield surface also. Figure 5.41a shows the yield surfaces of the four soils in $p' - q - e$ space and Fig. 5.42b shows the generalised yield surface in $p' - q - e/e_L$ space.

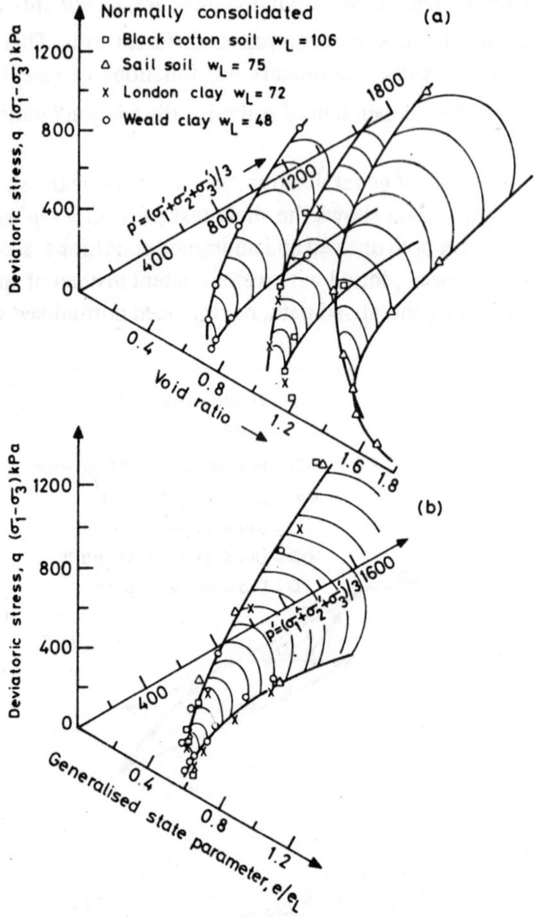

Fig. 5.41 Yield surfaces of differnt soils in $p' - q - e$ space and their generalisation in $p' - q - e/e_L$ space.

The equation for this generalised surface can be written interms of the normalised soil parameters as follows.

$$\lambda = C_c/2.303 = 0.234\, e_L/2.303 = 0.102 e_L$$

$$\kappa = C_r/2.303 = 0.046/2.303 = 0.02\, e_L$$

$$(\Gamma + \lambda - \kappa) = N = (0.99\, e_L + 1)$$

$$M = 0.80$$

Then

$$q = [Mp'/(\lambda - \kappa)]\,[(\Gamma + \lambda - \kappa) - v - \lambda \ln p')]$$

can be written as:

$$q = 10\,p'\,[0.99 - (e/e_L) - 0.102 \ln p'] \qquad 5.42$$

The values of the coefficients or the constants obtained in the above generalised equations may not be the exact values, and may be refined with a larger set of test results. The discussions presented are mainly to highlight the viability of the generalised approach in principle.

5.5 Permeability

It is needless to reiterate that permeability is one of the most important soil properties and its significance is increasingly realised with renewed interest in seeking acceptable solutions for environmental problems. For coarse grained soils, the coefficient of permeability has been related, with some success, to the characteristics of grain size distribution. Attempts to find similar approaches have been less successful with fine grained soils. Even with Kozeny-Carman equation, which takes into account the void ratio and through different parameters the fabric of soil, it has not been possible to predict the coefficient of permeability of fine grained soils within acceptable levels of accuracy needed for engineering purposes particularly as the stress state changes. Although mechanisms controlling the permeability of clays has drawn the attention as early as 1971 (Mesri and Olson, 1971) it has so far not been possible to identify the basic parameters by which the permeability of the clays be interpreted, generalised and possibly predicted.

Within the framework of generalisation of compressibility and strength of uncemented saturated soils let us analyse the permeability data of soil and examine the possibility of generalisation.

It is known that void ratio alone is not a sufficient parameter for quantifying the permeability of fine grained soils. Two soils with the same void ratio can have drastically different permeability values. The same soil under different conditions, but with the same void ratio can show different permeability coefficients like for example compacted and slurry consolidated states. Thus it is obvious that the microstructure or more clearly, the size, distribution and arrangement of voids also affects the permeability and not just the volume of voids.

The liquid limit water content, could serve as a reference parameter for generalising stress-strain behaviour because of the same order of effective stress existing at that state for all soils. If this parameter is to serve as a reference even for generalising permeability behaviour, all soils should posses the same pattern of microstructure at their liquid limit states. This fact has already been examined in Chapter 3 (see sec.3.7.2). To reiterate again, the reanalyses of data of Griffiths and Joshi (1989) on intruded volume per unit volume indicate that pore size distribution is of same pattern for the soils tested. It has been further reported that there exists a relation between ratio of pore volume V_t to that at liquid limit state V_{tL} and consolidation stress in the form

$$V_t/V_{tL} = 1.193 - 0.206 \log \sigma' \qquad 5.43$$

On the basis of the experimental evidence and the interpretation of microstructure data it can be inferred that

(i) the microstructure of fine grained soils at their liquid limit is of same pattern and

(ii) the change in micro-void volume with stress is proportional to its initial state.

Based on the above premises, the consolidation and permeability of soils compressed from slurry state to different consoldiation stress levels with variable head permeability test data has been examined (Nagaraj et al., 1993). In Figs. 5.42 and 5.43 the data of $e - \log \sigma_v'$ and $e - \log k$ for the soils considered in the series of experiments are shown. It can also be seen that upon normalisation of void ratios corresponding to water content at liquid limit, different $e - \log \sigma_v'$ and $e - \log k$

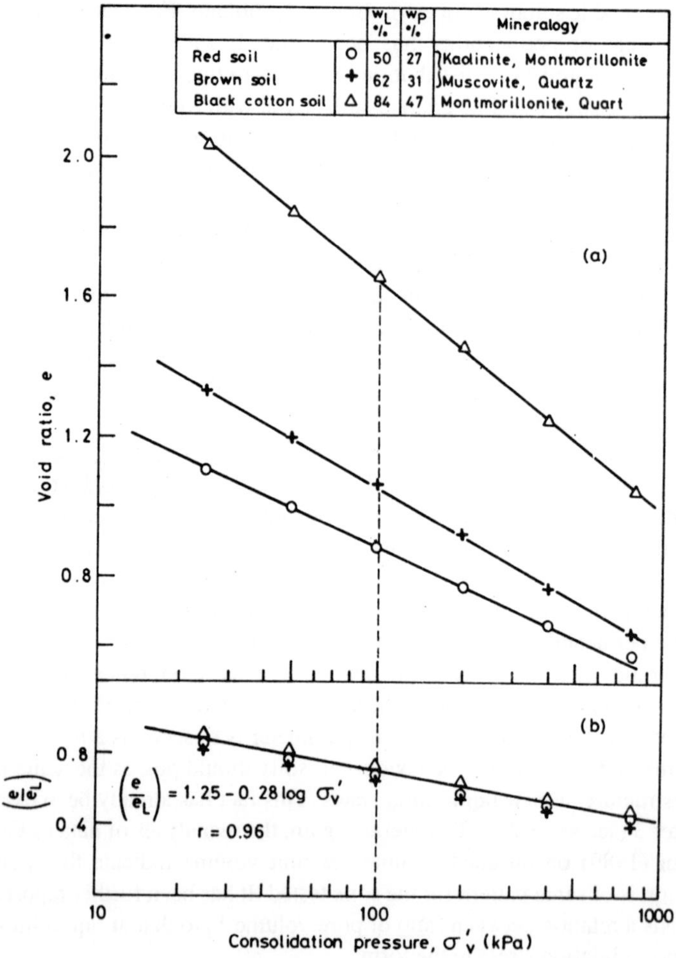

Fig. 5.42 Generalisated $e - \log \sigma_v'$ relation of three soils.

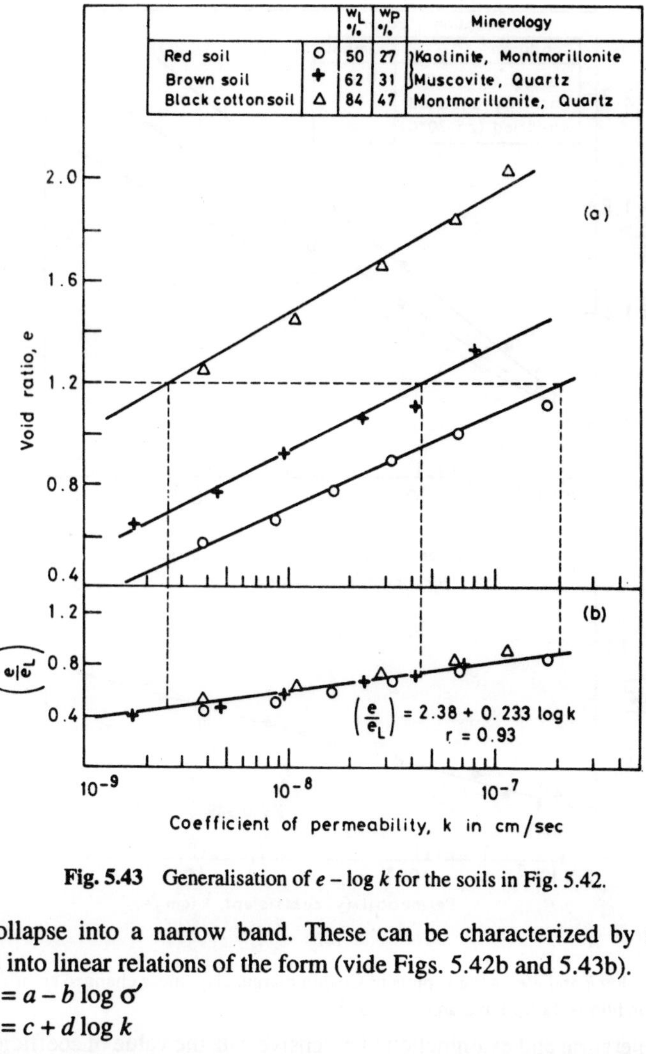

Fig. 5.43 Generalisation of $e - \log k$ for the soils in Fig. 5.42.

paths collapse into a narrow band. These can be characterized by regression analysis into linear relations of the form (vide Figs. 5.42b and 5.43b).

$$e/e_L = a - b \log \sigma'$$

$$e/e_L = c + d \log k$$

For the data analysed

$a = 1.25 \quad b = 0.280$ with $r = 0.96$

$c = 2.38 \quad d = 0.233$ with $r = 0.93$

Similar relationships have also been obtained for clays subjected to environmental changes. It has been observed that natural clays subjected to drying exhibited irreversible changes in its liquid limit water contents. When such clays are compressed from their corresponding liquid limit water contents and permeability characteristics determined at different consolidation stress levels e/e_L bears a relationship with consolidation stress and permeability (Pandian et al., 1993). Figures 5.9 and 5.44 show the $e - \log \sigma_v'$ and $e - \log k$ plots and their generalization in terms of generalized state parameter, e/e_L.

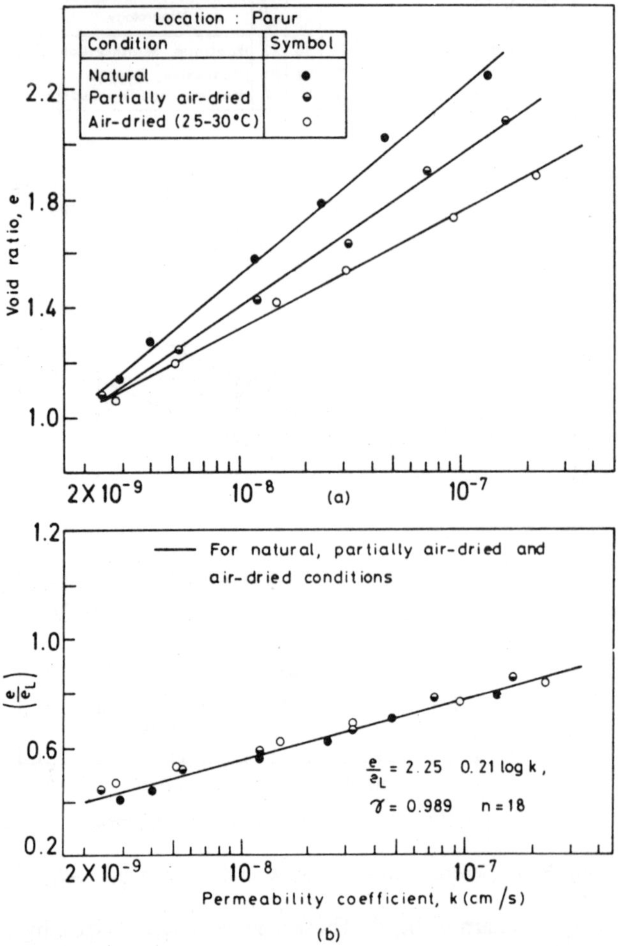

Fig. 5.44 $e - \log k$ and $e/e_L - \log k$ plots of Cochin marine clay under changed environmental conditions. (after Pandian et al., 1993)

With generation and examination of extensive data the value of coefficients for general usage can be established. Not withstanding this task another important aspect that can be highlighted is the possiblity of having permeability assessment as consolidation stress state changes.

Generally it is a misnomer that montmorillonitic soils have a low permeability than other soils of lower plasticity. But the fact is that all clayey soils will have the same permeability (at corresponding states) ranging between 10^{-7} cm/sec at high water contents close to their liquid limit state and 10^{-9} cm/sec at lower water contents of about 0.5 w_L.

5.5.1 Stress - state - permeability relations

Reverting to figures 5.42 and 5.43 the following inferences can be made. At the same pressure the void ratios reached are different but still e/e_L value is same and hence the value of k is same. On the other hand, if the void ratio is same for all three soils, since consolidation pressures are different, the e/e_L values are different and hence the permeability coefficients are different. For a soil of known liquid limit water content the change in permeability with consoidation stress can be assessed or for a required permeability the consolidation stress level can be computed from the relationships involving e, e_L, σ_v' and k. It is to be stressed again that these assessments are valid only for slurry consolidated soils and hence possessing similar microstructure. The permeability of compacted soils, subsequently saturated will not follow this relation because of their different microstructure and will be discussed in a later chapter.

5.5.2 Overconsolidated clays

Most often soils insitu might experience different stress histories. Even in the case of overconsolidated clays devoid of cementation, to characterize the compressibility response it is necessary to assess the preconsoldation pressure. With appropriate stress-state relationships discussed earlier (Eqns 5.5 and 5.6 and example cited) it would be possible to get the entire $e - \log \sigma_v'$ paths in the recompression, normally consoldiated and rebound stages.

It is interesting to find (Figs. 5.45 and 5.46) that permeability characteristics have been soley dependent upon the void ratio irrespective of the stress history

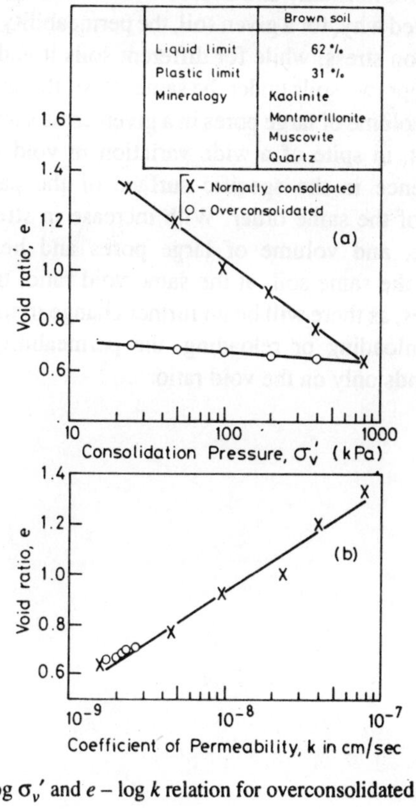

Fig. 5.45 $e - \log \sigma_v'$ and $e - \log k$ relation for overconsolidated states of Brown soil

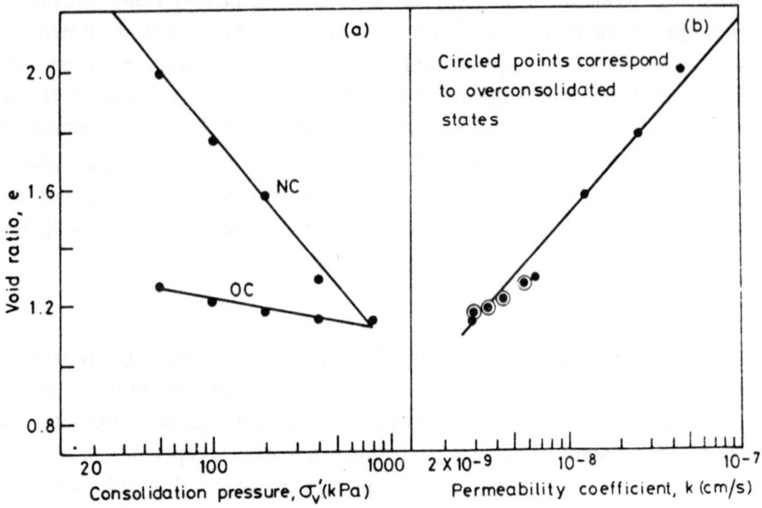

Fig. 5.46 Effect of overconsolidation on permeability coeffficient for Parur clay (after Pandian et al., 1993)

effects. With appropriate relationships stress - state - permeability realationships can be predicted both for normally and overconsolidated states of the clay.

It may be appreciated why, for a given soil, the permeability is maily dependent on void ratio and not on stress, while for different soils it is dependent on stress and not on void ratio. For two soils under the same stress, the size of the large pores will be same, and the volume of large pores in a given cross sectional area will also be not much different, in spite of a wide variation in void ratio, because of a corresponding difference in the specific surface of the particles. Thus their permeability will be of the same order. With increase in stress, there will be a reduction in the size, and volume of large pores and hence a decrease in permeability. But for the same soil, at the same void ratio, under normally and overconsolidated states, as there will be no further change in the size or volume of large pores during unloading or reloading, the permeability will be pressure independent but depends only on the void ratio.

6

CEMENTED SOILS

6.1 Introduction

Cemented soils are those in which additional bonding exists between particles through certain extraneous chemical substances present and hence exhibit higher resistance against deformation. Cementation bonds markedly influence the behaviour of soils. In general, cemented soils exhibit very low or negligible compressibility upto the yield stress and markedly high compressibility beyond it. These soils exhibit a definite cohesion component of shear strength. The serious practical implications with cemented soils are that they may lose their increased strength due to bonds either upon mechanical disturbance or due to chemical leaching which may lead to unexpected behaviour. The ratio of undisturbed strength to remoulded strength at the same void ratio, defined as the sensitivity S_t of the soil, is often used to denote the degree or intensity of cementation bonding.

It has not been possible, most often, to explain the behaviour of cemented soils, let alone predict it, in spite of concerted work by several researchers mainly from Canada, Japan and Norway. There are many anomalies in their behaviour when viewed from our knowledge of uncemented soils. All these may be because, the state of a cemented soil is not uniquely related with the effective stress as it is in the case of uncemented soils. Or in other words, the effective stress responsible for volume changes and shearing resistance is not known. In this chapter, we are proposing a totally different approach for analysing the behaviour of cemented soils. The approach is based on the finding that the load carrying capacity of a cemented soil is a superposition of two components: (a) resistance of the unbonded soil skeleton and (b) resistance of cementation bonds. Further it is brought out that the deformation is entirely due to changes in the stress on the unbonded skeleton, since the cementation bonds have a rigid response. The thoughts and the approaches presented are quite novel to the Geotechnical community and time alone has to testify their credibility.

Unfortunately there is only meagre data in literature, especially the information regarding the response of unbonded soil skeleton or the remoulded state, quite rightly so because no one would have thought it likely that the behaviour of remoulded states would be helpful in describing the behaviour of undisturbed states.

A greater emphasis is given to the analysis of behaviour since proper methods are not available till today and the possible generalisations and prediction methods are discussed later. The proposed mode of stress transfer with the superposition of two components is explained micromechanistically for the soil micro-structure described in Chapter 3.

Soils can aquire cementation bonding when they are in a normally consolidated state or overconsolidated state. In the former case, the soil will become soft sensitive and in the latter case, stiff cemented. Depending on the intensity of cementation and initial state of the soil, different soil states ranging from soft, highly sensitive to highly stiff are possible (Fig. 6.1). The stiff cemented soils are not generally problematic as they can mobilise sufficiently high strength and have low compressibility. Hence the discussions in this chapter will mostly pertain to soft sensitive soils although the same logic can be extended to stiff cemented soils. The proposed thoughts may not be applicable to insitu weathered soils from parent rocks, with considerable residual cementation bonding, since the micro-structure of remoulded soils as discussed earlier may not exist or this aspect has to be examined further.

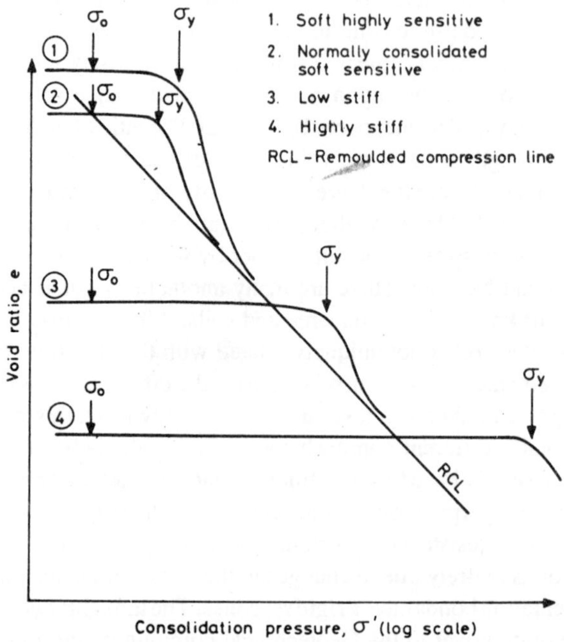

Fig. 6.1 Schematic representation of soils of different degrees of cementation.

6.2 Meta Stable State

The chemistry or mechanism involved in the formation of cemented soils is not within the scope of this book. This aspect has been studied by several researchers (Skempton and Northey, 1953; Kenney et al., 1968; Kenney, 1968; Rosenquist, 1955; Houston and Mitchell, 1969 and others). In brief and in very general terms, cementation bonds between particles / or their aggregates are developed under saline environments acting over long periods of time, as solid amorphous links of precipitates of calcites, iron oxides, alumina and other inorganic and organic matter. These bonds impart additional rigidity to the soil skeleton against deformation. As long as these cementation bonds are present, the initial void ratio prevails even under stresses greater than the initial overburden pressure. This results in higher equilibrium void ratios than those of uncemented soil under corresponding stresses, i.e, the void ratio - pressure relation of the soil gets altered due to the contribution of the bonds to the resistance. This state which is not a true reflection of the equilibrium stresses acting, is usually termed as the 'meta stable state', in the sence that under any given stress, if the bonds are leached or destroyed, the soil would collapse in an effort to reach the state of uncemented soil under the same stress.

6.3 Characteristics of Cemented Soils

A typical compression path of a soft cemented soil has an inverse - S shape with three distinct zones (Fig. 6.2). In zone 1, i.e., up to the yield stress, σ'_y, the compressions are negligible (almost zero), and beyond this stress i.e., zone 2, very high compressibility is seen upto certain high stress level, say σ'_t. The path in this zone is not linear even in semi-log plot as it would be in uncemented soils, but is of decreasing slope with increase in pressure. Finally in zone 3, beyond the stress σ'_t, the compressibility attains a normal value, which may be equal to that of the remoulded state of the soil. Usually, the third zone will be in the range of such high stresses that it may not be of much practical interest. Hence considering only the first two zones, i.e., negligible compressibility upto the yield stress, and relatively higher compressibility beyond, which is also typical of an overconsolidated soil, cemented soils are often treated or analysed as overconsolidated soils. The ratio of the yield stress to the current overburden pressure is generally referred to as the overconsolidation ratio treating the yield stress as the maximum past pressure. The shearing behaviour of soft cemented soils appears to be similar to that of overconsolidated soils in many respects. They also exhibit a high initial modulus, an early peak strength and then strain softening beyond the peak. But on close examination it can be found that the two are totally different. First of all, at any given stress, the state of a soft cemented soil always has an equilibrium void ratio higher than a remoulded normally consolidated soil, whereas an overconsolidated soil always has a lower void ratio (up to the preconsolidation pressure) at the same stress level (Fig 6.2). The compressibility of an overconsolidated soil is never greater than that of the normally consolidated state and it may just approach that value at high stress levels. Where as in the case of cemented soil, the compressibility in zone 2, beyond yield stress is always far greater than that of the uncemented normally consolidated soil. The same features are reflected indirectly even under shear. A cemented soil shows much higher volumetric compression and shear strains than the corresponding uncemented soil, even for confining pressures much less than the yield stress. The volume changes are compressive even during post peak stages. Whereas in overconsolidated soils, the post peak softening is associated with volumetric dilations (if not, at least reduced volumetric compressions in lightly overconsolidated states, in which case a clear softening in

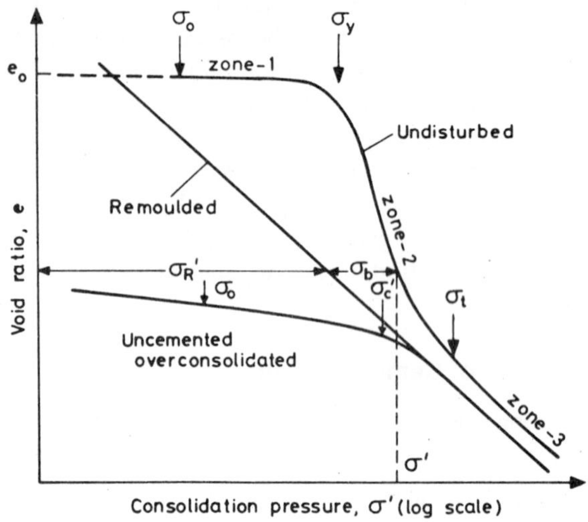

Fig. 6.2 Typical compression path of soft cemented soil.

strength may not be observed). In the case of undrained shearing, soft cemented soils would show continued positive pore pressures even beyond the peak strength, where as overconsolidated soils show negative or decrease in pore pressure. This is a very important point because strain softening associated with volumetric compression is a violation of the basic principles of thermodynamics and cannot be explained by the usual plasticity theories. But still, many researchers around the world are using plasticity based models to cemented soils. Another point of difference to be noted is that pronounced strain softening in soft cemented soils is observed in undrained shearing and generally not in drained shearing where as with overconsolidated soils a pronounced peak in strength is never observed in undrained shearing but only in drained shearing of highly overconsolidated states. Cemented soils show softening in undrained shearing even under confining pressures greater than the yield stress. This is a clear indication that the behaviour is not the same as that of a normally consolidated state even beyond the level of yield stress which is the case with overconsolidated soils. This is so because in cemented soils, there is an additional component of resistance from cementation bonds. All these specific characteristics can be clearly seen from the undrained test results of Gloucester clay (Bozozuk, 1984) and drained test results of Leda clay (Raymond et al., 1979) presented in Figs. 6.3 and 6.4.

In undrained test on Gloucester clay, strain softening is observed even for the test with a confining pressure of 2.2 kg/cm^2, which is much higher than the yield stress for the soil which is about 0.6 kg/cm^2 (Fig.6.5). Also, for all the tests the pore pressure continues to be increasing even beyond the peak strength. In the drained tests of Leda clay, we can see high volumetric compression even for the test with confining pressure of 20 kPa, which is less than the yield stress of about 40 kPa. For a similar case with uncemented soil say for example Weald clay, for

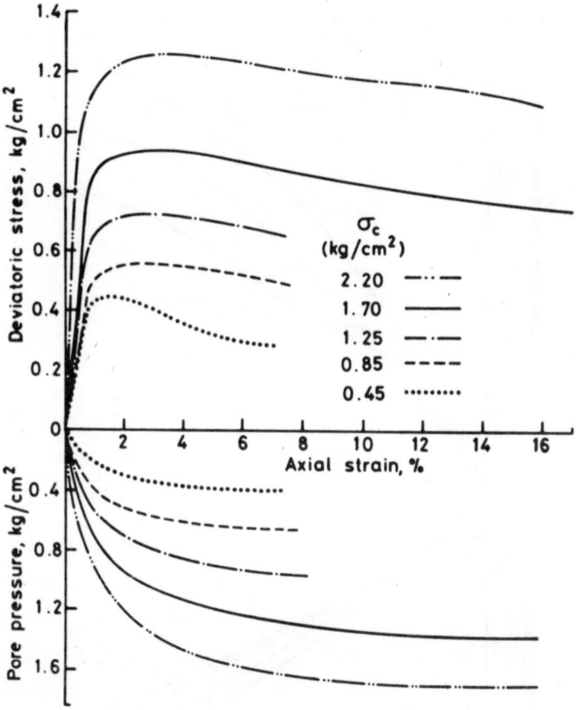

Fig. 6.3 Undrained triaxial test results of Gloucester clay (Data Bozozzuk, 1984).

an OCR of 2, i.e., at 60 psi confining pressure unloaded from a maximum pressure of 120 psi, the volumetric strain is nearly one third of (= 1.6%) what it is for the normally consolidated state (4.5%) (data Parry, 1960).

In this chapter, we are advancing a new mechanism of stress transfer which can consistently explain all the above observations and which provides a basis for the model to predict their stress-strain response without violating the basic physical laws.

6.4 Mechanisms of Stress Transfer

Before studying the mode of stress transfer at microlevel, let us make note of a very useful observation from the usual compression curves which infact has led to the formation of a systematic proposition of the micro model.

When the compression path of a soft cemented soil is examined in relation to that of its remoulded state (Fig. 6.2), it can be seen that, at any void ratio, e the pressure, σ' is in excess of the stress, σ'_R on the remoulded state by an amount $\sigma_b = (\sigma' - \sigma'_R)$ which may be regarded as the contribution from cementation bonds. Infact $(\sigma - \sigma'_R)$ is the minimum value of the stress on the bonds, since the stress on the uncemented soil skeleton can never be greater than σ'_R. Thus the load

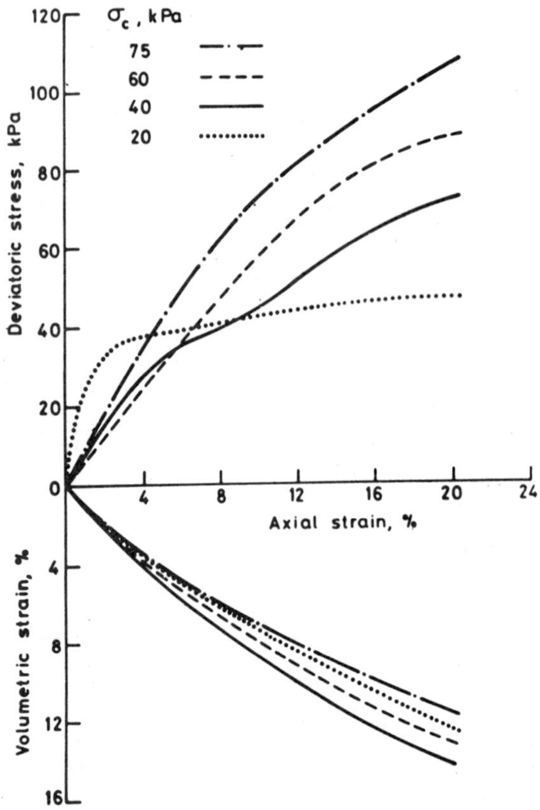

Fig. 6.4 Drained triaxial test results of Leda clay (Data Reymond et al., 1979).

carrying capacity of the soil is due to a combination of the resistance of the soil skeleton excluding bonds and of the cementation bonds. In zone 1 ($\sigma' < \sigma'_y$), the applied stress σ' is resisted totally by the bonds with a nearly rigid, non particulate response. Once the bonds are stressed to their maximum capacity σ'_b, further increments of load will have to be carried by the unbonded skeleton of the soil and this can happen only with compatible reductions in void ratio. It appears from the available data in literature (Quigley and Thomson, 1966; Yong and Nagaraj, 1977 and Lo, 1972) (Fig. 6.6), that this cementation component σ_b remains constant or even slightly increases but never decreases with increase in stress level within the working stress range of engineering interest. This is contrary to the usual notion that cementation bonds break down during loading and hence the compression path moves towards that of the remoulded state. There could be several possible reasons for this. The bonds may not break under compression loading, or though the bonds undergo a gradual breakdown during compression, the number of bonds per unit volume of the soil may remain the same or even more than what it was originally,

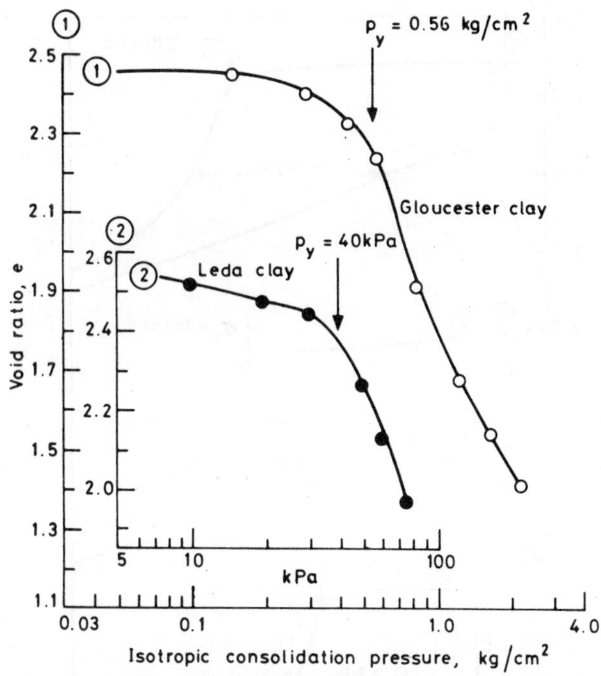

Fig. 6.5 Isotropic compression paths of Leda clay and Gloucester clay.

or the strength of each bond may itself increase with reduction in void ratio. Whatever may be the reason, it is an observed fact that the magnitude of cementation bond resistance, σ_b does not reduce during consolidation loading. However the sensitivity of the soil reduces with increase in pressure because of the drastic increase in the resistance of the unbonded skeleton with reduction in void ratio, while the cementation component being nearly constant.

Under any stress increment from σ_1' to σ_2', the soil would compress by such a magnitude as to increase the unbonded skeleton resistance to the required level from σ_{R1}' to σ_{R2}'. To enhance clarity let us consider the compression test data of Louiseville clay (Lapierre et al., 1989) (Fig. 6.7). The yield stress for this soil is about 180 kPa. The liquidity index of the insitu soil is greater than 1.0. Thus we can say that this yield stress of 180 kPa is entirely due to bonds since the uncemented skeleton resistance at such high water contents greater than liquid limit would be negligible (less than 5 kPa). Now, for a load increment ratio of 1, (from 180 to 360 kPa), i.e., at an applied stress of 360 kPa, the unbonded skeleton resistance is about 170 kPa, and the cementation component is still about 190 kPa. During this loading, though the applied load increment ratio was 1.0, it was actually much more (i.e., from less than 5 kPa to 170 kPa ≈ 34), on the soil skeleton excluding cementation bonds. This is responsible for the high compressions or the high values of C_c observed. These features are observed with many other sensitive

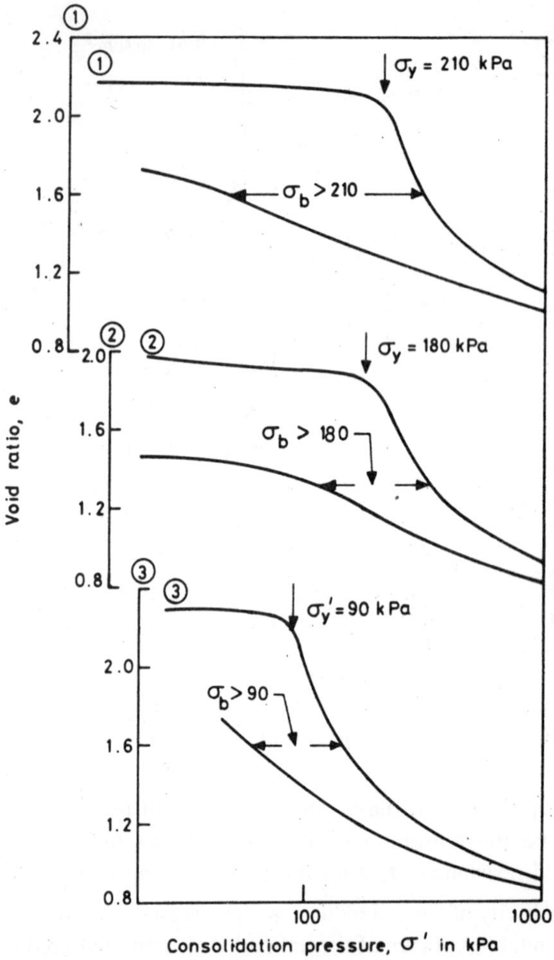

Fig. 6.6 Compression paths of three cemented soils [data from (1) Quigley and Thomson, (1966), (2) Yong and Nagaraj (1977) and (3) Lo (1972)].

clays (Quigley and Thomson, 1966; Yong and Nagaraj, 1977 and Lo, 1972), as can be seen from the Fig. 6.6.

We can restate the above as: the resistance of a cemented soil at any state is the sum of the resistance of the unbonded soil skeleton and the resistance of cementation bonds and that the volume changes are essentially due to changes in the stresses on the unbonded skeleton, the cementation bonds providing additional strength with rigid response. We have to examine the validity of this proposition micromechanistically, especially because, it has been emphatically brought out (Bloch, 1978) that in an interacting system such as clay-water system, consideration of independent stress fields for different phases and their superposition to get the total response is not physically tenable.

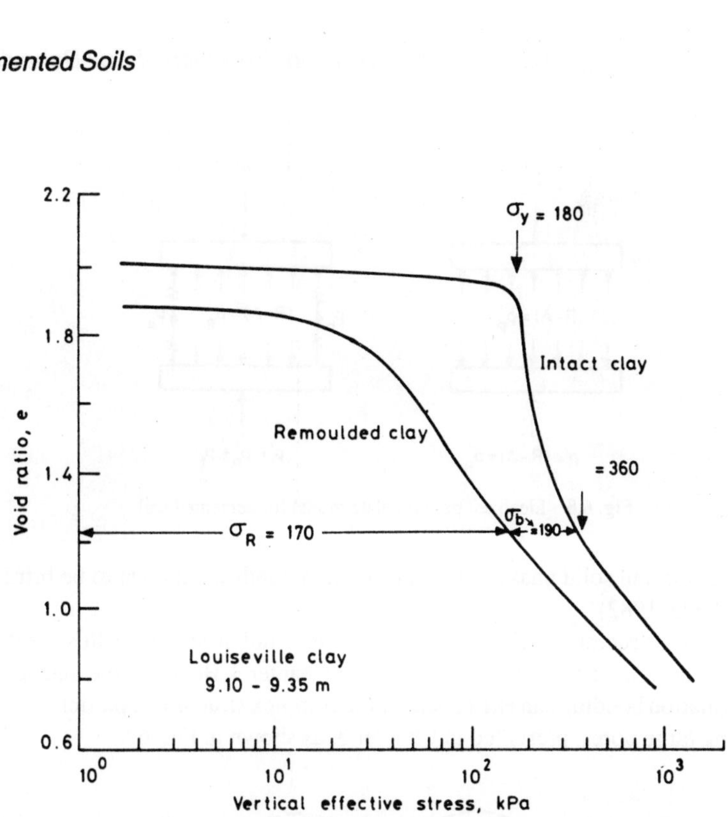

Fig. 6.7 Compression paths of Louiseville clay (after Lapierre et al., 1989).

6.4.1 Micro-structural considerations

As already discussed in Chapter 3, in clays, because of their plate like structure and high surface activity, the stress transfer between particles is essentially through long range forces, which are due to physico-chemical interactions with the fluid phase. The heights of micro-aspirities on their surface being quite small (10 to 100 Å units as against 1000 to 10000 Å units in sands), the possibility of direct contacts at the aspirities with short range forces acting is remote (Feda, 1982). Thus the applied pressure p is resisted by the dominant long range force which is the net osmotic repulsive pressure $p'_R = (R - A)$. This force is a function of the half space distance d between the particles.

Now cementation in soils is due to the presence of extraneous chemical substances precipitated between soil particles. It is quite possible that these substances result in contact junctions with particle surfaces at which short range forces, p_b operate as shown in Fig. 6.8.

There is nothing in principle to bar the coexistnace of the long range forces between the two particle surfaces and the short range forces at the sites of cementation, and the two together resisting the externally applied loading so that $p' = p'_R + p_b$. It should be noted that p_b is the average pressure obtained from the contribution of the bond forces at all junctions over a cross section, considered to be acting over the entire cross sectional area, as is usually done in defining the

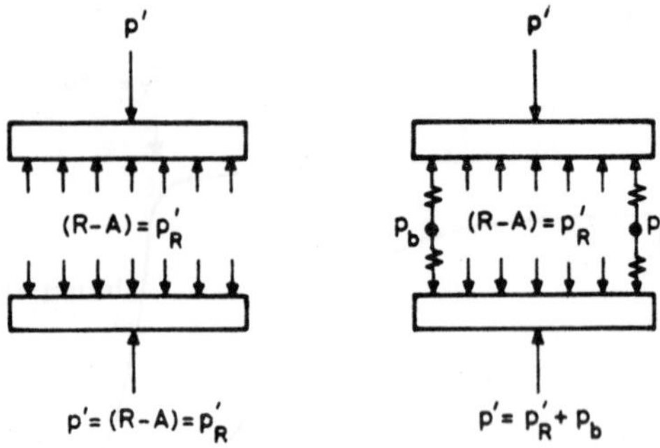

Fig. 6.8 Idealized parallel plate model for cemented soil.

effective stress in solid phase. Such cementation bonds are known to be brittle in nature (Feda, 1982).

The microstructure of clayey soils is of course not as simple as this idealized parallel plate structure as already discussed in Chapter 3. But the same mechanism of cementation bonding can exist even for the complex structure of particle clusters enclosing large pores, presented in Chapter 3, as shown in Fig. 6.9.

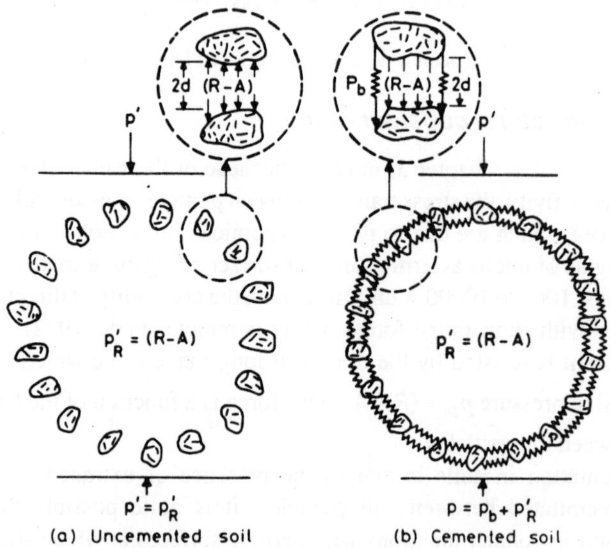

Fig. 6.9 Possible microstructure for cemented soil.

This mode of superposition is not a contradiction to Bloch's statement. It is still true that in interacting system, consideration of independent stress fields is not tenable in the sense that the equilibrium of fluid phase has to be always related to the force field in the solid phase, i.e., between particles. But it is only the long range forces $(R - A)$ which are interacting with the fluid phase, and the short range cementation forces do not play any role in the fluid retention. These cementation bond forces are similar to the contact forces in sands which are not capable of retaining fluid. Hence the osmotic suction on the fluid phase is only $(R - A) = p_R'$ and the total equilibrium can be written as $p' = p_b + p_R'$

Researchers in the past generally thought that cemented soils have a different microstructure, of more open and flocculated type, than uncemented soils. But the recent studies on permeability and pore size distribution with scanning electron microscopy (Delage and Lefebvre, 1984) or mercury porosimetry (Lapierre et al., 1989) show that the pore size distribution as well as permeability of cemented and remoulded states are identical at the same void ratio. This implies that the basic microstructure is the same in the two states although the equilibrium stresses are different because of bond resistance.

Now reverting to the deformation behaviour, the soil may be expected to show a rigid response until the yield stress of the bonds is reached, since the bonds are rigid and brittle. Beyond the yield stress, the specific feature of the two phase particulate system comes into play. In metals or concrete or any other continuum, a rupture of bonds would simply mean a corresponding reduction in strength on that plane. But a particulate system can compress and harden until it gains the resistance required to carry the stresses which were being carried by these ruptured bonds, provided there is scope for such compressions by way of drainage. If the soil is loaded beyond the yield strength of the bonds, some of the bonds may yield. A grouping of particles, which is responsible for plastic compressions, is now possible at such yielded junctions. The interacting system would reach a changed equilibrium with reduced intra- and inter-cluster spacing of clusters of particles and reduced size of large enclosed pores, there by increasing the unbonded skeleton resistance by such a magnitude that the additional applied pressure can be resisted.

During such compression under increased loading, it is quite possible that the density of bonds per unit volume may either remain constant or increase even though some of the bonds are really broken down, there by not reducing the magnitude of cementation resistance, which is an observed fact as seen in the earlier section. Or, the bond strength not decreasing with increase in pressure could be even due to the fact that with compression of the soil and decrease in the inter cluster spacing, the short range forces at the cementation bonds may themselves increase in magnitude. Whatever may be the reason, the fact remains that cementation bond resistance does not decrease with consolidation loading i.e., along constant q/p' path. However, during shearing, i.e., loading with q/p' ratio increasing, the yielding of bonds does result in a decrease of cementation component, both in undrained and drained shearing because of breakdown of bonds in excess of the corresponding volumetric compressions unlike in consolidation loading.

6.5 Effective Stress in Cemented Soils

With this understanding of the micro mechanism, we may have to redefine effective stress in cemented soils. By definition, effective stress is that stress which is transfered through solid particle contacts and which is responsible for volume changes and mobilisation of shearing resistance. Then for cemented soils, since the volume changes are due to the uncemented component of stresses p'_R, this component is the true effective stress. i.e. $p' = p'_R = p - u - p_b$, and the mobilisation of uncemented skeleton resistance to shearing q_R is also directly related to this effective stress p'_R, but there is an additional shearing resistance q_b due to bonds.

6.6 Analysis of Stress-Strain Response

Based on the premise that the uncemented component of resistance mobilizes at corresponding deformations as in a remoulded soil, and cementation bonds providing an additional resistance with rigid response, the following analysis and mode of describing the stress- strain behaviour of cemented soils are elucidated.

The uncemented component of resistance (q_R) mobilizes hyperbolically with strain as in a normally consolidated uncemented soil, drained or undrained as the case may be. The cementation component q_b mobilizes within negligible or very low strains. Upon further shearing, the bonds may undergo a gradual breakdown thus reducing the bond strength q_b. The total resistance q at any instant, being a sum of these two components may or may not show peak in the stress-strain behaviour depending on whether the rate of increase of uncemented skeleton component is lesser or not than the rate of decrease of cementation component. Now we can see why we generally have a peak and softening only in undrained shearing and not in drained shearing. This is because the component q_R does not increase considerably beyond 4 to 6% strain in undrained shearing whereas it continues to increase appreciably even upto 15% in drained shearing, the rate of decrease of q_b being the same in both the cases.

To obtain the stress-strain response we have to estimate the two components independently at any strain level and add the two.

As we can see from Fig. 6.10, for any given state $C\ (e, p')$ of the soil, we have the corresponding uncemented state $E\ (e, p'_R)$. Mobilisation of uncemented skeleton component of resistance with shear strain will be for this initial state (e, p'_R). Simultaneously the bond strength, q_b, mobilizes to its peak value, q_{bmax} for a particular (very small) strain and decreases there after due to breakdown of bonds. With the increase in the deviatoric component q_b, we would also have an associated increase in the mean principal stress equal to $q_b/3$. The bonds are incapable of resisting this additional mean principal stress $q_b/3$ if they are already stressed to their maximum capacity p_b (which will be true for confining pressures greater than the yield stress). In addition, the stress carrying capacity of the bonds itself will reduce from its initial value of $p_0\ (= p_{bmax})$ to zero at very large strains due to break down of bonds. These two stress components, $q_c/3$ and $(p_{b0} - p_{b\varepsilon})$ will have to be transferred either on to the remoulded state (if drainage is permitted) or to the pore water pressure (if drainage is not permitted) which is in effect equivalent to an external loading of the same magnitude on the uncemented soil state. At very large strains, the bonds may be completely broken when both p_b and q_b would reduce to zero and the whole of p_{b0} would have been transferred to the pore pressure or to the soil skeleton as effective stress. The soil would ultimately reach the same critical state F (e, p'_{fu}) in undrained shearing, as an uncemented soil at the same void ratio would reach (Fig. 6.10). We can now see why we get greater pore pressures $(= p - p'_{fu})$ in cemented soil than for uncemented soil $(p'_R - p'_{fu})$ at the same void ratio. Figure 6.11a shows the undrained stress path for a soft sensitive soil and the corresponding stress path on the unbonded soil skeleton (i.e. excluding bonds) explicitly showing the two components of resistance. At any instant, X_i is the applied total stress state, X_i' is the effective stress state and X_i'' is the corresponding stress state on the unbonded soil skeleton. The total resistance X_i' $(q, p')_i$ is the sum of unbonded skeleton resistance X_i'' $(q_R, p'_R)_i$ and the bond resistance $(q_b, p_b)_i$.

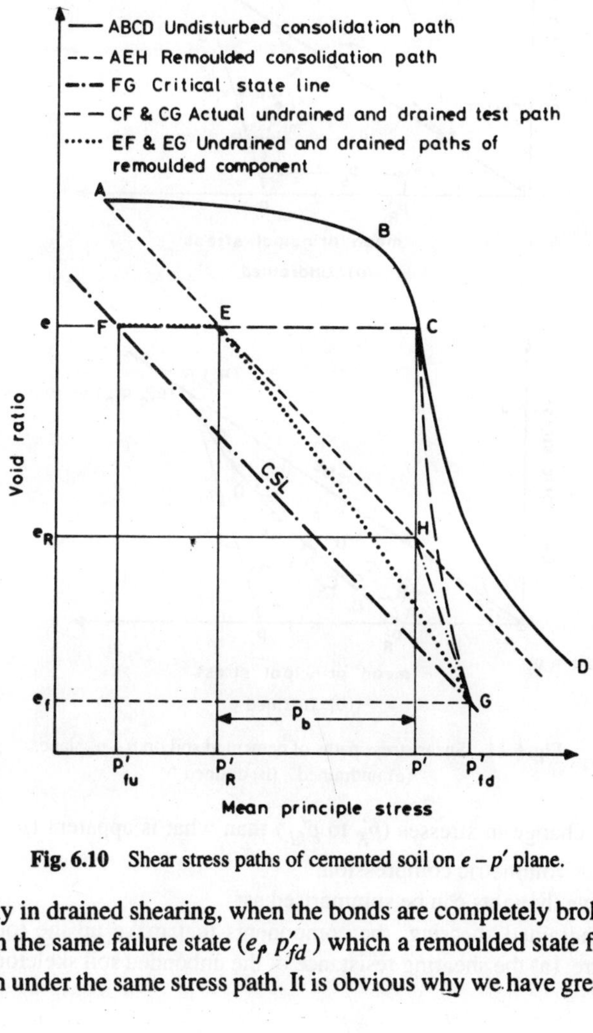

Fig. 6.10 Shear stress paths of cemented soil on $e - p'$ plane.

Similarly in drained shearing, when the bonds are completely broken, the soil would reach the same failure state (e_f, p'_{fd}) which a remoulded state from (e_R, p') would reach under the same stress path. It is obvious why we have greater volume compression for the cemented soil than for the remoulded soil at the same confining pressure. During drained shearing, the stress components $q_c/3$ and u_b $(= p_{b0} - p_{b\varepsilon})$ will be transferred to the uncemented skeleton as effective stress at each stage of shearing. Hence the initial uncemented state at (e, p'_R) is subjected to a loading path different from (with $\delta q/\delta p$ less than that for) the actual applied path (Fig. 6.11b). i.e., the initial soil skeleton excluding bonds (e, p'_R) is subjected

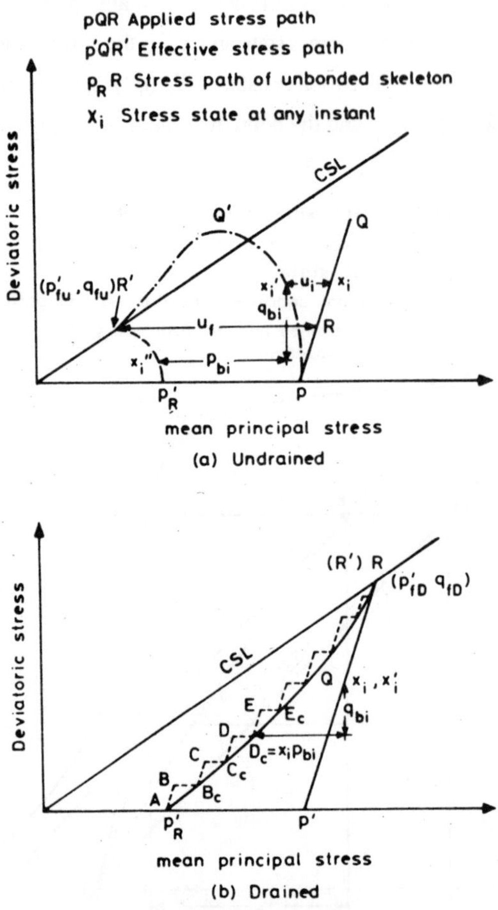

Fig. 6.11 Shear stress paths of cemented soil on $q - p'$ plane.
(a) undrained (b) drained

to a greater change in stresses (p'_R to p'_{fd}) than what is apparent (p' to p'_{fd}) and hence greater volumetric compression.

The above thoughts can be summarised as:

In an undrained shearing, the components that make up the total shearing resistance are: (a) the shearing resistance of the unbonded soil skeleton, q_R for the initial state (e, p'_R) and (b) the bond resistance, q_b together giving the total strength q at any given instant. The components that make up the total pore pressure, u are (a) an uncemented skeleton component of pore pressure, u_R corresponding to q_R, (b) a component $q_b/3$ equal to the increase in mean principle stress due to increase in the deviatoric stress, q_b and (c) a component u_b equal to the decrease in p_b caused by breakdown of bonds (Fig. 6.12a). i.e.

$$q = q_R + q_b \quad 6.1$$

$$u = u_R + u_b + q_b/3 \quad 6.2$$

In drained shearing, the total shear strength q at any instant is the sum of the cementation component, q_b and the uncemented skeleton component, q_R with its distorted stress path to include the stresses $q_b/3$ and u_b transferred from the broken bonds incrementally throughout the shearing process. There is only one component of volumetric strain ε_{vR} i.e., due to the changes in the stresses on the uncemented soil skeleton, since the cementation component has a rigid response (Fig. 6.12b).

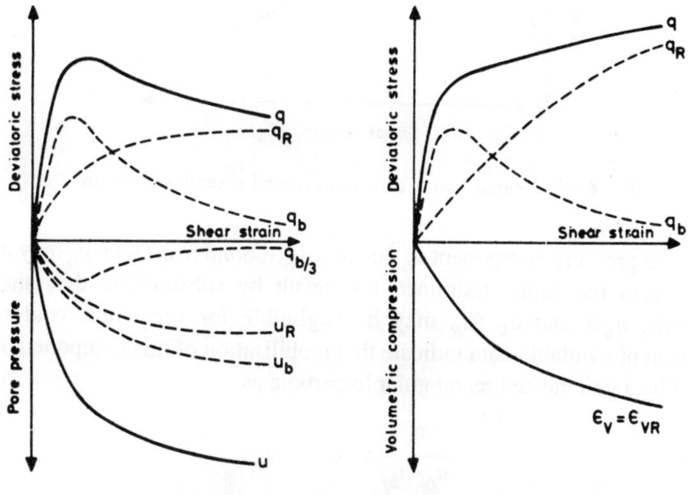

Fig. 6.12 Cemented, uncemented and net components of shear strength with strain – (a) undrained (b) drained (Nagaraj et al., 1991).

We can obtain the response of the uncemented soil skeleton at any state under any loading path using any suitable constitutive relations as discussed in Chapter 5. To obtain the characteristics of the cementation component we may have to conduct typical shear tests to get the maximum bond strength and also the rate of degradation of q_b with strain due to breakdown of bonds. An undrained test at a confining pressure less than the yield stress would be preferable because for such cases the uncemeted skeleton component would be negligible due to the very high water content, and hence the observed strength itself is the q_b component. From the available data in literature it appears that the bond strength mobilises to a peak value at strain level of about 1 to 2 %, and the reduction thereafter nearly follows an inverse hyperbola as shown in Fig. 6.13. Very large strains (greater than 35 to 40%) may be required for causing complete breakdown of bonds. Please note, that Fig.13 does not indicate a stress vs strain relation for the bond resistance. It only indicates the magnitude of available or intact bonds at any stage of shearing. The bonds' response is infact rigid-brittle. The magnitude of peak strength, q_{bmax} may itself increase with increase in confining pressure in the same way as the variation of p_{bmax} with confining pressure as discussed earlier.

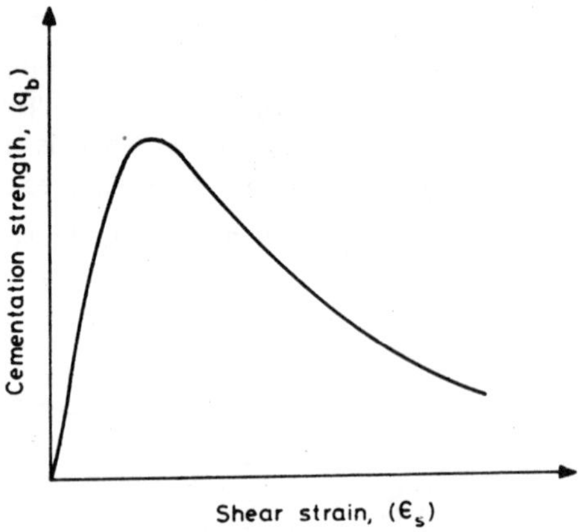

Fig. 6.13 Degradation of cementation bond strength with strain.

The pore pressure component u_b due to loss of bond resistance p_b may also be obtained from the same undrained test result by substracting the other two components, $q_b/3$ and u_R (u_R may be negligible for very high void ratios). Observation of available data indicate that mobilisation of this component can be described by a generalised rectangular hyperbola as

$$\frac{\varepsilon_s}{u_b/u_{bf}} = a + b\varepsilon_s \qquad 6.3$$

where u_{bf} is the final value of u_b equal to p_{b0}.

The scheme of working using a constitutive model is quite simple for the undrained case. It is exactly the same as for an uncemented normally consolidated soil at (e, p'_R). We only have to compute q_R and u_R with strain and then add q_b and ($q_b/3 + u_b$) respectively to get the total strength and pore pressure at any strain.

For the drained case the procedure is slightly involved. For the initial state (e, p'_R) the uncemented component has to be worked out for the applied stress path (say $\delta q/\delta p' = 3/1$ for the usual triaxial test) incrementally as for the uncemented soil. But at the end of each increment, the additional stress component $\delta p_{ad} = (\delta q_b/3 + \delta u_b)$ transferred from the broken bonds during that strain increment, has to be applied and the corresponding strains computed. To get the total resistance at any stage the cementation resistance q_b has to be added to q_R as before.

Summarising the above, we can obtain the stress-strain response of a soft cemented soil under any stress path, as the combined effect of resistances of the uncemented skeleton and of the cementation bonds if we have the following information.

(a) Consolidation data on the undisturbed soil to know the state of the soil at any pressure.
(b) Consolidation data on the remoulded soil to provide compressibility coefficients λ and κ (or C_c & C_r) and to estimate p_{bmax} or u_{bf} at any void ratio (along with (a) above).
(c) Undrained shear test on the undisturbed sample to get the magnitude and rate of decrease of q_b (and also of u_b) with shear strain
(d) Shear test on the remoulded soil to know the friction factor M (or ϕ).

The above discussed analysis or prediction is for a given soil knowing its basic properties. In the following sections we will examine if further generalization and simplification are possible for different cemented soils as we could do in the case of uncemented soils.

6.7 Generalisation and Prediction of Behaviour

It is obvious that we cannot simply extend the generalisation of behaviour presened in Chapter 5 for different cemented soils by normalising with their respective liquid limit values. This is because liquid limit being a parameter determined on a completely remoulded soil, cannot, in any way reflect the effect of cementation bonds. However, since the response of soil skeleton excluding bonds is exactly the same as that of an uncemented soil, the generalisations brought out in chapter 5 are definitely possible for this component exclusively. That is, if we know the liquid limit of the soil we can get its compressibility parameters λ and κ (or C_c and C_r) with respect to the component of stresses on the uncemented soil skeleton, and hence we can generate the compression path in e vs log σ'_R plot for the uncemented component. i.e., the relation of the form

$$e/e_L = a - b \log \sigma'_R \qquad 6.4$$

is still valid for cemented soils where σ'_R is the component of the total stress σ carried by the uncemented soil skeleton. These soil parameters can further be used to get the yield surface for the uncemented component and strain computations for any stress path.

The bond strength q_{bmax} or p_{bmax} is not likely to be a generalisable quantity for different soils because the extent or degree of cementation in any soil is the choice of the nature and can be determined only by experimentation. However, the rate of degradation of q_b and p_b with shearing may follow a similar pattern for any cemented soil, and it may be possible to have generalised relations for q_b or p_b with shear strain for all cemented soils.

6.7.1 Prediction of yield stress

The bond strength p_{bmax} in compression can be determined from the usual isotropic consolidation test on the undisturbed sample. But the problem with soft cemented soils is that they are highly susceptible for disturbance during sampling, handling and testing and as such the laboratory testing may not give realistic values. Hence it is necessary to estimate the probable yield stress in field, σ_{ct} from the laboratory test results or if possible, by alternate methods. We are presenting here two simple methods (Nagaraj et al., 1990) for estimating the field yield stress.

By modifying the laboratory compression curve

It has been shown (Milovic, 1970; Lo, 1972 and Holtz et al., 1986) that with increase in degree of sample disturbance, the magnitude of experimentally determined yield stress decreases while the strain upto this stress increases. Nagaraj et al. (1990) have shown that the locus of σ_y for different degrees of sample disturbance is a straight line perpendicular to each of the $e - \log \sigma'$ curves at their points of maximum curvature (Fig.6.14). In the limiting case this line would be perpendicualar to the remoulded path which corresponds to 100% disturbance.

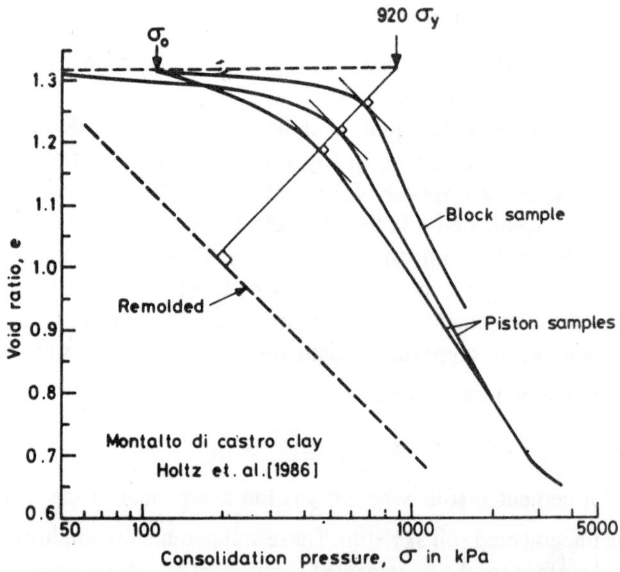

Fig. 6.14 Effect of disturbance on e vs log σ curves of Montalto di Castro clay (after Holtz et al., 1986).

Secondly, there are indications (Fig. 6.15) from the insitu load settlement test results (Pelletier et al., 1979 and Folks and Crooks, 1985) that the settlement is negligible upto the yield stress (the small strains that we observe in laboratory tests is due to sample disturbance). With these two findings, the most probable value of the yeild stress in field can now be predicted from the laboratory curve of field

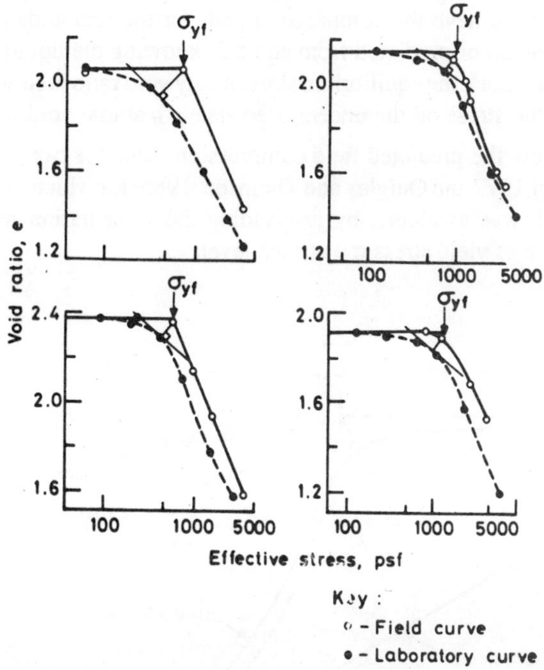

Fig. 6.15 Comparison of predicted σ'_{ct} with field values (after Pelletier et al., 1979).

sample as the point of intersection of a horizontal line from the point σ'_o, e_o and the normal to the e vs $\log \sigma'$ curve at the point of maximum curvature (Fig. 6.14). The validity of this approach is shown in Fig. 6.15 wherein the field yield stress values have been predicted for four soils.

Using field vane strength

Alternatively, in the absence of a laboratory compression curve, the field yield stress can be determined using the field vane strength data. This is an useful method since it is a routine practice to determine the vane strength of the soil during site investigations and hence this data would be always available. Since both the yield stress σ_y (= σ_b) and the undrained strength s_u (= q_b) are reflections of the same bond strength, it is logical to relate these two parameters. Detailed analysis of published data of many Canadian sensitive soils has resulted in a functional relationship of the form

$$\sigma_{y,\,\text{field}} = 3.78\, s_u + 7 \qquad 6.5$$

with a high degree of correlation of 0.98 (Nagaraj et al., 1990) s_u being in kPa.

6.7.2 Prediction of field compression curve

Once the bond resistance σ_b is known, the prediction of the field compression curve is quite simple. If we have the compression path for the remoulded soil (either laboratory determined or predicted from eqn 5.2, knowing the liquid limit of the soil as discussed earlier), the equilibrium stress at any void ratio is simply obtianed by adding σ_b to the stress on the uncemented state σ'_R at that void ratio. Figures 6.16 and 6.17 show the predicted field compression paths for two soils (data by Yong and Nagaraj, 1977 and Quigley and Thomson, 1966) for which the remoulded compression path was available, by just adding the cementation resistance σ_b (which is the same as yield stress σ_y) at each level.

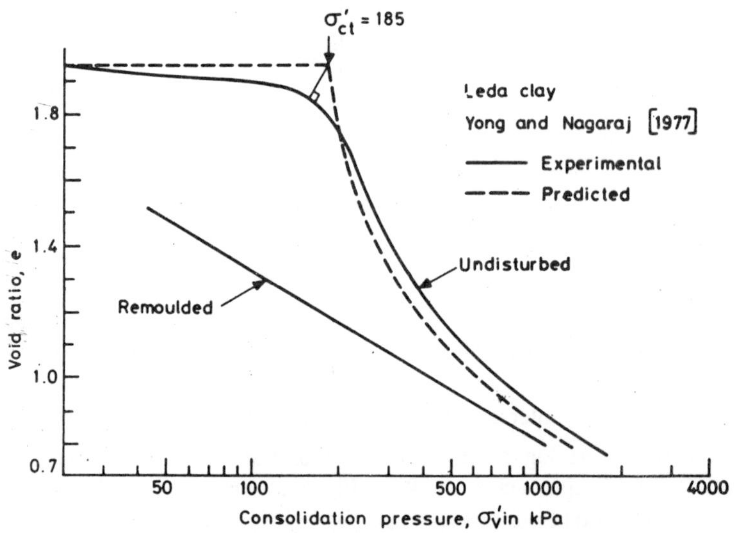

Fig. 6.16 Prediction of σ_{ct} and probable field e vs log σ'_v path using laboratory curve (Data, Yong and Nagaraj, 1977).

6.7.3 Estimation of sampling disturbance

As we know it is almost impossible to get a truly undisturbed sample in sensitive soils. Mechanical disturbance may be caused at various stages i.e., during advancement of bore hole, insertion of sampling tools, transporting, handling and testing of specimen. It is necessary to estimate the degree of such disturbance in order to assess the reliability of laboratory test results and if possible to account for the effects of disturbance on the laboratory measured properties.

For cemented soils, with high liquidity index, we have seen that the response of the soil, for stress levels less than the yield stress, is essentially due to cementation component (the uncemented skeleton resistance being negligible for such cases) and hence is directly affected by sample disturbance. As such, properties like undrained shear strength, tangent modulus etc. for tests with

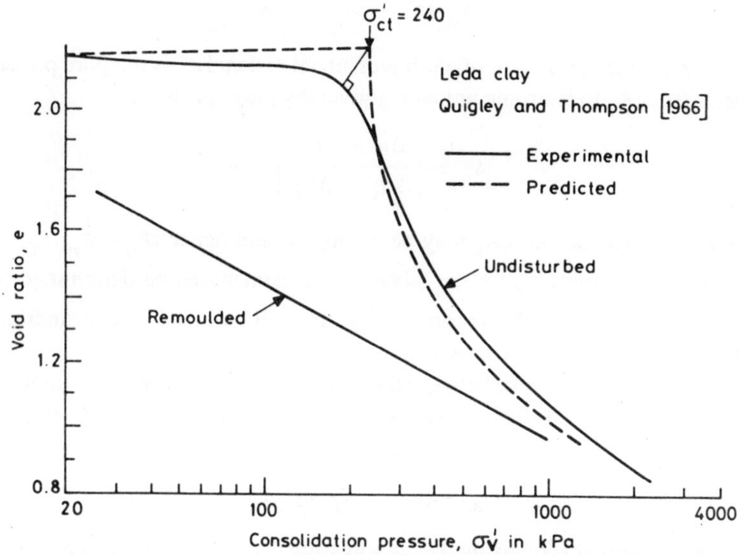

Fig. 6.17 Prediction of σ_{ct} and probable field *e* vs log σ'_v path using laboratory curves (Data, Quigley and Thomson, 1966).

confining pressures less than the yield stress can be corrected if the degree of disturbance is known. Where as for confining pressures higher than yield stress, or for drained shear, or for soil states with I_L much less than 1, the mechnaical properties are not affected one to one by the sample disturbance since in these cases, the uncemented skeleton resistance, which is not affected by disturbance, is not negligible. Hence it may not be possible to apply reasonable corrections to the measured values in such cases.

Evaluation of sample quality by numerical rating can only be done in terms of the ratio of a particular property measured in the test specimen to that of the soil in-situ. To do this, there is always the problem of determining the absolute value of the property of the in-situ soil. A few methods have been tried to evaluate the sample quality. Comparison of field vane test results and the laboratory cone or unconfined compression test results provide an indication of the sample quality although with the limitation that the type of tests are different. Stress- strain curves of unconfined compression tests provide a relative assessment of the degree of disturbance. Failure strains less than 2-4% indicate good quality samples and failure strains greater than 10% indicate significant disturbance (Andresen and Kolstad, 1979). The method is known to be applicable only to soft cemented soils.

A better quantitative evaluation of mechanical disturbance is obtained by residual effective stresses i.e., the negative pore pressure developed upon extraction of the sample. The residual effective stress for a perfect sample, for conditions of complete saturation and no access to water, can be theoretically calculated (Ladd and Lambe, 1963) by

$$\sigma'_{ps} = \sigma'_{v0}\{K_0 + A_u(1 - K_0)\} \tag{6.6}$$

where K_0 is the coefficient of earth pressure at rest and A_u is the pore pressure parameter from K_0 to isotropic unloading given by the expression:

$$A_u = \left(\frac{\Delta u - \Delta\sigma_h}{\Delta\sigma_v - \Delta\sigma_h}\right)$$

For most practical cases A_u may be negligible and hence $\sigma'_{ps} = \sigma_{v0} K_0$. The residual effective stress σ'_R for the laboratory specimen can be determined in a triaxial setup by increasing incrementally the cell pressure under undrained conditions. The cell pressure at which the pore pressure becomes zero gives the residual stress. The difference between the residual stress in the perfect sample and that in the laboratory sample gives a measure of the degree of disturbance (Okumura, 1971 and Nelson et al., 1971)

$$D_d = (\sigma'_{ps} - \sigma'_R)/\sigma'_{ps} \tag{6.7}$$

A simpler method is by using the compression test data (Nagaraj et al., 1990). If we know the true undisturbed value of yield stress in the field σ_{yf}, and the laboratory determined value σ_y, then

$$SD\% = [(\sigma_{yf} - \sigma_y)/\sigma_{yf}] \times 100 \tag{6.8}$$

σ_{yf} can be estimated either from the laboratory compression curve itself or from field vane strength data as already discussed in the earlier section.

The laboratory measured values of undrained shear strength or tangent modulus can then be corrected to obtain the possible undisturbed values as:

$$q_{uf} = \frac{q_u}{(1 - SD/100)}; \quad E_{uf} = \frac{E_u}{(1 - SD/100)} \tag{6.9}$$

The above proposition has been verified for Milovic's (1970) data on two sensitive soils, for samples obtained from three different techniques resulting in different degrees of disturbances. The validity of the approach is revealed by the fact that the same undisturbed value has been obtaibed for all the three samples (Table 6.1).

6.8 Stiff Cemented Clays

So far we have discussed only soft sensitive soils i.e., where the uncemented soil skeleton is in a normally consolidated state. The concepts are applicable even to stiff cemented soils. i.e., when the cementation bonds have been developed on a soil which was already in an overconsolidated state. The only difference is that to compute the uncemented skeleton resistance, we have to consider an initially overconsolidated state and appropriate stress-strain relations. But the

Table 6.1 Soil properties modified for sample disturbance
(after Nagaraj et al., 1990)

Soil type	Type of sample	Disturbance SD (%)	Unconfined Compression Test				CIU test			
			q_u expt.	q_u in-situ (predicted)	E expt.	E in-situ (predicted)	q expt.	q in-situ (predicted)	E_u expt.	E_u in-situ (predicted)
St. Simon clay	Block	21.9	0.60	0.768	75	76	0.60	0.768	69	88
	Piston	31.4	0.41	0.598	54	78	0.52	0.758	58	84.5
	Shelby	43.8	0.25	0.445	30	53.6	0.35	0.765	50	88.96
Nicolet clay	Block	24.7	0.32	0.425	42	56	0.51	0.677	67	89
	Piston	36.5	0.28	0.441	32	50	0.46	0.724	54	85
	Shelby	47.0	0.22	0.415	11	21	0.35	0.660	30	56.6

determination of the bond strength σ_b or the separation of the two components is not very easy in this case. This is because for the normally consolidated state, there is a unique stress related to each void ratio, but not so for overconsolidated state, i.e., for a given void ratio we cannot know the equilibrium pressure unless we know the σ'_p of the soil. For the field soil we would not be knowing the σ'_p for the uncemented skeleton unless we know the complete stress history and cementation bond developement, which is not a possibility. And unless the component of uncemented skeleton resistance is known at a given equilibrium state, the cementation component σ_b cannot be known from the observed total stress. For example, the current state E (Fig. 6.18) of a cemented soil might have been acquired in several ways (Nagaraj et al., 1991).

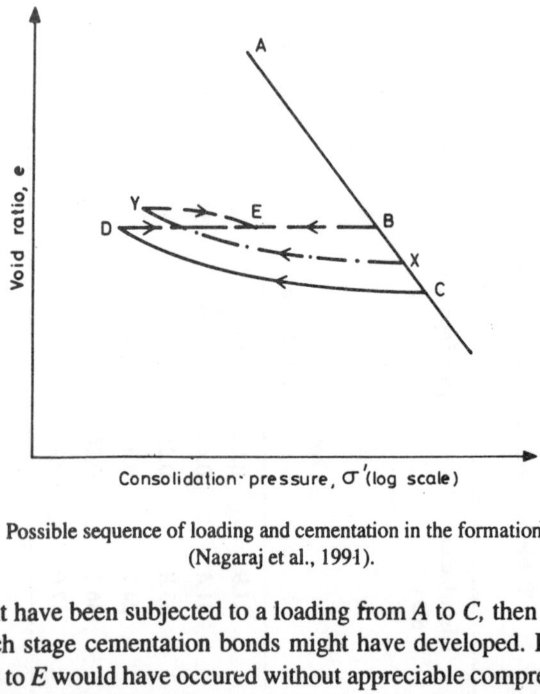

Fig. 6.18 Possible sequence of loading and cementation in the formation of stiff clays (Nagaraj et al., 1991).

(a) It might have been subjected to a loading from A to C, then unloaded to D, at which stage cementation bonds might have developed. Further loading from D to E would have occured without appreciable compression since the bonds resist compression. In this case, stress dependent σ'_p of the soil would be the pressure at C.

(b) The soil might have developed bonding after a loading from A to B. A stress release to E at this stage takes place without much change in void ratio, since the bonds resist dilation also. The σ'_p in this case would be the stress at B itself.

(c) Alternately, the soil might have undergone a loading from A to X, an unloading from X to Y, a reloading from Y to E and then the bonds might have developed, in which case, σ'_p would correspond to the stress at X.

The compression path for each of the above cases would follow different paths depending on the value of maximuim past pressure σ'_p on the unbonded soil skeleton, as per equation,

$$e/e_L = 1.122 - 0.188 \log \sigma'_p - 0.046 \log \sigma'_R \qquad 6.10$$

and $\sigma = \sigma'_R + \sigma_b$

Thus, it is obvious that the uncemented component cannot be assessed with just index properties and insitu condition unless the complete stress history is known. We may have to run an actual compression test to arrive at the unknown σ'_p and σ_b. However the complete compression curve with a load increment ratio of one at the end of 24 hours may not be required but only the data of equilibrium states at two sufficiently high stress levels (*X* and *Y* as shown in Fig. 6.19.) will be sufficient, i.e., with stress level *X* high enough to ensure that the bond strength is fully utilised and the uncemented state is still in the overconsolidated range, and stress level *Y* much higher so that the uncemented component σ'_R has crossed the overconsolidated range or has reached the normally consolidated range.

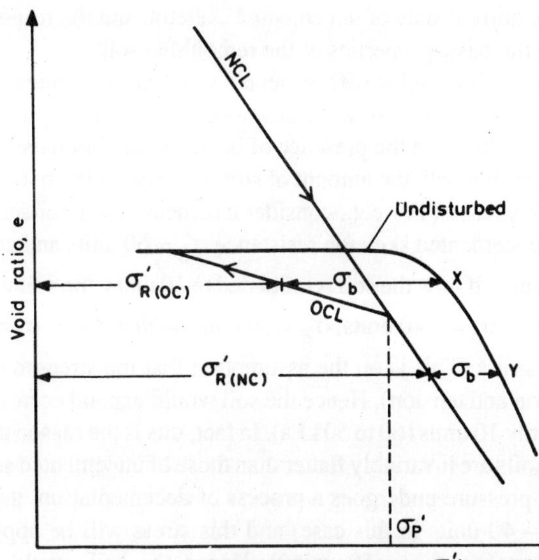

Fig. 6.19 Schematic representation of suggested procedure to predict stress dependent σ'_p and cementation bond strength σ_b (Nagaraj et al., 1991).

From the data of void ratio and pressure at *Y*, σ_b can be computed since the uncemented component at this void ratio is known from eqn. 5.2 i.e., $e/e_L = 1.122 - 0.234 \log \sigma'_R$ and subtracting this from the equilibrium stress at *Y*

$$\sigma_b = (\sigma' - \sigma'_R)_{@Y} \qquad 6.11$$

Once σ_b is known, the uncemented component (σ'_R) at the stress level X can be computed

$$(\sigma'_R)_{@X} = (\sigma)'_{@X} - \sigma_b \qquad 6.12$$

Then the stress dependent σ'_p on the unbonded soil skeleton can also be computed from equation 5.5. With both σ_b and σ'_c known we have all the required data and the behaviour under compression or shear can be predicted by superposing the two components as in the case of soft cemented soils.

6.8.1 Effect of decementation

Another important aspect with cemented soils is the effect of decementation. If, during the process of decementation, the physico-chemical environment of the soil does not alter considerably as to affect the uncemented state of the soil, it would be possible to predict the changes in the soil state. The soil loses that part of the stress carrying capacity (σ_b or q_b) which was the contribution of bonds. If the soil is soft sensitive, the soil will compress until it reaches a state of equilibrium with the increased skeleton resistance now balancing the applied stresses. This is equivalent to applying a load increment of σ_b, q_b (which was being carried by the bonds so far) on the current state of uncemented skeleton and the response can be estimated knowing the basic properties of the remoulded soil.

On the other hand, if the soil is stiff cemented where a stress release has taken place after the cementation bonds were developed, the soil may expand upon decementation. This is because the presence of bonds would not have allowed the soil to expand compatible with the amount of stress release as the bonds can resist (most likely) tensile stresses. For e.g., consider a cemented state under a stress of 100 units with the uncemented skeleton resistance $\sigma'_R = 60$ units and cementation resistance $\sigma_b = 40$ units. If now the soil is unloaded to 10 units, the soil would come to an equilibrium with $\sigma_b = -40$ units, $\sigma'_R = 50$ units so that $\sigma'_R + \sigma_b = 50 - 40 =$ 10 units = applied stress (This is on the assumption that the strength of bonds is equal in compression and tension). Hence the soil would expand corresponding to a stress release of only 10 units (60 to 50 kPa). In fact, this is the reason the rebound paths in cemented soils are invariably flatter than those of uncemented soils. If now the soil at 10 units pressure undergoes a process of decementation, it would lose its bond strength (– 40 units in this case) and this stress will be applied on the uncemented skeleton (*e*, σ'_R (= 50 units)). Hence the soil would now swell corresponding to a stress release of 40 units (i.e., from 50 to 10 units) on the uncemented skeleton. Once the equilibrium has been established after the removal of bonds the soil behaviour would be the same as that of the remoulded soil which can be easily predicted. Experimental data by Griffiths et al., (1988) where the cementation bonds in a stiff clay have been removed by a process of EDTA flushing, very well confirm these behaviour in general although further experimental verification is required to establish quantitative predictions.

6.9 Permeability of Cemented Soils

The available data in literature on permeability of cemented soils indicate a similar relationship between void ratio and permeability as for remoulded clays. The permeability values are of the same range i.e., $k \approx 10^{-7}$ cm/sec, at high water contents close to liquid limit and reducing to $k \approx 10^{-9}$ cm/sec at lower water contents (close to 0.5 or 0.45 times w_L). These values are also in the same range as for remoulded soils. This has been observed for several cemented soils (Tavenas et. al 1983). More recently Lapierre et al., (1989) have presented void ratio versus permeability relations for the same clay both for intact and remoulded states, and have shown that the relations are identical for both the states. This indicates that the presence of cementation bonds does not affect the flow through the soil which is essentially through the enclosed large inter-aggregate pores, and hence at any given state the permeability is the same as that of a remoulded soil at the same void ratio and can be predicted from the same generalised relation

$$e/e_L = a + b \log k \qquad 6.13$$

Infact the permeability of intact clay and remoulded clay being the same is the most convincing evidence that the micro structure of both the states is identical and hence the logical consequence of possible superposition of uncemented and cemented components of stresses as discussed earlier.

Slope b in equation 6.13 can be regarded as permeability index (C_k) similar to compression index, C_c. Tavenas et al., (1983) and Lapierre et al., (1989) analyzed extensive data of permeability of naturally sensitive soils. As a first approximation the following relationship has been suggested.

$$C_k = 0.5\, e_o \qquad 6.14$$

where C_k is the permeability index and e_o is the initial void ratio.

Reexamination of permeability index of clays by Sivakumar Babu et al., (1993) on the basis of the generalised state parameter (e/e_L) approach discussed earlier, reveals that the form of relationship cited in equation (6.14) is tenable both for uncemented and cemented soils. But for normally consolidated soils the normalising parameter is void ratio at liquid limit e_L instead of the initial void ratio. Infact, e_L is the appropriate parameter even for cemented soils, but for most cemented soils with $I_L > 1$, e_o will be close to the e_L at the time of deposition and hence e_o itself can be used.

7

PARTLY SATURATED SOILS

7.1 Introduction

Most of the natural soils above ground water table (or above capillary zone) and invariably all the compacted soils in their placement condition are in a state of partial saturation. Partly saturated soils exist all over the world where the weather conditions are such that the annual evaporation potential is in excess of the precipitation. The heave or collapse of partly saturated soils upon saturation can cause considerable distress or damage to buildings founded on them. It is estimated that the damages world wide due to expansive soils exceed those due to floods, hurricanes, earth quakes and tornedos put together (Jones and Holtz, 1973).

Some of the problems encountered with partly saturated soils are:

(i) compression due to applied stresses,
(ii) heave or collapse and development of swelling pressure upon wetting,
(iii) shrinkage due to drying,
(iv) shear strength of partly saturated state and its variation upon soaking and
(v) compaction and its control.

7.2 Basic Aspects and Approaches

The mechanisms of volume change and shear strength behaviour of partly saturated soils are not very well understood even to this day, although there has been intensive research in this area since 1960's. The early efforts (Atchison and Donald, 1956; Croney et al., 1958; Bishop 1961; Jennings, 1961 etc) were in general attempts to define or arrive at the effective stress which controls the mechanical behaviour, mostly by modifying Terzaghi's effective stress equations to account for partial saturation in the form:

$$\sigma' = \sigma - u_a + \chi(u_a - u_w) \qquad 7.1$$

where u_a and u_w are the pore air and water pressures respectively and χ is a constant ranging between 0 and 1 and is a function of degree of saturation. But such effective stress equations failed to answer satisfactorily the volume changes under different loading conditions.

The difficulties in finding a single effective stress led to the acceptance of two independent stress fields as a necessary frame work (Coleman, 1962; Fredlund and

Morgenstren, 1977; Josa et al., 1987; Lloret and Alonso, 1985 and Mathyas and Radhakrishna, 1968) and is the most widely used approach today. The suitable stress fields are shown to be any two of the three stress fields i.e., $(\sigma_{ij} - u_a\delta_{ij})$, $(\sigma_{ij} - u_w\sigma_{ij})$ and $(u_a - u_w)$. Further, stress strain relations have been developed by considering different elastic modulii for each phase, independently for the two stress fields.

With the same concepts, the relation between shearing resistance and normal stress has been defined (Fredlund et al., 1978 and Tolls, 1990) independently for each of the two stress fields or alternately the contribution of pore water suction to shear strength is considered to be an increased cohesion (Alonso et al., 1990).

Reiterating again the statement by Bloch (1978), that in an interacting system such as clay-water -air system, independent stress fields for different phases are not valid and the overall stress field is the only acceptable stress field. The pore water and pore air pressures and the physico-chemical interparticle forces are all interdependent and are only the consequences of the total stress variations.

In this Chapter, a fundamental search into the mechanisms governing the behaviour of partly saturated soils is attempted considering the forces at the interfaces of different phases from basic laws of physics with the concept of the microstructure similar to that presented for saturated soils in Chapter 3. The study essentially brings out that the net osmotic stress $(R - A)$ between particles is the effective stress but this effective stress is not uniquely related with the void ratio as in saturated systems but is related with an involved state parameter $e\sqrt{S_r}$. This is because although the effective stress is the same for all the phases, the volumetric response of the air phase is different from that of the water phase as rightly proposed by the earlier workers.

7.3 Microstructure

To understand the mechanism of stress transfer in partly saturated state it is necessary to visualize the probable microstructure. Generally three broad classifications of partly saturated states have been recognized. For soils with very low degree of saturation (< 20%) the gas phase is continuous and the water phase forming menisci at the grain contacts is discontinuous. More precisely, even if the adsorbed water is continuous, it is held to the surface and not free or permeable. At very high degrees of saturation (> 85%), water is continuous but the gas, in the form of discrete bubbles within the water phase, is discontinuous. At intermediate saturations both water and air phases are continuous.

The occluded bubbles, in the case of high saturation, which just exist within the liquid phase, are found not to alter the structure of the saturated matrix, and hence their presence does not affect the basic nature of saturated soil behaviour except that the soil tends to be more compressible. The case of very low saturation is not of practical interest as compacted soils cannot be placed at such low saturation conditions. Such states can be attained in nature only in the process of drying and below shrinkage limit void ratios (which may vary with the initial structure and moisture content in case of compacted soils). Hence only the intermediate case, which is of greater concern will be discussed here.

It was discussed in Chapter 3 that for saturated soils starting from a slurry state without prestress, the particles are likely to be grouped together to form clusters or aggregates which get arranged enclosing large pores within themselves as shown in Fig. 3.13. Diffuse double layer interactions and stress transfer through osmotic repulsions prevail between two such clusters. Under an isotropic state of stress, the net osmotic pressure $(R - A)$ between interacting particle units, which is a function of the distance between them and is equal to the applied pressure p, and the fluid in the large pore is held by this osmotic suction, u_m provided by the double layer interactions between particle units all along the periphery of the pore. The size R of the large pores was shown to be decided by surface tension considerations given by $u_m = p = 2T/R$. For subsequent changes in the stress system, the distance $2d$ between clusters and the size R of the large pores get altered compatibly. The probable changes in the microstructure of the soil such as grouping or dismembering of particle units to accommodate the required changes in the pore sizes during loading or shearing was also discussed.

The structure of partly saturated soils is likely to be very much similar, except that some or all of the large pores may be filled with air and not water, as shown in Fig. 7.1. The spaces between and within the clusters are likely to be completely saturated because of the very high affinity of particle surfaces to water and also the amount of water required to fill all these spaces is quite small (less than 5 to 10% of the total void volume, even at liquid limit water content as shown by pore size distribution data). The air-water interphase is formed by the menisci which bridge the space between two particle units around the air pore (Fig. 7.1). A capillary suction $(u_a - u_w)$ is sustained by these menisci, the magnitude of which is a function of the curvature of the menisci, R_c, following the equation:

$$(u_a - u_w) = 2T/R_c \qquad 7.2$$

The radius of curvature, R_c of the menisci need not be equal to the pore radius R, but can be much smaller, the limiting value being half the spacing between two particle units which is controlled by diffuse double layer interactions between them.

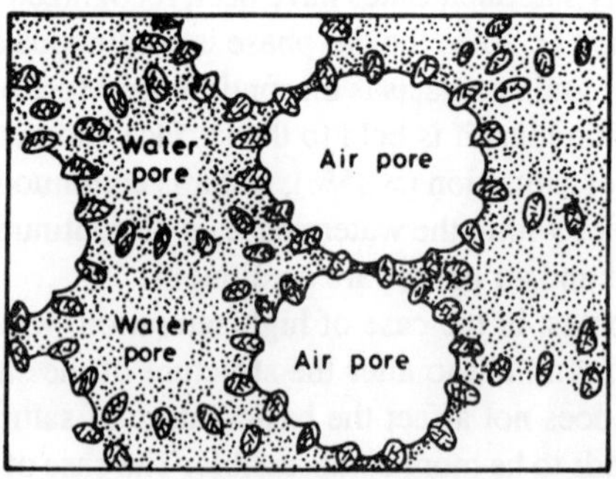

Fig. 7.1 Micro structure of partly saturated soils.

The size of the water pores and their compression with subsequent loading would be exactly the same as in the case of saturated soils. The size R_a of the air pores is not necessarily related to the stress state because it is the curvature R_c of the menisci which controls the stress. The air pores can be much larger than the water pores. The pore size distribution data by Sridharan (1968) of compacted and remoulded kaolin samples, consolidated to the same stress level indicate that there are greater amounts of pores of larger size in the compacted sample. In general, for a partly saturated soil, freshly mixed and loaded from an unstressed state, the size and number of air pores may get decided by the initial mixing water content. The subsequent compressions may be controlled by the possible extent of cluster grouping or breakdown under the available energy as in the case of saturated soils discussed earlier in Chapter 3.

Now for a monotonic loading from an unstressed state, the net osmotic pressure $(R-A)$ between particle units is equal to the applied spherical pressure p, and this same osmotic suction is transformed into capillary suction, $-u_w$ at the air water interface, and hence the total stress p is the equilibrium stress in the entire system and hence p alone is the effective stress. At any stage, the deformation of the soil depends on the type of stress probe which may be a change either in the applied stress p, or in the osmotic suction $(R-A)$ or the capillary suction u_w.

Before examining the response under different probes there is one more point to be discussed. There appears to be a basic difference in the overall structure of a compacted soil from that of a remoulded saturated soil, although the geometry of large pores will be similar in both the cases, i.e., a compacted soil, even upon saturation tends to be different from a remoulded saturated soil. The permeability results by Mitchell et al., (1965) (Fig. 7.2), which show a high influence of initial moulding water content on the permeability values, strongly support this point. For the same void ratio, a specimen compacted dry of optimum and saturated, results in a permeability in the range of 10^{-5} cm/sec, where as the specimen wet of optimum has a value in the range of 10^{-8} cm/sec. Remoulded saturated fine grained soils, even in their liquid limit sate have a maximum permeability in the range 10^{-7} cm/sec. This indicates that compaction dry of optimum renders a more open structure with much larger air voids, and with a greater fraction of particles locked in clusters which are not available for enclosing large pores (Fig. 7.3). This sort of grouping of particles results due to their attempt to share the available but insufficient water. To make it clear, for the same volume of large pores through which water can permeate, a compacted soil would have greater solid content or a lower void ratio. In other words, for the same void ratio, a compacted soil would give a higher permeability than a remoulded soil. On the wet side of optimum, the permeability values are quite close to those obtained for remoulded saturated samples indicating that the structure is similar to that of remoulded soils i.e., without such locked in groups of clusters. In fact the change in the structure with water content appears to be drastic around the moisture state of OMC, permeability suddenly falling from 10^{-5} to 10^{-7}. The magic number of OMC may perhaps be just the water content required to disperse the clusters. This structure for water content dry of optimum, is unlikely to be disrupted during mere consolidation loadings or by saturation but may be broken during shearing to very large strains. Thus although the compressibility or strain responses of compacted soils are similar to that of remoulded soils in principle or mechanism wise, the numerical values could differ even upon saturation.

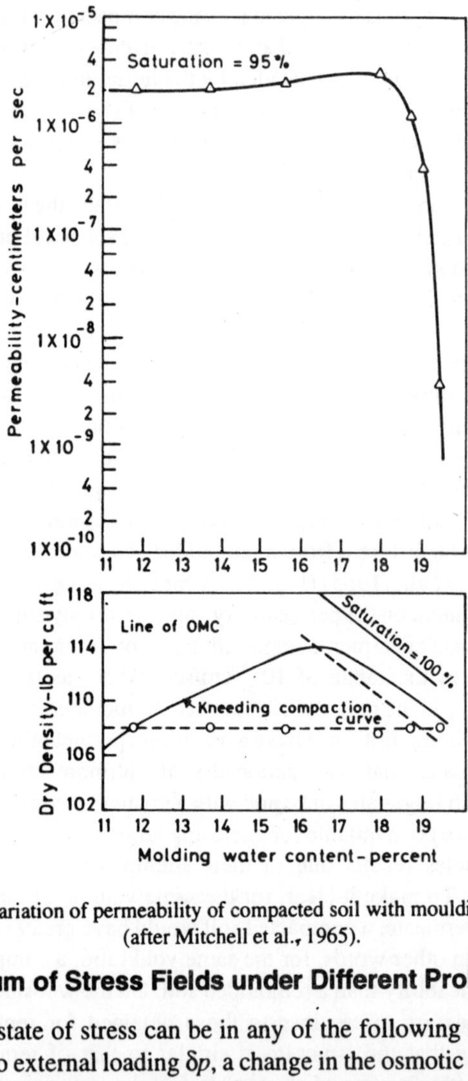

Fig. 7.2(a) Variation of permeability of compacted soil with moulding water content (after Mitchell et al., 1965).

7.4 Equilibrium of Stress Fields under Different Probes

A change in the state of stress can be in any of the following forms: a change in the stresses due to external loading δp, a change in the osmotic suction $\pm\, \delta(R - A)$, and a change in the capillary suction, $\pm\, \delta(u_a - u_w)$ brought about either by change in pore air pressure, $\pm\, \delta u_a$ or change in the pore water pressure, $\pm\, \delta u_w$. Of these, changes in osmotic suction and in pore air pressure are not of much practical significance although osmotic suction changes are possible under rare conditions of changing pore fluid characteristics. Changes in pore air pressure may occur in oil bearing areas where methane and other gases are produced. Otherwise under normal circumstances u_a is always atmospheric. Of more practical relevance are the applied stress changes, $\pm\, \delta p$ at constant water content condition as during usual

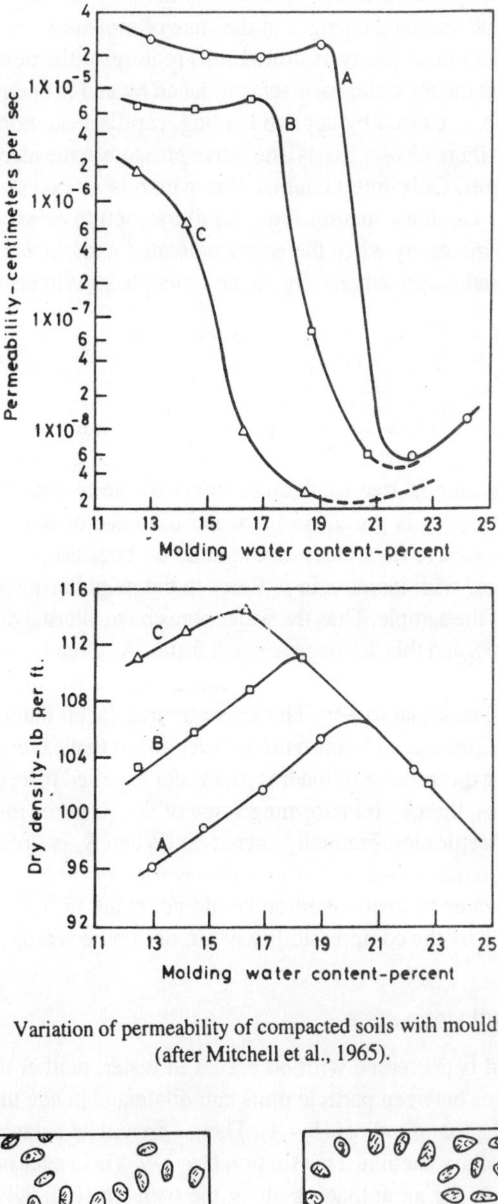

Fig. 7.2(b) Variation of permeability of compacted soils with moulding water content (after Mitchell et al., 1965).

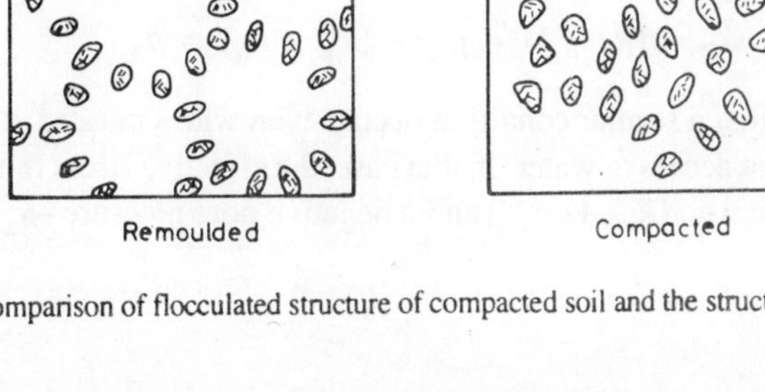

Fig. 7.3 Comparison of flocculated structure of compacted soil and the structure of remoulded soil.

construction loading, and changes in pore water pressure, $\pm\delta u_w$ under constant loading, $+\delta u_w$ corresponding to the case of inundation and $-\delta u_w$ to the case of drying in nature with seasonal changes in the state of moisture.

The capillary suction in partly saturated soils requires little more clarification. When the suction at the air water menisci is induced by and is in equilibrium with the osmotic suction as caused by applied loading, capillary suction is considered to be zero although there always exists a negative pressure at the menisci and hence has a thirst for water. Only interfacial suction which is in excess of the applied loading is noted as capillary suction. This capillary suction or excess pore water tension is induced internally when the soil is unloaded without access to water or when a negative head is applied directly on the water phase as in the case of drying.

7.4.1. Loading under constant water content

(a) Virgin state

For a partly saturated soil, a constant water content condition does not mean a constant volume or an undrained condition as is the case with saturated soils because the air volume is free to change. Infact for soils with $S_r < 85\%$, it is immaterial whether there is provision for water drainage or not. It is found that even if water is allowed to flow out, it does not do so, because the water squeezed out from water pores with increase in p, flows to the neighboring air pores rather than flowing out of the sample. Thus the water pores have internal drainage as long as there are air pores, and this drainage is much faster. As already stated, under this condition, $\delta(R - A)$ between particles associated with a change in the spacing between them, will be equal to $+\delta p$. The air-water interfacial menisci correspond to or will be in equilibrium with this osmotic suction, so that excess pore pressure will be zero. During this process of loading, the water expelled from the water pores flow to the air pores, thereby transforming some of the air pores into water pores, and the degree of saturation gradually increases. When S_r is greater than about 85%, the air fraction may loose continuity and may form occluded bubbles. In such a case a constant water content condition would not result in $\delta(R - A)$ equal to δp and may be decided by the compressibility of air, or in otherwords possible extent of drainage.

(b) unloading from virgin state

When applied load is decreased with no access to water, neither the water filled pores nor the spaces between particle units can dilate and hence there is unlikely to be a change in the magnitude of $(R - A)$. Then, $-\delta p$ will be balanced by an equal increase in the capillary suction. This suction $\delta(u_a - u_w)$ is in excess of the applied stress $(p - u_a)$. Thus for an unloading along the path AB (Fig. 7.4), the effective stress at any point B, remains the same as it was at A so that at B

$$p'_B = (R - A) = p_B + (u_a - u_w) = p'_A \tag{7.3}$$

We may note that a similar condition occurs even with saturated soils when it is unloaded without access to water. In that case, the effective stress remains equal to the past pressure, i.e., $(R - A) = p'_p$ and a negative pore pressure $-u_w = (p'_p - p)$

will mobilize to keep the systems in equilibrium. It is for such unloaded states (as during a compaction process) that the expression for effective stress i.e., $p' = (p - u_a) + (u_a - u_w)$ is applicable. Of course this expression is valid even for virgin loading in cases where $(u_a - u_w)$ is an applied suction, where as the suction observed with unloaded states is an induced suction, and the effect of the two on deformation will be just the opposite (which will be discussed in a later section).

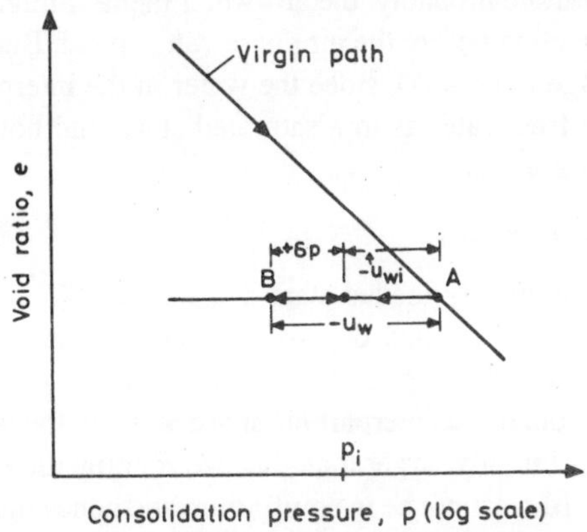

Fig. 7.4 State of stress for partly saturated soil.

Again for reloading i.e., for $+\delta p$ on an unloaded state *B* under constant water content, since the micro structure is already in a compressed state corresponding to a higher stress at *A*, there will not be further increase in effective stress, only the excess capillary suction, $-u_w$ reduces to balance δp so that at any instant *i*,

$$p'_i = (R - A) = p'_A = p_i - (-u_{wi}) \tag{7.4}$$

On the other hand, if there was access to water during unloading, just sufficient to permit the increase in '*d*' spacing between particles and compensation of capillary suction, but not for inundating the soil, then the effective stress would reduce compatible with the stress decrement so that at any stage the effective stress, $p' = (R - A) = p_B$, with capillary stress always equal to zero, similar to the case of usual unloading of saturated states with access to water. This may be possible only under controlled laboratory testing and may be rare in nature, except when the soil can absorb moisture from atmosphere. In such a case the large water or air pores may not undergo any swelling. Such overconsolidated states in total equilibrium, if reloaded i.e., $+\delta p$, the effective stress would again increase with increase in *p*.

7.4.2 Suction Probes under Constant Load

(a) Increase in suction, $-\delta u_w$

It is well known that for saturated states, an applied pore water suction has the same effect as that of an applied loading, *p*. Even for an unsaturated soil, for the continuous water filled pores, an applied suction will be equivalent to an applied loading and the pores would compress equally in both the cases. But for the air phase, which is compressible in contrast to the incompressible water phase, the two processes are different.

An externally applied stress increment $+\delta p$ would instantaneously increase the air pressure through the excess pore water pressure, and hence the air would have a tendency to compress, whereas an applied suction $-\delta u_w$ would depress the air pressure instantaneously, and hence the air pores should if at all, have a tendency to expand. However more probably, the air- water menisci might deepen to adjust to the new state of suction before the air phase can expand. But both would bring about an equal change in (R – A), since the water in the interparticle spacings is continuous with the free water as in a saturated state, and both probes have the same effect on the water phase.

(b) Suction reduction, $+\delta u_w$

If the soil is in monotonically loaded state i.e., with $p = (R - A)$ with no excess capillary suction, then an applied δu_w to the water phase, would immediately dissipate by way of water flowing into the air pores. The air - water menisci would initially start moving out of the interparticle space and into the large air pore. If the size of the air pore is initially larger than the size R compatible with the pressure p, the intruded water film cannot be in equilibrium and hence the air pore will have to collapse to that size R. And with continued flow of water, the air pores gradually get saturated. In the whole process, the effective stress $(R - A)$ between particle units remains unaffected and is in equilibrium with the applied stress p. Thus only the state of the soil changes without a change in stresses until the soil is completely saturated, beyond which a positive pore water pressure increment, $+\delta u_w$ would decrease the effective stress as in a saturated system.

On the other hand, if the soil is initially in an unloaded state (as at B) with excess capillary suction compensating the stress removal, an applied $+u_w$ would now compensate the existing suction, thereby effective stress $(R - A) = p_B + (u_a - u_w)$ would reduce until $(R - A) = p_B$ when $(u_a - u_w) = 0$. During the process, the applied positive pore water pressure, δu_w if it is greater than the external stress, p_B may force the large water pores to dilate by an amount possibly greater than what it would be in the case of a saturated soil. For eg., consider a partly saturated state monotonically loaded to about 8 kg/cm^2 (Fig. 7.5) and then unloaded to say 2 kg/cm^2. Under that state, the capillary suction would be –6 kg/cm^2. If now the soil is inundated, it amounts to applying an internal $\delta u_w = 6$ kg/cm^2 which is much greater than the applied external stress 2 kg/cm^2, hence water pores are forced to dilate. If instead the soil was unloaded to only 4.5 kg/cm^2, in which case $+\delta u_w$ upon inundation would be 3.5 kg/cm^2 which is less than applied stress of 4.5 kg/cm^2, the soil may not heave upon inundation more than the usual rebound for the saturated soil.

7.5 Deformation Under Different Probes

As already stated earlier, though under any applied probe, the entire mass of soil reaches an unique state of stress in equilibrium with the applied stress, with the osmotic suction $(R - A)$ between particles being the effective stress, the volumetric

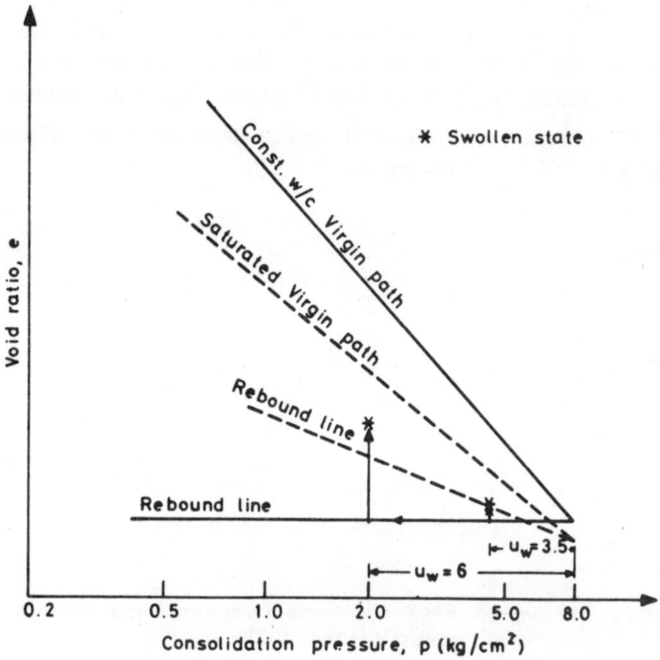

Fig. 7.5 State of stress defining the heave potential upon inundation.

or the deformation response of the air phase is different from that of the water phase for a given probe and also the two are different for different probes as such. That is the reason why we do not get a unique relation between the overall void ratio and the effective stress, as in the case of saturated systems, but varies with relative amounts of air and water pores and with the type of probe. Hence the constitutive relations have to be defined for different cases. However, the shearing resistance would be a function of the effective stress $(\bar{R} - A)$ whatever be the type of probe. It is here that the authors differ from the currently accepted concepts which propose different friction angles ϕ and ϕ_b for the two probes of applied load and pore water suction, which will be discussed in a later section.

7.5.1 Loading at constant water content for virgin states

Under this probe of incremental loading, the mobilised osmotic suction $\delta(R - A)$, will be equal to δp. The volume change of the water filled pores will be exactly the same as that of a saturated soil, the reduced radius of the pores being compatible with the changes in the state of stress as per equation 7.3. The air pores, subjected to an instantaneous increase in pressure through the water pressure, will have a tendency to compress, may be governed by the usual gas laws. But such a compression is again restricted by the possible extent of micro-strucutural changes such as grouping or dismembering of particle units to accommodate the changes in pore perimeter, before the air pressure dissipates. Since the air pores are initially larger than the water pores, a specified amount of microstructural changes or reduction in perimeter would lead to a greater reduction in the volume of air pores than that of water pores. That is the reason why the e vs log p curves are steeper with increase in air fraction or decrease in S_r. This is clearly reflected in Toll's (1990) experimental results, λ_a being higher for lower degrees of saturation (Fig. 7.6). (Here the subscript 'a' refers to the applied stress probe, to differentiate it from the subscript 'w' for suction probe).

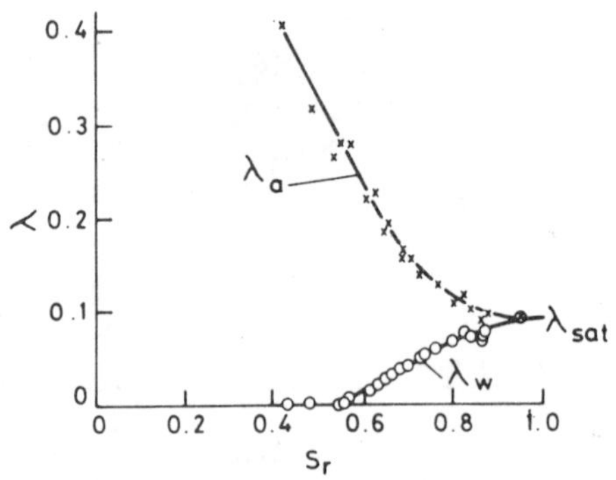

Fig. 7.6 Variation of compression index with degree of saturation for total stress and suction probes (after Tolls, 1990).

The compressibility of a compacted but subsequently saturated soil, may not be equal to that of a remoulded saturated soil, because of the basic differences in the microstructure, compacted soils having a greater volume of large pores for the same given void ratio when compared to the remoulded soils.

During unloading, i.e. for $-\delta p$, if there is no access to water, neither the interparticle spaces nor the large water pores can dilate. The air is momentarily subjected to a depression in pressure through the water phase and may have a tendency to expand but before that the air-water menisci may simply deepen to resist this suction, not allowing the air volume to increase. Thus the unloading e vs log p paths with no access to moisture, is likely to be horizontal.

For reloading, under constant water content, since there is no change in the osmotic suction, $(R - A)$ but only the existing excess capillary stress, $-u_w$ being nullified with changes in p, there need not be any compression of the water phase. But for the air phase, which would have already equilibrated with the atmospheric pressure under the unloaded state B, a reloading of $+\delta p$ is as good as a virgin loading for the current volume of air and hence the air pores would have a tendency to compress. The compression of air volume for subsequent load application may of course be restricted again by the extent of possible microstructural changes. The recompression paths of a partly saturated soil may be relatively steeper than that of its remoulded state.

7.5.2 Applied suction under constant loading

As already discussed, the compression of the water filled pores under an applied suction will be exactly the same as for the case of applied $+\delta p$ and also will be the same as what it is for saturated soils. But for the air pores the applied suction $-\delta u_w$ in turn causes a pressure depression which is infact similar to the case of $-\delta p$ and hence does not induce any compression. Thus the compressibility parameter λ_w of a partly saturated soil for a given change in suction, $-\delta u_w$, is less than the compressibility λ_a for the same magnitude of change in applied loading, δp and is always less than λ for a saturated soil. This is clearly reflected in Toll's (1990) experiments which show an increasing λ_w with increase in saturation and always less than that for the case of applied $+\delta p$ for the corresponding degree of saturation (Fig. 7.6). The fact that air pores do not compress as much as the water pores for applied suction is supported by the higher shrinkage limit values for soils dried from an initially unsaturated compacted state than those obtained from drying the standard initial slurry state, shrinkage void ratio increasing with decreasing initial S_r (Table. 7.1– Authors data).

Table 7.1 Variation of shrinkage void ratio for different initial conditions

e_o	w (%)	S_r (%)	e_s
a. Black cotton soil			
$w_L = 81\%$, $G = 2.73$, $w_s = 13.4$			
0.84	21.9	71.2	0.53
0.93	18.0	52.8	0.71
1.00	12.3	33.6	0.90
1.23	33.9	75.2	0.51
1.22	24.9	55.7	0.72
1.20	17.9	48.9	0.91
1.64	45.47	75.6	0.50
1.63	34.0	56.9	0.72
1.61	22.2	37.6	1.10
b. Red soil			
$w_L = 38.4\%$, $G = 2.67$, $w_s = 15.4$			
0.73	20.9	76.4	0.57
0.73	16.4	60.0	0.63
0.89	14.2	42.6	0.82
0.93	20.7	59.4	0.78
0.94	26.9	76.4	0.65

7.5.3 Suction reduction under constant load

For the above condition, for a soil in the virgin state, the applied back pressure $+\delta u_w$ immediately gets dissipated by the water flowing into the nearby air pores. There will be no change in the osmotic suction, $(R - A)$ but the air pores gradually get saturated and in the process get compressed also due to their initial larger size which is not compatible with the applied pressure for the water medium. Thus the final equilibrium state and hence the amount of collapse can be known by the corresponding saturated state at that stress level.

For an unloaded state as at B (Fig. 7.4), with an excess capillary suction of $-u_w$, such that $p' = (R - A) = p_B - (-u_w)$, with the application of $+\delta u_w$, the effective stress decreases as the magnitude of $-u_w$ decreases. If it is a process of inundation or rise in water table, which tries to bring the pore water pressure to that of atmospheric pressure (or even higher), the applied $+\delta u_w$ will be equal in magnitude to $-u_w$ at B. If this internal pressure is higher than the applied pressure p, it would force the water pores to swell. On the air pores, an applied positive pore water pressure, $+u_w$ would have a compressive tendency initially. But as the air pores get saturated, such pores may also be subjected to a swelling tendency. The magnitude of net swelling which is a combined effect of the above two processes may be difficult to theoretically estimate and one has to resort to experimental observations.

7.6 Phenomenological Models to Predict Constitutive Relations

7.6.1 Compressibility behaviour under constant water content

Now we know that a unique relation does not exist between e and p for partly saturated soils and that it depends on the degree of saturation. To arrive at the possible stress-state relations, one dimensional compression test results conducted under constant water content conditions with different initial water contents and starting from a stress free state (i.e., a state of loose filling) were examined (Nagaraj and Srinivasa Murthy, 1985 and Panidan et al., 1992). The data show that the e vs log σ'_v paths lie above that of the saturated e vs log σ'_v path, the deviation increasing with decreasing degree of saturation. But the same data when plotted in $e\sqrt{S_r}$ vs log σ_v plot, shows a unique relation irrespective of the initial water content. (This may not be valid for very low mixing water contents in which case the moisture is insufficient to develop double layers around the particles). Figures 7.7a & b show the compression data of three soils, following the above behaviour. Further, it has been shown that $e\sqrt{S_r}$ vs log σ'_v relations of different soils can be generalized in the form:

$$\{e/e_L\}\sqrt{S_r} = a - b\log\sigma'_v$$

This describes the compressibility behaviour of monotonically stressed partly saturated soils. It has been discussed earlier that for monotonically loaded virgin states, the applied stress p is the effective stress p'. This indicates that for partly saturated states, the effective stress p' is probably uniquely related with the involved state parameter $e\sqrt{S_r}$ and not with just e. The unconfined compression test results (Nagaraj and Tiwari, 1985) also support this thought wherein a unique relation between q_u and $\{e/e_L\}\sqrt{S_r}$ has been observed irrespective of the initial moulding water content and type of soil (Fig.7.8) which will be discussed later.

It is quite appropriate that the state parameter involves e and S_r but there is no rationale apparently for the particular form of $e\sqrt{S_r}$. The mechanism of volume change of the incompressible water phase being quite different from that of the

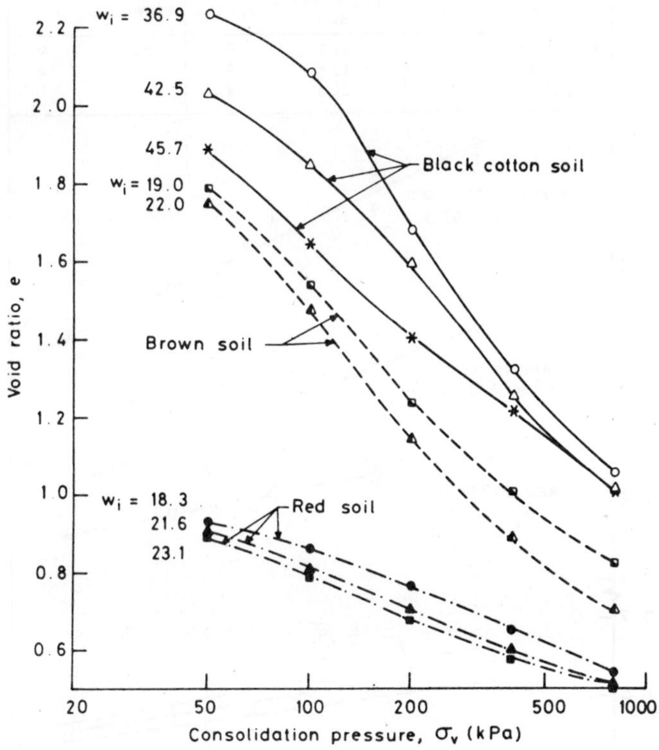

Fig. 7.7(a) *e* vs log σ_v plots for soils with different initial moisture contents.

compressible air phase (which could be both by expulsion and compression), it may not be possible to reflect the overall volume changes with a single parameter $e\sqrt{S_r}$. But the total compressibility may be governed on the whole by some law, as it is finally restricted by the possible extent of microstructural changes under the available energy input. Also, the compression processes of the two phases are coupled in the sense, as long as the pressure in the water phase is not dissipated, it would be acting on the air phase and on the other hand the compressible nature of the air phase would increase the rate of dissipation of pressure in the water phase. A rigorous theoretical analysis of the whole process may be too complicated and it is reasonable to resort to experimental results. Infact, even for saturated systems, the well known e *vs* log *p* relation is not a theoretically founded relation. From the laws of physics only the radius R of the pores is related to *p* and this relation is not in direct agreement with observed *e* vs *p* relations. Thus we need not look for a theoretical basis for the $e\sqrt{S_r}$ vs p' relation but can conveniently use the observed unique relation.

Now, with the known unique relation between $e\sqrt{S_r}$ and p', for isotropic loading, p' plotted in natural log scale

i.e., $$e\sqrt{S_r} = N' - \lambda' \ln p' \qquad 7.5$$

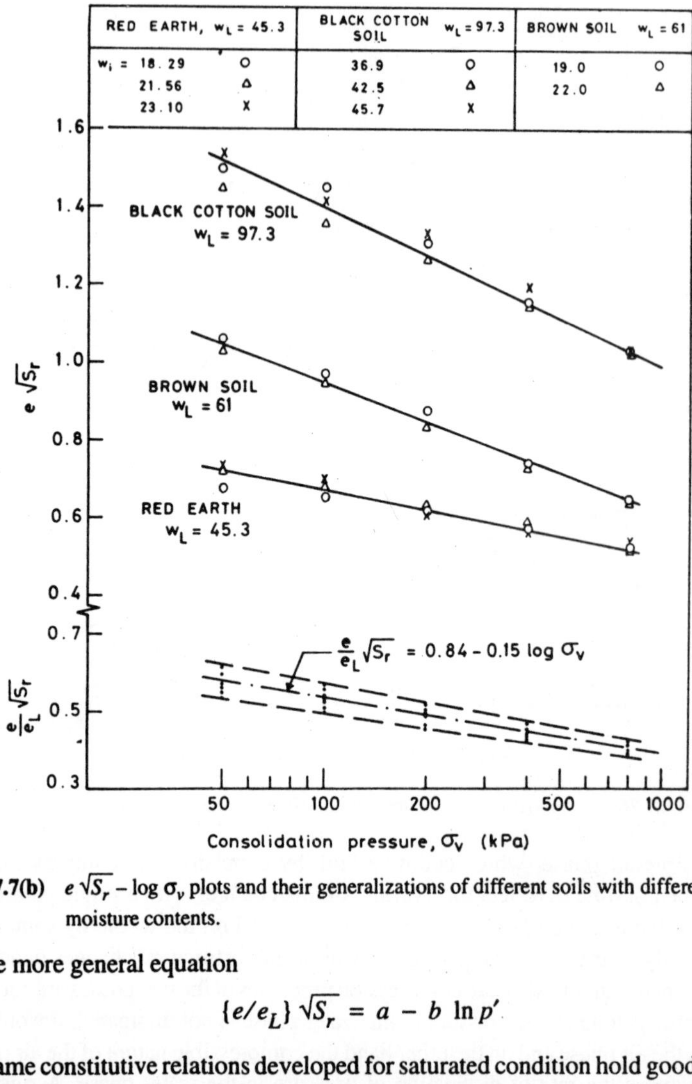

Fig. 7.7(b) $e\sqrt{S_r} - \log \sigma_v$ plots and their generalizations of different soils with different initial moisture contents.

or the more general equation

$$\{e/e_L\}\sqrt{S_r} = a - b \ln p' \qquad 7.6$$

the same constitutive relations developed for saturated condition hold good for any general loading under constant water content except that the compressibility coefficient is to be suitably computed from the above unique relations (equations 7.5, 7.6).

Differentiating eqn (7.5) the compressibility coefficient, λ will be

$$\lambda = 2\lambda'/\sqrt{S_r} \qquad 7.7$$

Similarly from equation 7.6,

$$\lambda = 2be_L/\sqrt{S_r} \qquad 7.8$$

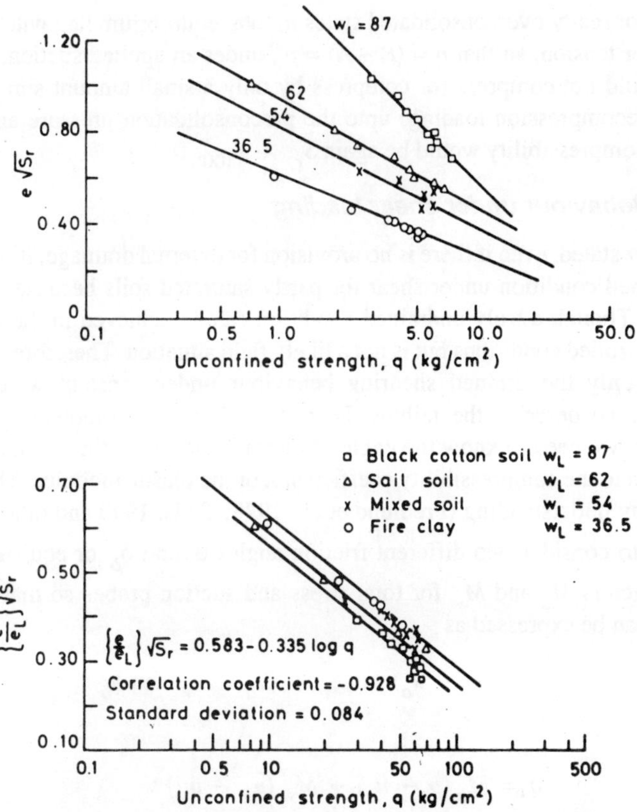

Fig. 7.8 Generalization of unconfined compression strength of different soils with differnt compaction conditions (Nagaraj and Tiwari, 1985).

For the set of soils indicated in Figure 7.7, $b = 0.15$. Then

$$\lambda = 0.13\, e_L/\sqrt{S_r}$$

The elastic compressibility coefficient κ for a partly saturated state cannot be experimentally evaluated. It is because in the usual unloading process, unconstrained states cannot be achieved, i.e., in the constant water content condition the effective stress does not change while with access of water the state of the soil itself gets changed. It is likely that the value of κ will be in the same ratio with λ as it is for a saturated case.

7.6.2 Drying or Suction ($-\delta u_w$) probes under constant load

For virgin states as at A (Fig. 7.4) and also for states unloaded under constant water content as at B with still an internal virgin effective stress as at state A, as already discussed only the water pores shrink but the air pores do not change much in volume. Hence the compressibility of the soil will be reduced to S_r times that of the saturated state so that

$$\lambda_s = S_r \lambda_{100}$$

But, for really overconsolidated states in total equilibrium i.e., with no excess pore water tension, so that $p = (R - A) = p'$, under an applied suction, the water pores would not compress (or compress by only a small amount similar to that under a recompression loading) upto the preconsolidation pressure and beyond this, the compressibility would be again $S_r \times \lambda_{100}$.

7.6.3 Behaviour under shear loading

As already stated, even if there is no provision for external drainage, it will always be a drained condition under shear for partly saturated soils because of internal drainage. Though a truly undrained condition can be achieved in the laboratory under controlled conditions but is not a likely field situation. Therefore we need to consider only the drained shearing behaviour under constant water content condition. To describe the failure shear strength or the complete stress-strain behaviour, we need to know the angle of internal friction or the friction factor M in addition to the compressibility coefficient λ or the plastic modulus. The general present day understanding (Fredlund et al., 1978; Tolls, 1990 and others) on this aspect is to consider two different friction angles ϕ' and ϕ_b' or equivalently the friction factors M_a and M_w for total stress and suction probes so that the shear strength can be expressed as

$$\tau = (\sigma - u_a) \tan \phi_a' + (u_a - u_w) \tan \phi_b' \qquad 7.9$$

or

$$q = M_a (p - u_a) + M_w (u_a - u_w) \qquad 7.10$$

However, a few researchers (Alonso et al., 1990) have considered an increased cohesion due to suction.

But, from the discussions presented in this book, since the effective stress which is the net interparticle repulsive pressure is the same for both total stress and suction probes, it may not be reasonable to have two different friction factors for the two probes. The friction coefficient would be a single-value with respect to the effective stress which may however be different from that for the remoulded saturated state because of the basic difference in the structure. As discussed earlier, an initially unsaturated soil, will have a structure with a greater number of soil pockets which are not enclosing large pores. Thus at a given void ratio, there may be a greater number of interacting particle units on a given cross section, which are actually the sites of shearing resistance. Hence the friction factor for partly saturated state M_{ps} may be higher than that for remoulded state.

The magnitude of this M_{ps} may turn out to be a function of initial moulding water content which determines the initial structure. With continued shearing the value of M_{ps} may decrease, and may reach the normal value of M for remoulded states at very large strain. These aspects are yet to be established with extensive experimental programme. Once the compressibility constants λ and κ, and the friction factor M_{ps} are established, the stress-strain behaviour can be modelled by the usual elasto- plastic models.

While analysing the behaviour of partly saturated soils we see that the degree of saturation is an additional state parameter over the usual void ratio which is directly related to water content for saturated states. Hence the equations can be solved only if one of them is kept constant, or can be expressed as a function of the other, as under a constant water content condition or known final degree of saturation which are infact the real practical situations. A monotonically loaded partly saturated sample without stress history as at A would follow the normal drained path AF in a standard triaxial test (Fig. 7.9) but the strains correspond to higher λ which varies with the progress of the test.

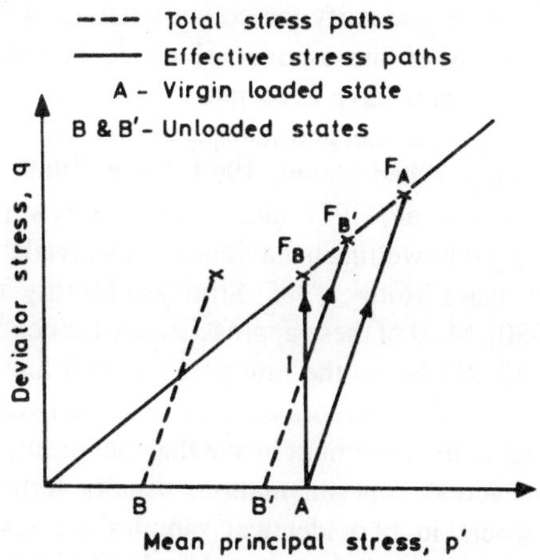

Fig. 7.9 Schematic representation of effective stress paths under triaxial shearing for different initial states.

For unloaded or compacted sample as at B, (Fig. 7.9) since the effective stress is still equal to that at A ($= p_A$) with negative pore water pressure $u_w = p_A - p_B$, the effective stress path would follow a constant p' path as if starting from A, at each level the applied stress increment Δp_i compensating equal magnitude of the existing negative pore pressure. If the initial negative pore water pressure is small as for the sample B', the soil may take the effective stress path $IF_B{}'$ after the total stress path intersects the constant p' path at I. If during shearing, the sample gets saturated (or $S_r > 85$ - 90% for practical purposes), further shearing may follow an undrained path, if there is no drainage. The corresponding strain computations for each stress path can be done using the appropriate parameters. From the discussions presented above, it may be logically inferred that the unconfined compressive strength of a compacted soil is unique for a given compactive effort, since the effective stress induced is same. Its magnitude can be related to the parameter $e\sqrt{S_r}$ since, this parameter is uniquely related with p or the compaction effort.

Figure 7.8 typically shows the variations of unconfined compressive strength with $e\sqrt{S_r}$ for four compacted soils with different initial moisture contents. The relationship can be linearised in semilog plot. Similar observation was reported by Da Cruz (1963). Further in Figure 7.8, the possibility of generalization using their respective void ratios at liquid limit has been presented. This relationship can be used to predict the shear strength of compacted partly saturated soils.

7.6.4 Collapse or heave upon inundation

In literature, attempts to estimate swell or collapse potentials are mostly qualitative. For example, it is known that type of soil, initial dry density and water content and the applied stress level all affect the magnitude of heave or collapse, upon saturation. Low initial dry density and high stress levels may show a collapsing behaviour, while a higher initial dry density and lower over burden stress conditions lead to a swelling. Higher liquid limit soils show a higher potential of swelling. Lutenegger and Saber (1988) present a comprehensive information of the various attempts made to qualitatively identify the collapse nature of the soils. Methods for quantitative predictions of the amount of heave or collapse are not yet satisfactorily defined. Attempts have been made to evolve statistical/ empirical relations between swelling behaviour and state parameters (Gibbs and Holtz, 1956; Seed et al., 1962; Komarnik and David, 1969; Saito, 1969; Vijayavergia and Ghazzaly, 1973; Snethan et al., 1977 and others). A few phenomenological approaches for predicting the swelling behaviour are also available (Yoshida et al., 1983; Nagaraj and Srinivasa Muthy, 1985; Srinivasa Murthy and Nagaraj, 1987 and Fredlund et al., 1980). Most of these approaches use the condition that the final swollen state of the soil will be on the saturated rebound line. But experiments clearly indicate that the swelling is more than the predicted values. It can be said that these methods predict the minimum heave that can occur. However the best known method is by actual experimentation, usually termed as the double oedometer method, where in two identical samples are tested in oedometer compression, one sample saturated at a nominal load and then compressed and the other compressed at constant water content throughout. Figure 7.10 shows the compression paths of two samples, one initially of low dry density and the other of high dry density, which would collapse and heave respectively. The difference between the void ratio on the constant water content compression path and that on the saturated compression path at any pressure gives the magnitude of heave or collapse.

The method works well for prediction of collapse i.e., if the compression path of the saturated sample is known, the possible collapse at any load can be precisely predicted. But it is known that this method overpredicts the amount of heave at higher stress levels.

A closely related aspect with the phenomenon of heave upon inundation is the swelling pressure, which is the pressure exerted by the soil upon inundation, if it is not allowed to swell. The magnitude of swelling pressure is quite often required in the design of structures in expansive soils. The swelling pressure can be experimentally evaluated. The three widely used methods for evaluating the swelling pressure are: (1) free swell and load, (2) swell under different surcharge loads and (3) no swell-loading test. Figure 7.11 illustrates the different methods. The three methods are found to yield considerably differing values of swelling

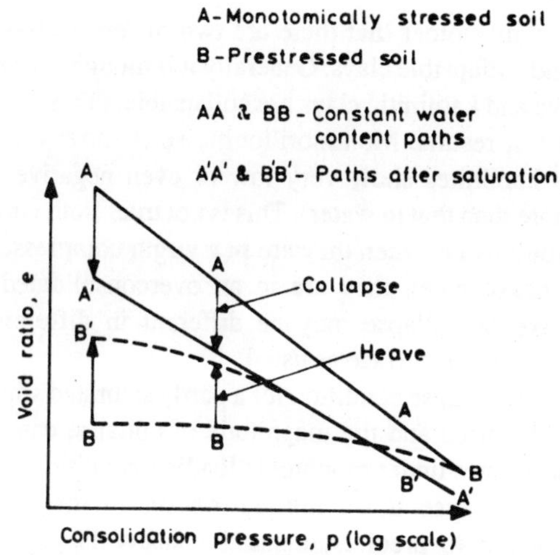

Fig. 7.10 Prediction of heave and collapse from double oedometer test.

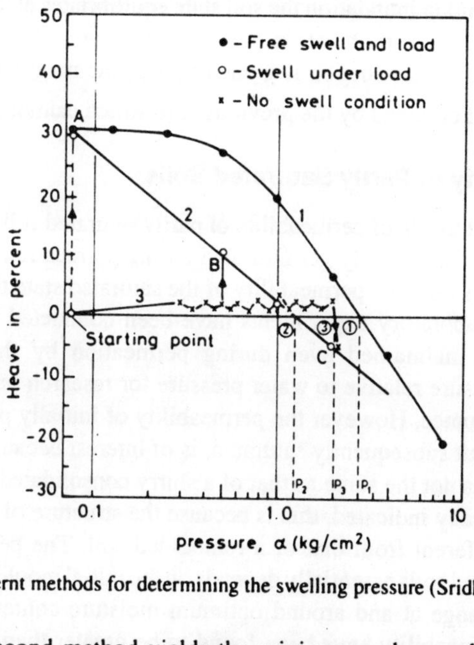

Fig. 7.11 Differnt methods for determining the swelling pressure (Sridharan et al., 1986).

pressure. The second method yields the minimum value, and the method with inundation under nominal load and subsequent compression loading, yields the maximum value (Sridharan et al., 1986). It is recommended that the swelling pressure determined from either second or third method could be used in designs.

For computation of heave under a given load, the second approach gives satisfactory results.

It is generally a misnomer that there are two different classes of clays - the expansive clays and collapsible clays. Generally it is thought that montmorillonitic clays are expansive and kaolinitic clays are collapsible. (This is a feeling gathered by the free swell test results. Montmorillonitic soils show very high free swell volumes whereas kaolinites show very low or even negative values, i.e. their volume in air is more than that in water). This is not true. Both clays are collapsible under certain conditions i.e. when they are in a virgin compressed state, and both the clays would heave when they are in an overconsolidated state. Only the magnitude of heave or collapse may be different in different type of clays, depending on their specific surface as usual.

In conclusion, the collapse condition for a partly saturated soil upon inundation can be precisely identified and the magnitude of collapse can be predicted. All partly saturated soil states under monotonic loading conditions without the effect of prestress will collapse upon inundation on to the saturated state. A comprehensive method to predict quantitative heave is yet to be formulated. However, based on the discussion on the microstructure and responses of the soil under different stress probes it can be said that, the soils with excess negative pore pressure will heave upon inundation. The magnitude of heave will be dictated by the relative values of external load and the excess suction. If the suction is less than the applied load p, upon inundation the soil state equilibriates at the corresponding saturated rebound line from the same maximum past pressure which is equal to $(p - u_w)$. When the excess suction is greater than p, the magnitude of heave will be greater than that obtained by the previous case which cannot be quantified.

7.7 Permeability of Partly Saturated Soils

It is infact incorrect to talk of permeability of partly saturated soils because a partly saturated soil would eventually get saturated during permeation. Once the steady state is reached it is only the permeability of the saturated state that is of concern. Of course some laboratory experiments have been conducted with the state of partial saturation maintained even during permeation by the application of controlled air pressure relative to water pressure for research purposes but is not of practical importance. However the permeability of initially partly saturated or compacted soils but subsequently saturated, is of interest because it is found that this permeability is not the same as that of a slurry consolidated state at the same void ratio. As already indicated, this is because the structure of a compacted soil is going to be different from that of a remoulded soil. The permeability of the compacted saturated soils essentially depends on the initial moulding water content with a drastic change at and around optimum moisture content. The values of coefficient of permeability have been found to be greater than 10^{-5} cms/sec for soils compacted dry of optimum while they are less than 10^{-7} cm/sec when compacted wet of optimum. It is likely that generalisations are possible with permeability relations of different compacted soils, which is yet to be examined.

7.8 Compaction Characteristics

The value of compaction as a practical method to control strength, compressibility and permeability characteristics of fine grained soils has been amply attested by the fact that it is in use since the time when the early earth structures were built. Every day, millions of cubic meters of soils are compacted all over the world since compaction is the only simple viable means of effecting marked changes in engineering properties of soils economically. Properly placed and compacted soil mass has strength and stability that are as good or even better than many natural formations. Soils in their as compacted condition are partly saturated.

Compaction is a simple process of densification by expulsion of air using mechanical manipulation. This is in contrast to densification by consolidation by expulsion of water under an external load. The consolidation process is very much slower, because of low permeability to water than the compaction process which is almost instantaneous because of high permeability to air and possible internal drainage for water.

For a given soil, the dry density that can be realized by compaction depends on the compaction effort and the moulding water content. Figure 7.12 shows typical dry density - water content relationships which is normally referred as compaction curves. The dry density - moisture content relationship of the soil obtained from laboratory tests forms the basis for specification and field compaction control.

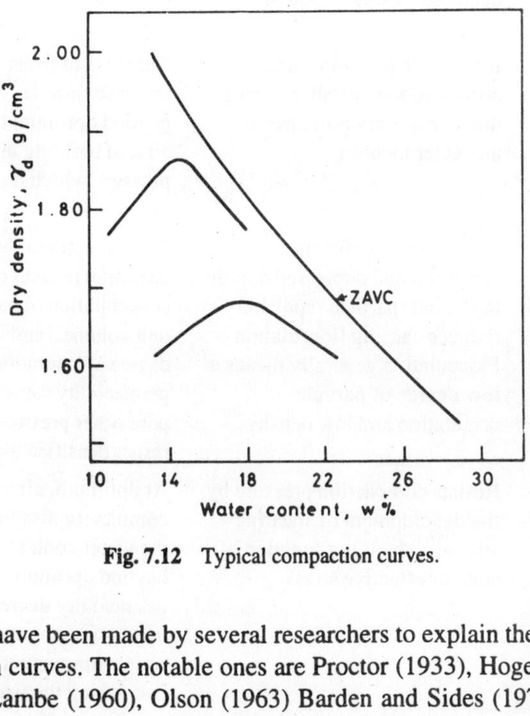

Fig. 7.12 Typical compaction curves.

Attempts have been made by several researchers to explain the characteristics of compaction curves. The notable ones are Proctor (1933), Hogentogler (1936), Hilf (1956), Lambe (1960), Olson (1963) Barden and Sides (1970), Hodek and Lovell (1979). In table 7.2, a critical examination of the theories is provided. In all the theories except the last two, the distinct phases involved in the process of compaction, viz.,

(i) initial phase of mixing the soil with water and
(ii) subsequent phase of application of compactive effort to densify the same

were not considered separately. Even the explanations provided by Barden and Sides (1970) and Hodek and Lovell (1979) with regard to these two phases were only qualitative and did not enhance the predictive capability of the compaction behaviour. The cluster theory of soil compaction (Srinivasa Murthy et al., 1985) attempts to formulate an approach to encompass the effects of both compaction effort and soil type.

Table 7.2 Comparison of different theories of compaction

Theory	Densification	
	Dry of optimum	Wet of optimum
Proctor (1933) Lubrication theory	Capillary effects at different water contents and the resulting resistance to compactive effort	Increased lubrication due to higher water contents and associated particle rearrangement
Hogentogler (1936) - Viscous water theory	Viscous effects due to absorption of water and its decrease as water content increases	Predominantly lubricating effects and rearrangement of particles
Hilf (1956) - capillary theory	Resistance to compaction attributed to resistance arising due to capillary pressures at air-water menisci	Decrease in density with increase in water content beyond optimum to trapping of air and building up of pore air pressure which reduces effectiveness of compaction
Lambe (1960) - physico-chemical theory	Due to water deficiency condition and depressed double layer, interparticle repulsion reduces causing flocculation. Flocculation generally means a low degree of particle orientation and low density	Beyond optimum level of water, attraction reduces causing lesser concentration of particles per unit volume. Further increased degree of saturation and low permeability causes positive pore water pressures which resists densification
Olson (1963) - effective stress theory	Resists compaction pressure by the development of shearing stresses which is a function of state of effective stress	At optimum, air voids become completely discontinous. As the water content is increased beyond optimum, air permeabilty decreases resulting in positive pressures which resist compaction consequently decreasing the density

Table 7.2 contd.

Barden and Sides (1970) - cluster characteristics theory	At low moisture contents, the dry density is attributed to high strength of air filled pores of macropeds. As water content increases the macropeds become wetter and consequently weaker with the possibility of greater distortion. Density increases till the optimum level	Further increase in compaction moisture content beyond optimum as the water layers between the soil particles increase in thickness decreases the drydensity.
Hodek and Lovell (1979) - Soil aggregate theory	Compaction is accomplished by the reduction of void between aggregates with very little change in the volume of aggregates at constant water content. With batches of different water contents density increases to its optimum value	Air permeability progressively decreases with increase in density and water content upto optimum level beyond which air voids become discontinuous. At this stage very little further densification will occur regardless of the input effort. A common water content versus unit weight relationship develops wet of optimum that is independent of input energy

7.8.1 Cluster hypothesis for soil compaction

We can observe a well defined pattern of compaction curves. For a given soil, as compactive effort increases the maximum dry density increases together with a decrease in optimum moisture content (Fig. 7.13a). For a specified compactive effort, as the liquid limit of soil increases, the optimum moisture content increases and the dry density decreases (Fig. 7.13b).

We can also notice that compaction curves dry of optimum cut across the constant degree of saturation paths, whereas the paths beyond optimum moisture content decend rapidly as the degree of saturation further increases with a tendency to run parallel to zero air void curve. We can imply from these observations that the dry density realized is due to a combined influence of the parameters, viz., moisture content and liquid limit water content and the external compactive effort.

Cluster theory of compaction (Srinivasa Murthy et al., 1985), enables to identify the role of above parameters on the mechanism of compaction.

When water is added to fine grained soils which is far less than the maximum water holding capacity of soils (liquid limit), due to inadequacy of available water, clay particles share the available water resulting in the formation of clusters. These clusters by themselves are in equilibrium under the influence of physico-chemical forces and pore water tension. The rigidity of such clusters against deformation, coalescence or breakdown depends upon the physico-chemical potential of soils (reflected in liquid limit water contents) and the mixing moisture content. In terms of the defined parameters (moisture content and liquid limit) the rigidity, R for a

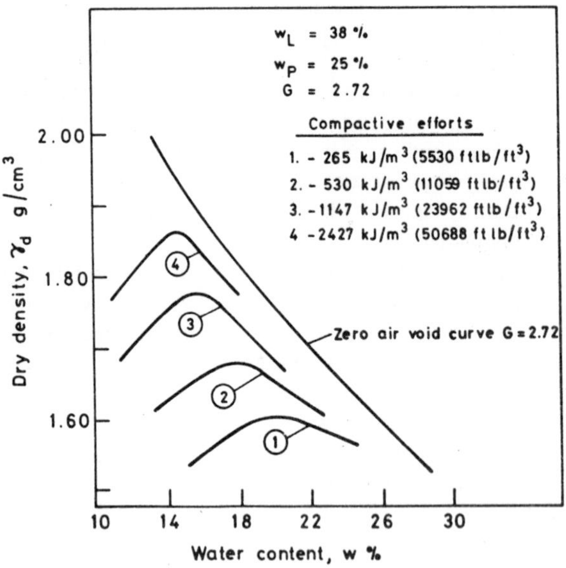

Fig. 7.13(a) Compaction curves of a soil for different efforts.

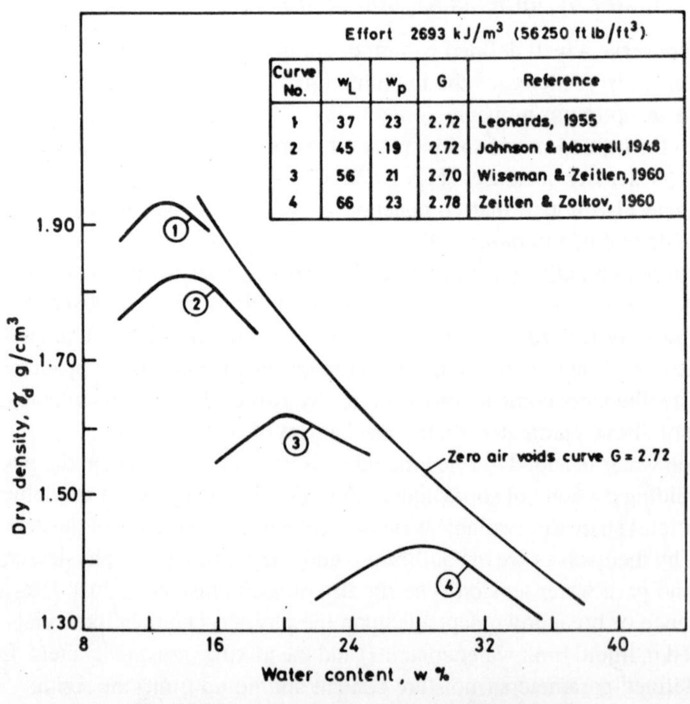

Fig. 7.13(b) Compaction curves for different soils.

soil of liquid limit w_L, is inversely proportional to water content. With water content remaining same the rigidity is a direct function of liquid limit of soil. Hence the rigidity, R is a function of w_L/w. The density that can be realized for a compactive effort is inversely proportional to rigidity of clusters, viz.,

$$\gamma_d = f(1/R) = f(w/w_L) \qquad 7.11$$

7.8.2 Analysis and prediction

The possibility of employing the above hypothesis to analyse and predict compaction characteristics merits examination. Let us consider the compaction data, from literature which is shown in Fig. 7.13. The soils are identified by their liquid limit water contents. The corresponding optimum moisture content and dry density are indicated in the inset table of Fig. 7.13. For a given dry density, the degree of saturation realized can be linked with moisture content. It is interesting to note that, for a given compactive effort there exists a unique linear relation between water content and liquid limit for a specified degree of saturation for all soils (Fig. 7.14) (Nagaraj and Bindumadhava, 1992). It implies that the value of degree of saturation is not arbitrary, but depends upon the intrinsic cluster characteristics dictated by water content and liquid limit of soil, compactive effort being the external factor in densification. Each experimental point on a compaction curve reflects the equilibrium state realized imparting a definite dry density to the soil system and in turn a certain value of degree of saturation. Although definite functional relationship between γ_d, w and w_L is not apparent, the possibility of a phenomenological approach to predict the compaction curve of a soil knowing its liquid limit water content is presented.

For a given soil with liquid limit, w_L a set of water content points can be identified on the lines representing different degrees of saturation in Fig.7.15. Now, for each water content, the corresponding dry density can be computed knowing the corresponding degree of saturation for the known value of specific gravity of the soils. The method can be adopted upto a degree of saturation of 85-90%, which corresponds to the optimum moisture content. Obviously different set of such relations are required for each compactive effort. The predicted compaction curves using this method are shown by dashed lines in Fig. 7.16.

Analysis of extensive compaction data available in literature sometimes shows compaction curves not following the defined pattern, some criss-crossing have also been reported. Nagaraj and Bindumadhava (1992) have identified that of the several possible experimental errors causing such deviations, the major one could be the final height of compacted soil being more than that of the standard mould. Figure 7.16 shows the results of standard compaction tests wherein the heights of the compacted soil was deliberately varied. The range of deviation indicated in Fig. 7.17 is quite large as much as 40% in dry density and over two fold in optimum moisture content. This may explain the cause for criss-crossing of compaction curves.

A complete development of prediction methods and theories of compaction based on fundamentals of soil behaviour is still not amenable. As such only the phenomenological approach presented in this section may be gainfully used.

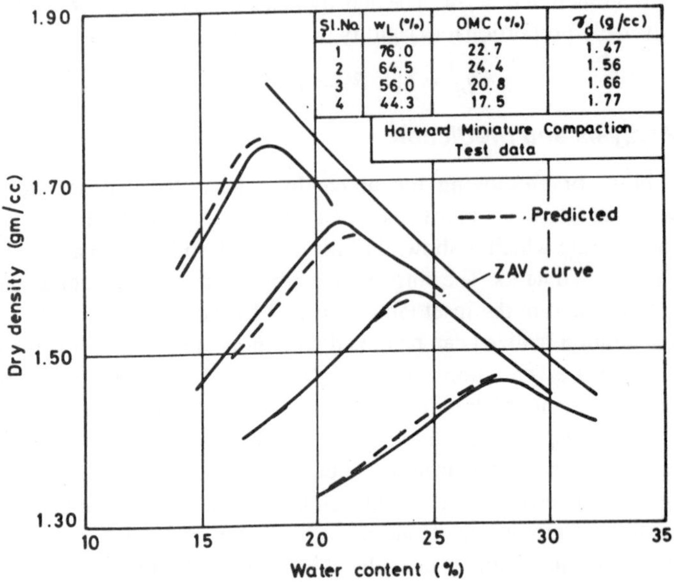

Fig. 7.14 Standard Proctor compaction curves for four soils (Nagaraj and Bindumadhava, 1992).

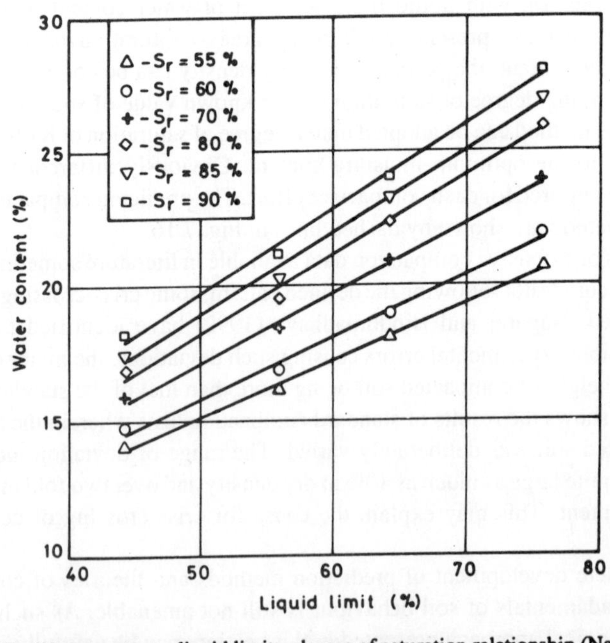

Fig. 7.15 Water content – liquid limit – degree of saturation relationship (Nagaraj and Bindumadhava, 1992).

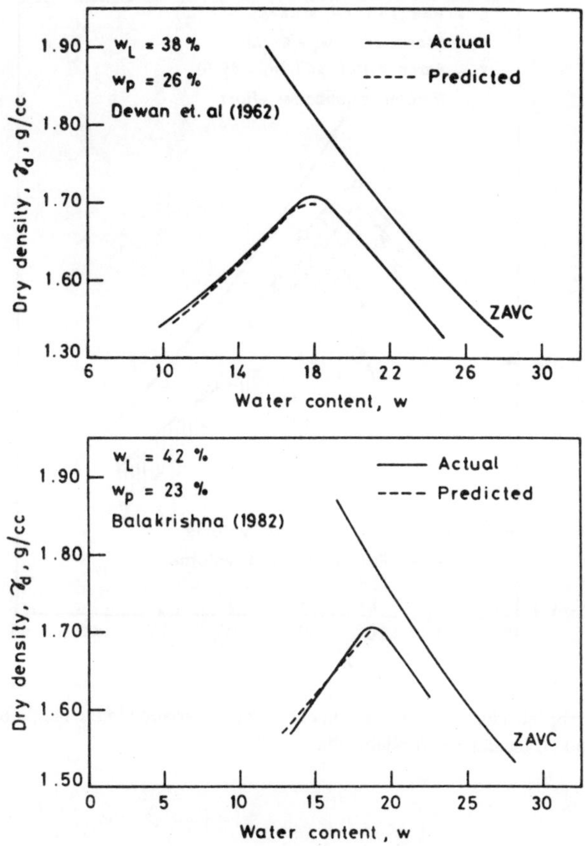

Fig. 7.16 Prediction of compaction curve.

In retrospect, it has been attempted in this Chapter to describe the behavior of partly saturated soils. The limitations in the present day understanding have been brought out. A generalized approach to define the compressibility of partly saturated soil under total stress probe has been presented. The behaviour under suction probe has been discussed. The possibility of having a constitutive model for describing the shear behaviour has been examined. The prediction of heave and collapse of partly saturated soils upon inundation has been attempted. Possibility of generalizing the compaction behaviour has been discussed. It is the belief of the authors that most of these discussions are applicable to compacted soils and a restraint has to be exercised while applying to desiccated soils.

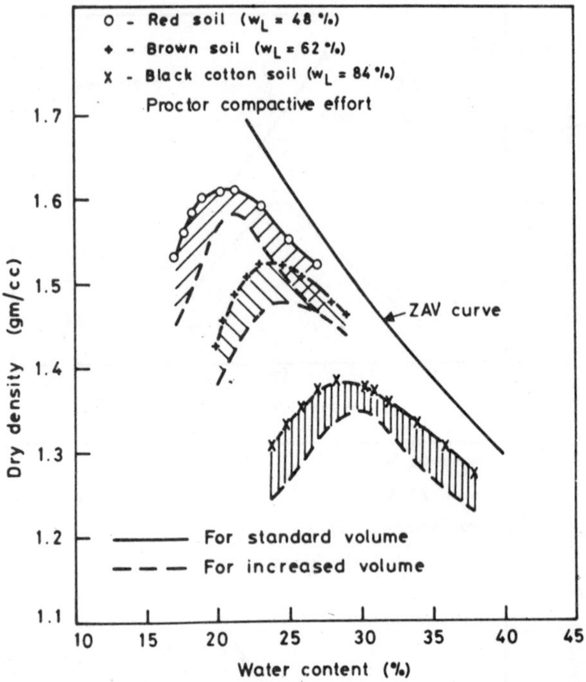

Fig. 7.17 Error introduced in compaction curve due to variation in the height of compacted soil (Nagaraj and Bindumadhava, 1992).

8

COARSE GRAINED SOILS

8.1 Introduction

Sands and clays have much in common as members of the same family of particulate materials. For e.g., dense sands show volumetric dilation and strain softening under shearing as do overconsolidated clays and loose sands show strain hardening with volumetric compression as do normally consolidated clays. Still their differences are as important as their similarities. There is a total difference in the mechanism of stress transfer and deformation of the two classes of soils as already discussed in Chapter 3. The distinction between coarse and fine grained soils is generally by the particle size. i.e., soils containing particles larger than 0.075mm are classified as coarse grained soils. Sands and gravel fall under the category of coarse grained soils.

In these soils because of their larger size and hence smaller specific surface, gravitational forces dominate their behaviour rather than the surface physico-chemical forces. Hence generalizations such as those proposed in the previous chapters (for which the Gouy-chapman diffuse double layer theory for colloidal particles was the basis) are simply not applicable to coarse grained soils. As of now, each given soil, not only that, sometimes even a given soil under each initial relative density, is treated as a different material with its own strength and deformation characteristics (except for the recent attempts by Been and Jefferies, 1985 where it has been shown that many mechanical properties are similar for all soils under identical states).

Since the thrust of this book is on possible generalizations and predictions of soil behaviour, the coverage in this chapter will be limited to such attempts of generalisations (or failures?) and to some special behavioural features of sands in comparison with those of fine grained soils, although there are a large number of investigations reported in literature on various other aspects. We will examine some of the general behaviour in the same sequence, if possible, as we did for fine grained soils, i.e., compressibility, shear strength and stress-strain behaviour and permeability. Before that, we will just recall in a very broad sense, the mechanism of stress transfer and deformation in such soils which may help us understand better the observed behavior.

8.2 Mode of Stress Transfer

As we have already seen in Chapter 3, the stress transfer in coarse grained granular medium is essentially through actual physical contacts between grains by friction bonds (Feda, 1982). Deformation of the medium is due to sliding or rolling of the grains to newer positions rather than due to the strain in the actual solid particles. Under an applied load increment, the particles slide to finally reach an arrangement that is most stable under this new set of loading. It can be shown that the most stable arrangement is one where the force at each of the contacts is normal to the contact area. But there would be resistance to such sliding of particles. This resistance to sliding may be the result of two components;

(a) friction at the contacts and
(b) interlocking of particles.

It can be seen from Fig.8.1 that if an initially dense assembly of particles is to deform, the particles will have to ride over the neighbouring one with an increase in the volume. Additional energy will be required to lift the particles against the applied normal load. Whereas, if an initially loose assembly is to deform, it can easily slip into the adjacent voids, resulting in a volume decrease (which may release energy, i.e., a negative component of dilation energy).

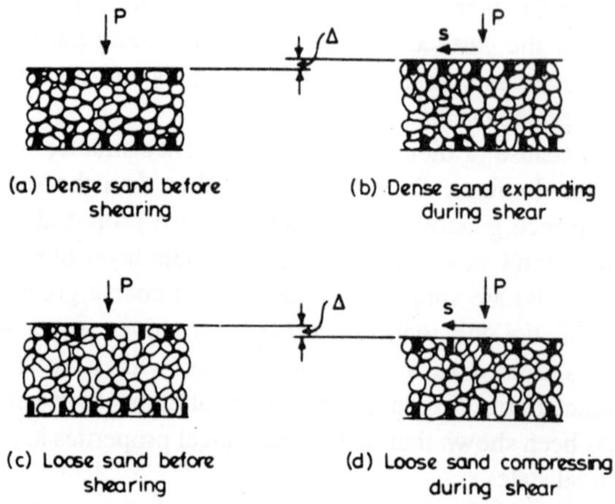

Fig. 8.1 Volume changes in dense and loose sands during shear.

The frictional component is a function of the number of contacts and the strength of bond at each contact. The dilation component is dependent on the density which in turn may depend on the size, shape and grading of the grains. Well graded angular grains result in a denser packing than poorly graded rounded ones. Thus the strength and deformation of the soil is controlled by the arrangement of grains or the fabric. In response to an applied load increment, the internal fabric of the system changes. Attempts have been made to define a suitable fabric factor which can be measured and can yield quantitative estimates of the external behaviour. (Nemat Nasser and Mehrabadi, 1982; Satake, 1982 and Shahipoor, 1982). The fabric factor includes a statistical picture of the number of contacts, inclination of contact normals, area of contacts, distance between contacts, deviation of local porosities from the mean porosity etc. There have been extensive.

studies all over the world, (Rothenberg and Bathrust, 1990; Cundall and Strack, 1982; Matsuoka, 1982 and others) both theoretical and experimental, to understand and quantify the macrolevel response in terms of the behaviour at microlevel i.e., as the net effect of response of each grain at each contact junction. In this chapter, an attempt is made only to discuss the various observed behavior in the light of the findings of such fundamental studies.

8.3 State and Compressibility Behaviour

Cohesionless soils can exist in nature or can be placed under different degrees of initial compaction. The degree of compaction is usually defined as dense or loose or more quantitatively, in terms of the 'relative density' D_r,

$$D_r = (e_{max} - e)/(e_{max} - e_{min})$$

where e_{min} and e_{max} are the minimum and maximum void ratios that the given soil can be placed in the laboratory by standard procedures. But generally, the range of densities that can be achieved is small, and are such that most of them (sometimes even the loosest possible state) are denser than critical (Fig.8.2) (critical state as defined in earlier chapter). Of course, some very fine sands are exceptions which exist in unusually loose states under certain conditions. In other words, the state of a coarse grained soil does not reflect the current equilibrium stresses as in normally consolidated clays and the stresses compatible with the current void ratio may be far too high and may never be applied on the soil in the normal working stress levels of geotechnical structures.

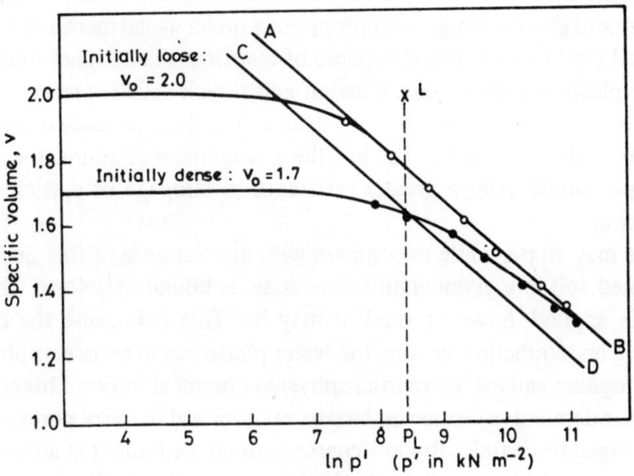

Fig. 8.2 v vs $\ln p'$ paths for initially loose and dense sands (Data by Vesic and Clough, 1968 reported by Atkinson and Bransby, 1978).

The existence of coarse grained soils in relatively denser states is quite understandable because nothing prevents the grains from physically being placed closely. You can actually pick and place grain by grain, to see that the voids are minimum. Whereas in clays, the surface physico chemical repulsive forces prevent the particles from being placed closer without the aid of an external force. Thus under an isotropic compression test, a sandy soil may withstand a high pressure, may be 1000 kPa or even more, before undergoing appreciable compressions. As such it may be difficult to obtain the linear portion (if at all there exists one, which we will detail little later) of the compression path similar to that of virgin or normally consolidated clays, within the usual testing ranges.

The occurrence of exceptionally loose state (such as point L in Fig. 8.2) is rather difficult to comprehend. The behaviour of such soils is somewhat similar to that of cemented quick clays. The soil in its loose state, can withstand pressures greater than what could be expected (p_L) for that density as per the extended virgin path, but beyond a certain stress level, undergoes sudden and drastic compressions. We may note here that the critical state model (or others) cannot accommodate the existence of a state outside or above the normally consolidated line and hence fail to predict the behavior of such soils. It merits examination as to what makes the soil exist in such loose states.

The term relative density is not very well defined. The standard method for determining e_{min} and e_{max} is to weigh the mould of known volume filled with air dropped soil, without (for e_{max}) or with a speficied amount of vibration (for e_{min}). But it has been found that the density achieved is very much dependent on the method of placement like air deposition, hydraulic pluviation or moist tamping. It is generally seen that moist tamping results in such loose fabric, may be because of capillary or surface tension forces, where as the other two methods result in relatively denser states. (Been & Jefferies, 1985).

An initially loose fabric at deposition would not have been a problem by itself, if the soil were to exhibit the expected deformation behaviour under further loading. i.e., the loose soil should compress appropriately under a load increment. But many times the soil (or fabric) is found capable of resisting much higher loads without appreciable plastic compressions. Cundall and Strack (1982) present interesting sketches (Fig. 8.3) to show the effect of structure on the resistance to deformation of a granular medium. It can be seen that the arrangement of grains is such that the looser sample would offer a greater resistance to slippage of particles than the denser packing.

Here we may stop a while to compare with the response of fine grained soils. A fine grained soil in a given equilibrium state, is bound to deform when a load increment is applied, however small it may be. This is because the clay-water system being an interacting system, the water phase has to be in equilibrium with the applied loading and the interparticle physico-chemical forces. This equilibrium of water phase demands a change in the size, number and shape of pores compatible with the changed force field. But in a coarse grained medium, if at all water is held in the pores it is only by gravity and it has no role to play in the process of stress transfer. Hence, the soil in a given state can resist deformation under load increments as long as the induced shear stresses are less than the limiting shearing resistance at each of the contacts.

Now it is quite possible that a coarse grained soil exhibits such a rigidity against deformation upto some magnitude of load increments not only at the initial

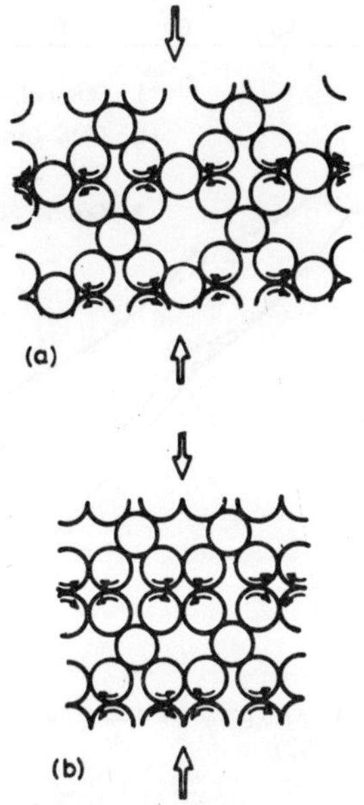

Fig. 8.3 Effect of structure on strength: (a) low density - high strength (b) higher density-lower strength (After Cundall and Strack, 1982).

depositional void ratio but also at subsequent equilibrium states during a long stepwise loading process, i.e., it is quite likely that the soil shows a compression path with broken lines or with stretches of flat and steep sections.

In addition, there is the phenomenon of particle degradation, i.e., the particles may undergo different degrees of actual crushing under sufficiently high stresses, which may alter the size, shape, particle size distribution which may further affect its basic strength and deformation characteristics. This is very well reflected by the data of compression test result of St. Peter sand (Fig. 8.4) (Borg et al., 1960) reported by Feda (1982) where the break points indicate increased compressibility due to particle crushing.

With all the above discussions it is quite clear that it may not be possible to have a defined unique virgin compression path for coarse grained soils as we had for fine grained soils.

8.4 Shear Strength

The most common observation about the strength parameters of cohesionless soils is that they have higher friction angles than clays and generally negligible or zero cohesion intercept (Fig. 8.5). Grading, mean size and angularity of particles are the factors accounting for different values of ϕ for different soil types; well graded very coarse, angular particle types yielding higher values. And for a given soil, relative density plays an important role, denser samples showing higher friction angles. As discussed in Chapter 3, the basic resistance due to sliding alone is probably the

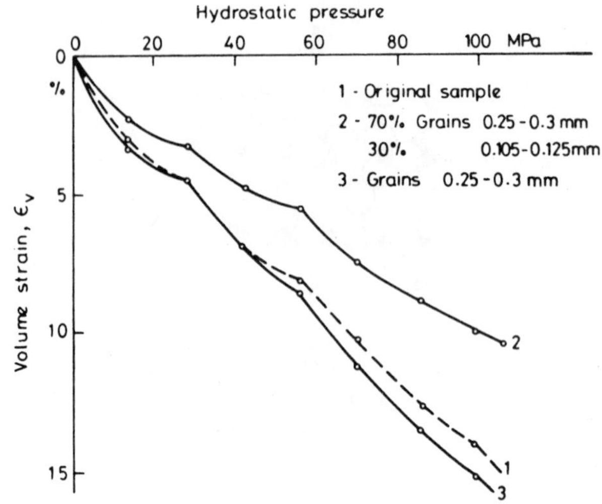

Fig. 8.4 Change in compressibility with pressure due to particle crushing (Data by Borg et al., 1960 reported by Feda, 1982).

same for most types of soil minerals, since they all basically belong to the same silicate group, and is found to be of the order of 26° – 28°. (Feda, 1982 and Mitchell, 1976). Therefore, the friction angle, greater than 27° observed in any case is due to the interlocking effect and hence is increased by all such factors which increase the interlocking.

For a given coarse grained soil, the failure envelop is not always a perfect straight line for the entire range of stresses but actually droops down or flattens with increase in stress level as shown in Fig. 8.6 (Lambe and Whitman, 1969 data of Holtz and Gibbs, 1956). The magnitude of drooping may vary for different relative densities until at very large stresses all of the failure envelops tend towards that for the loosest density. But this may occur at very high stress levels (greater

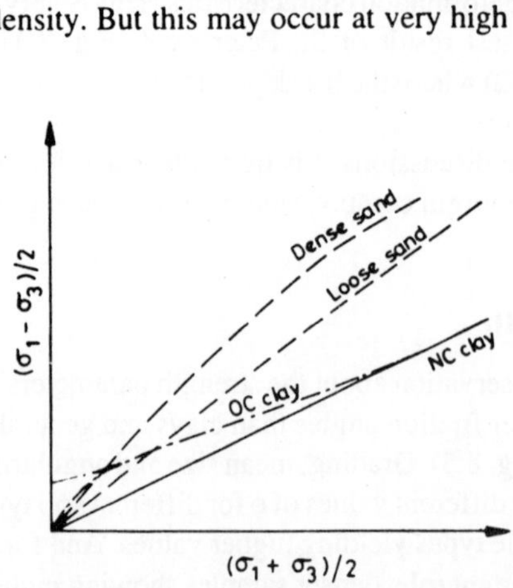

Fig. 8.5 Mohr diagram for sands and clays under different states.

than 4–5 MPa) and at stress levels of engineering interest the envelops may be approximated to be straight lines passing through the origin.

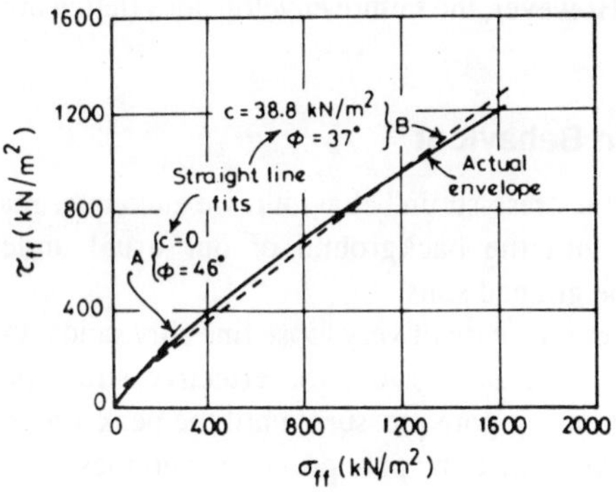

Fig. 8.6 Non-linear Mohr envelope for sands.
(After Lambe and Whitman, 1969)

It may be worthwhile at this stage to compare the shear behaviour with that of fine grained soils. The increase in strength for an overconsolidated soil is reflected more as an increase in cohesion than in friction angle, ϕ' or rather even a slight reduction in ϕ'. It is for this reason that coarse and fine grained soils are often referred to as cohesionless and cohesive soils respectively. But it is now understood (Lambe and Whitman, 1969) that if shear tests are really conducted at zero or very low confining pressures, we do get a zero cohesion even for overconsolidated fine grained soils, so that the whole of available strength is always frictional for any soil. i.e., the failure envelops may pass through the origin and curve very steeply as shown by dotted lines.

The reduction in the value of ϕ' for overconsolidated fine grained soils may be explained as follows. On the basis of postulations advanced regarding shearing of overconsolidated soils, which involves a process of breakdown of particle clusters, it is likely that the additional energy required to separate the particles is not so much a function of the normal stress but is rather dependant on the strength or magnitude of intrinsic attractive forces, '*A*' acting between particles, (the range of σ'_3 values that the soil is subjected to, being far less than the '*A*'). Ofcourse, the amount of breakdown and in turn the increased resistance itself is dependent on the level of σ'_3 or more precisely, decreases with decrease in the OCR. That is the reason the failure envelops are flatter, gradually approaching the envelop for normally consolidated state.

In contrast, the additional energy required to overcome interlocking in dense sands is likely to be a function of the normal stress acting. Both the sliding friction and the interlocking or dilatancy components being functions of normal stress, the total increased strength is reflected as a higher ϕ' with no cohesion component. Ofcourse the relative effect of dilation component should decrease with increase in σ'_3 or decrease in OCR as in the case of finegrained soils but the level of σ'_p (the equivalent prestress) being very high in the case of dense sands, this effect may not be significant for the usual low range of confining pressures that the soil is subjected to. (It may be noted here that σ'_p need not mean the actual maximum preconsolidation pressure, but is only the equivalent pressure compatible with the present void ratio). However, the failure envelop does flatten at much higher stress levels.

8.5 Stress-strain Behaviour

When we consider the stress-strain behaviour, we encounter many uncomprehendable phenomena with the background of our usual understanding and in comparison with fine grained soils.

Let us first consider the case of very loose fine silty sands. In a static undrained test (Fig. 8.7) (data of Castro, 1969) the effective stress path would show a decreasing p (or increasing pore pressure) until the peak strength as expected for a loose state. But beyond that, the pore pressure continues to increase with both p and q decreasing very rapidly to finally reach nearly zero strength. Similar test results have been reported by several other researchers (Been and Jefferies, 1985 and Poorooshasb and Pietruszezac, 1985). These are special class of soils called liquefying soils. (Under dynamic loads, these soils would liquefy which will be explained in a later section). This mode of strain softening is never observed with the counterpart in fine grained soils i.e., normally consolidated clays. A fine grained soil under any given void ratio would yield indefinitely upon reaching the failure strength, q_f but would continue to possess that strength. Or, if the void ratio of the soil were to be so high as to reach such a reduced ultimate strength, q_u at large strains, there is no means by which it can be in equilibrium under an initial high stress of p'_o or can mobilize such a high peak strength of q_f. As discussed in the earlier section, the increased load carrying capacity in the case of sands even in a very loose state is due to the special 'structure' or arrangement of particles. Once it is loaded beyond the capacity of that state, the structure collapses and the soil looses its additional strength if the void ratio is held constant at the loose state. If it is a drained condition, the soil would suffer unusually high compressions beyond the peak strength. The behaviour is very much analogous to that of cemented sensitive soils. This perhaps is the reason for the often used common terminology of quick sands and quick clays for these two types of soils.

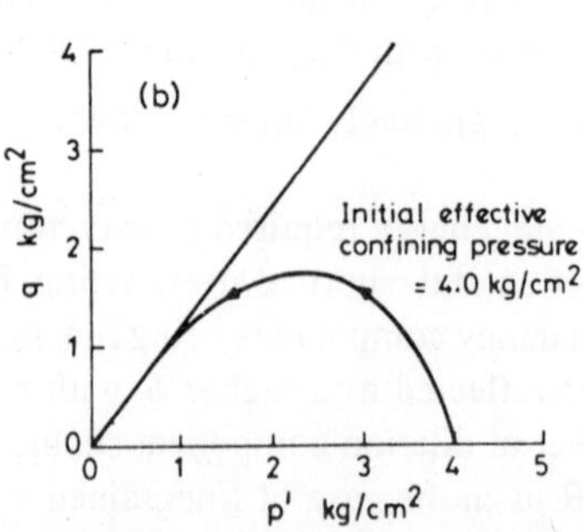

Fig. 8.7 Undrained stress path of loose sand (data by Castro, 1969 reported by Pietruszezac and Poorooshasb, 1985).

The other typically observed behaviour, perhaps for medium dense sands, is as shown in Fig. 8.8 (Ishihara and Okada, 1978; Castro, 1969 and Luong, 1980). The stress path in an undrained test proceeds leftwards as for a loose sand (i.e., compressive or u increasing) upto a particular strength level. But beyond this, it reverses its direction following a path similar to that of dense sands, showing a dilating tendency or negative pore pressures. Again this mode of a reversing in stress path is never observed with fine grained soils either for lightly or for heavily overconsolidated soils. Of course, for normally consolidated clays a dilating behaviour is never seen at any stage. The compression observed for heavily overconsolidated clays at very early strain levels is only of elastic nature and hence does not show a decreasing p in an undrained test. On the other hand, a lightly overconsolidated clay never exhibits plastic dilation and hence increasing p cannot occur in an undrained test. In other words, for fine grained soils, there is a clear demarcation as to whether the plastic volume change of a soil in a given state is dilation or compression for a specific p, and there is no overshooting beyond the compatible void ratio from either side. In contrast, the decrease in p for sands in Fig. 8.8 is due to an initial plastic compression even beyond or crossing the critical void ratio compatible with the stress, and then plastic dilation resulting in a stress path reversal.

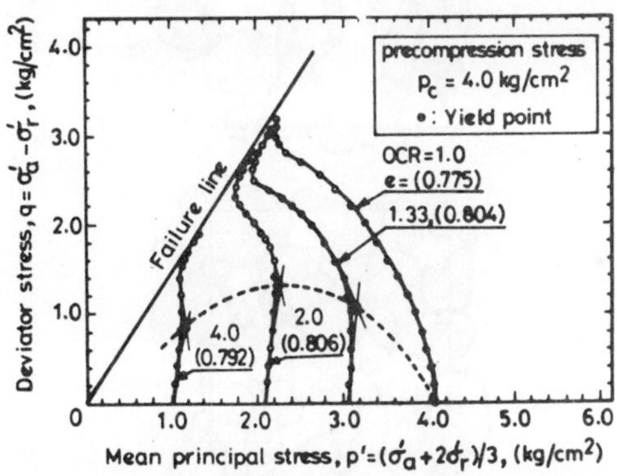

Fig. 8.8 Undrained stress path reversals in medium dense sands (After Ishihara and Okada, 1978).

The reason for this appears to be the distribution of local porosities in granular soils. It has been experimentally shown (Mogami, 1965) that there can be a large variation in porosity locally from the overall or average observed porosity (Fig. 8.9). Infact this aspect has been studied in great depth with regular and irregular arrays of uniform and nonuniform particle sizes. A statistical quantification of the porosity distribution has also been possible to some extent (Shahinpoor, 1982). The variation in local porosity is in general, found to be more for loose states, gradually tending towards more uniform fabric with increasing densities. Thus, at some stage, though the average void ratio may correspond to a relatively dense state, there may be considerable local loose pockets surrounded by very dense groups of grains. These dense groups are generally found to move as single units, so that the behaviour is largely controlled by the loose pockets and hence the initial observed behaviour which is similar to that of loose soil. Once this fabric is disturbed, leading to a more uniform arrangement, the soil behaviour may correspond to that of the denser state.

Sometime a combination of the two factors, i.e., the variation in local porosity and the special 'structure' effect (responsible for liquefaction discussed for loose sands) may result in a stress-strain curve which even shows a slight decrease in strength and then again appreciable increase (Fig. 8.10). The behaviour is sometimes called partial liquefaction.

The behaviour of very dense sands is in general more close to the expected one, since such irregularity or non-uniformities in fabric are known to be less pronounced at higher densities. i.e., a strain softening, dilating behavior in drained test or a corresponding negative pore pressure in undrained condition is observed (Fig. 8.11) (data of Baladi and Rohani, 1979 reported by Pietruszezac and Poorooshasb, 1985).

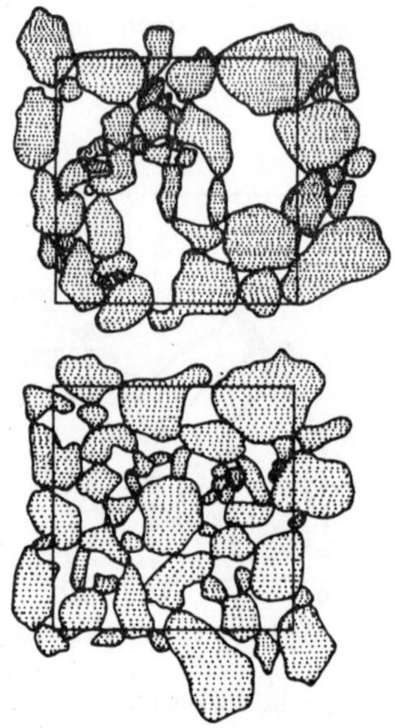

Fig. 8.9 Variation of local porosity with the same overall porosity (After Holtz and Kovacs, 1981).

An important observation from the general stress-strain behaviour of coarse grained soils is that there is a definite strength ratio, q/p' at which the stress paths decide their further course i.e., they either change from moving towards the origin to away from the origin, i.e. from compressive to dilative (medium dense sands), or they droop drastically towards the origin (very loose sands) (Fig. 8.12). The angle corresponding to this stress ratio is sometimes called the phase transformation angle (Ishihara et al., 1975) which may be the same as the characteristic angle (Habib and Luong, 1978 and Wood, 1982) which corresponds to the zero dilatancy state, i.e., the soil shows a compressing behaviour for stress ratios less

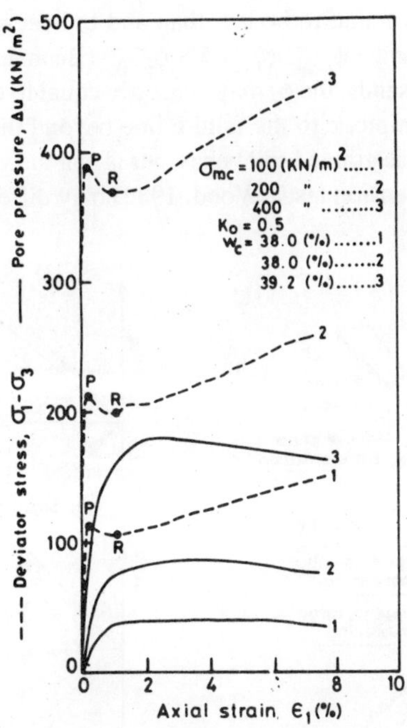

Fig. 8.10 Partial liquefaction as reflected in transitory loss in strength (PR) during undrained shearing (After Hanzawa et al., 1979).

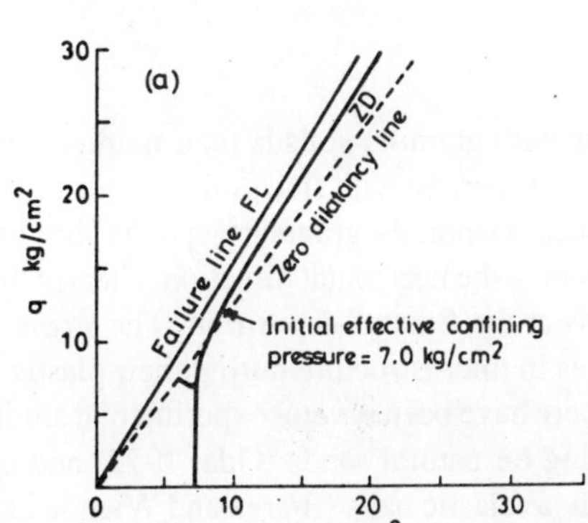

Fig. 8.11 Undrained stress path of dense sand (data by Baladi and Rohani, 1979 reported by Pietruszezac and Poorooshasb, 1985).

than this, and a dilating nature above this line. Perhaps this stress ratio corresponds to the basic friction angle for that initial specific structure. Once the stress ratio exceeds this ratio, the fabric gets disturbed. The soil shows considerable plastic strains only beyond this stress ratio. It is also observed that once the soil is loaded beyond this level it becomes stiffer in the direction of loading but softens considerably for reverse loading (Ishihara et al., 1975). This is due to the stress induced anisotropic fabric which will be discussed in the next section. The phase transformation angle is found to be less than and to bear nearly a constant ratio with the peak friction angle ϕ_{peak}, ($\phi_s = 5/8\ \phi_{peak}$) (Been and Jefferies, 1985) for dense sands. For loose sands, the ϕ_s may be nearly equal to ϕ_{peak}. The stress paths are found to move asymptotic to the failure line beyond the transformation line. Another feature of coarse grained soil behaviour is that the values of friction angle in compression and extension tests (Wood, 1982) may differ by as much as 10°.

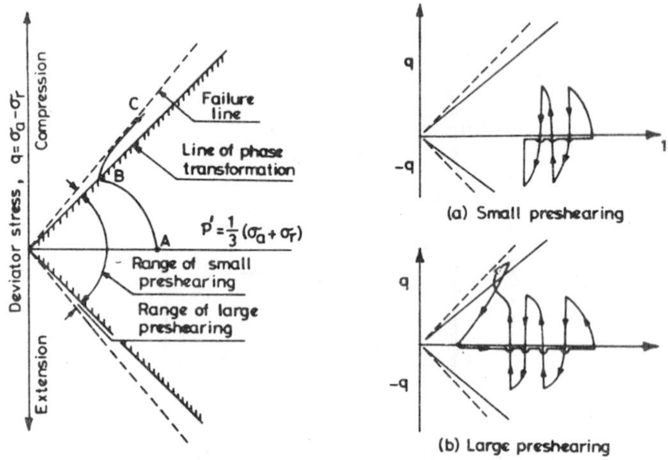

Fig. 8.12 Phase transformation during shearing - onset of large plastic strains (After Ishihara, et al., 1975).

8.6 Anisotropy

Another special feature with granular soils is their marked anisotropic behaviour which is probably not so important with fine grained soils. The anisotropy may be inherent or stress induced. Generally granular deposits show a higher stiffness in the vertical direction than in the horizontal direction. Clearly this initial anisotropy is due to the conditions at the time of deposition. The stress induced anisotropy results from the changes in microstructure during their plastic flow in response to the applied loading. There have been several experimental studies on such changes in fabric during shearing on natural sands (Oda, 1972) and on two dimensional model assemblies of photo elastic balls (Biarez and Wiendieck, 1963). It has been concluded that during the course of rearrangement of particles upon loading, the distribution of contact normals changes in such a manner as to produce greater concentration of contact normals along the direction of the maximum principal stress. This is also closely related to, or this results in, the formation of column like load paths leading to induced anisotropy. i.e., the material will have greater stiffness in the direction of maximum principal stress and a reduced stiffness in a direction perpendicular to that.

In fine grained soils too, the microstructure changes with applied deviatoric stresses. But here the change is mainly with the shape of the large pores as discussed in Chapter 3 and perhaps not at the level of particles. The pores which were spherical under isotropic loading, deform to an elliptical shape with the major axis of the ellipse perpendicular to the direction of major principal stress. But once the deviatoric stress is removed or the stress is changed to a new hydrostatic stress, the pores are bound to get back to a spherical shape to satisfy equilibrium conditions. Hence anisotropy may not remain in fine grained soils, i.e., the microstructure is always compatible with the current state of stress. Whereas in granular soils the fabric which was oriented with the direction of major principal stress need not revert back to an isotropic fabric upon removal of the deviatoric stress because the particle contacts are capable of carrying some shear stresses and hence can retain an anisotrpic fabric. If the distribution of contact normals corresponds to an anisotropic stress system and the actual stress acting is isotropic, then many of the contacts will be required to resist shear stresses. Dafalias (1986) states that in soils stress induced anisotropy gets erased out as easily as it is formed. May be it is completely true in fine grained soils but sandy soils can exhibit considerable anisotropy.

With all the above discussions it can be concluded that the behaviour of granular materials is quite complicated and that there are many factors which control their behaviour. The fabric or the micro structure plays an important role in addition to the usual factors like relative density, particle size distribution, extent of angularity etc. Though attempts have been made to quantify some of these fabric factors, viz., contact normal distribution, variation of local porosity etc and to relate them with the actual behaviour, the study has not yet culminated to a stage where generalizations and predications of behaviour can be made. But fortunately, in practice, sands are less problematic than clays. For the normal densities that sandy soils exist in nature, settlements or deformations are quite small upto the peak strength. And even if the deformations are considerable, because of their high permeability, deformations are immediate and most of it will occur during the construction period itself. Hence in any stability problem shear strength is the only consideration, and if a sufficient factor of safety is used, a limit analysis without elaborate consitutive relations may be sufficient. Perhaps the only major problem is the liquefaction of loose sands under repeated loadings, a condition when the soil looses its strength completely leading to catastropic failures of structures. Hence this aspect will be discussed in greater detail though the basic feature has been explained earlier for the case of static loading. Before that, a brief review of the very few attempts towards generalization of behaviour of granular soils will be presented.

8.7 Generalization and Prediction of Behaviour of Sands

As we have discussed, predication of behaviour of coarse grained soils is quite difficult, so much so that even for a single soil, it may not be possible to predict its behaviour at some other state and for a different loading path because of the effect of fabric. For this reason, constitutive models have not been very successful or successful only over a small range of conditions for which the basic properties were calibrated. Generalizations of behaviour irrespective of soil type are bound to be all the more unlikely. The only attempt in literature (as far as we are aware of) towards generalization is by Been and Jefferies (1985), where it has been shown that the general engineering behaviour of a soil is uniquely related to a state parameter called $\psi(= e - e_{ss})$, which is a measure of the difference of the current void ratio from the steady state void ratio at the same stress level (Fig. 8.13). It is claimed that the parameter ψ is a better parameter than the relative density since it takes into account the state of stress also, and in addition it relates the current state to a steady reference state (which corresponds to a state devoid of fabric effects). Thus the parameter may indirectly account for the fabric factors. It has been shown that there are unique relations between some soil properties and the state parameter such as, ϕ'_{peak} vs ψ, rate of dilation $d\varepsilon_v/d\varepsilon_a$ vs ψ, phase transformation angle ϕ_s vs ψ, for all soils (Figs. 8.14, 8.15 and 8.16).

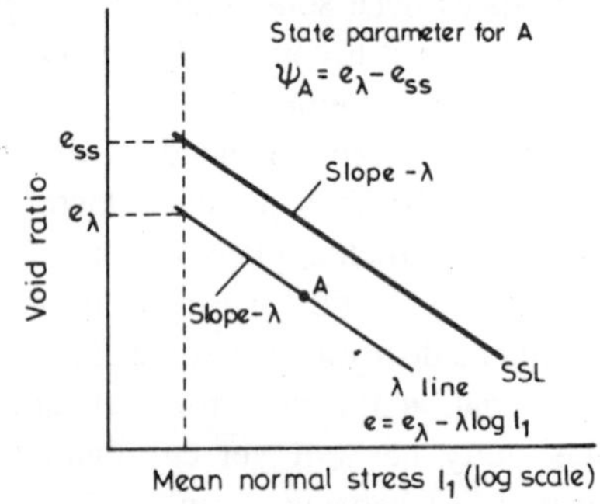

Fig. 8.13 Definition of state parameter, (After Been and Jefferies, 1985).

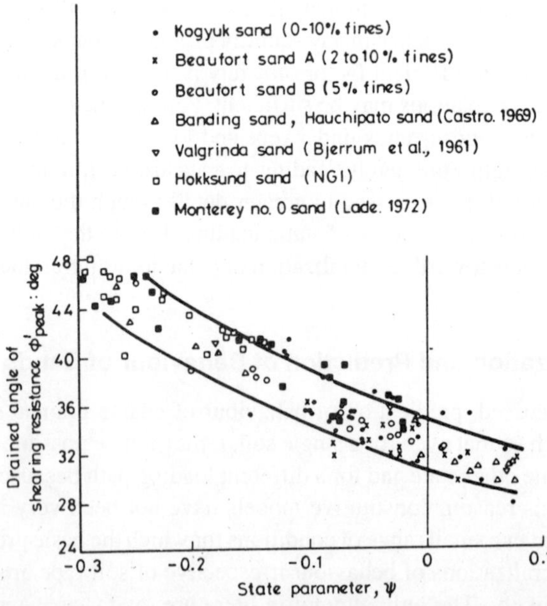

Fig. 8.14 Relation between drained angle of shear resistance and state parameter for several sands (After Been and Jefferies, 1985).

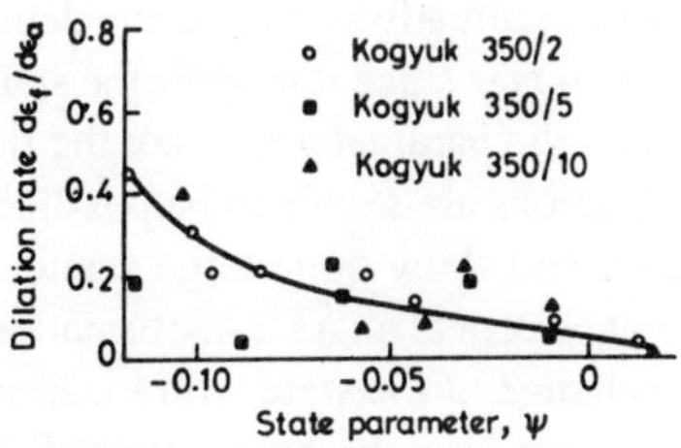

Fig. 8.15 Relation between dilation rate and state parameter for Kozyuk sand (After Been and Jefferies, 1985).

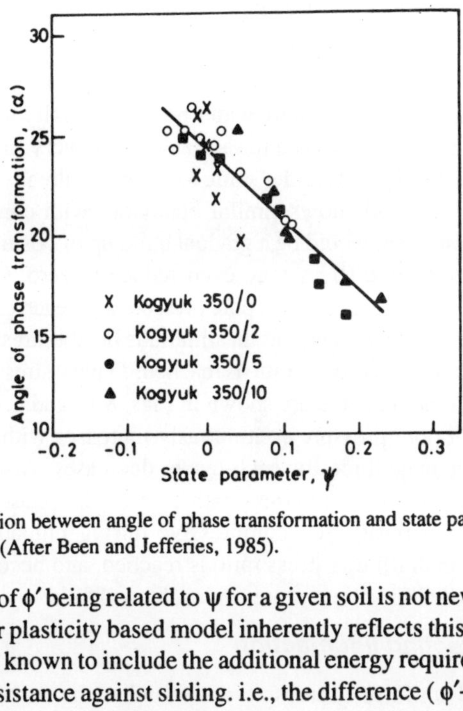

Fig. 8.16 Relation between angle of phase transformation and state parameter for Kozyuk sand (After Been and Jefferies, 1985).

The concept of ϕ' being related to ψ for a given soil is not new. Infact the critical state or any other plasticity based model inherently reflects this. The peak ϕ' at any state is very well known to include the additional energy required for dilation other than the basic resistance against sliding. i.e., the difference (ϕ'–ϕ_{ss}) is the dilation component, which would mean a $q/p > M$ for states dry of critical (dilating) or $q/p < M$ for states wet of critical (compressing).

The same thought was implied in Rowe's (1962) dilatancy equation $R = DK_{\mu}$ where R is the stress ratio (= σ_1/σ_3), D is the dilatancy factor ($D = 1 - d\varepsilon_r/d\varepsilon_a$) and K_{μ} is a constant factor equal to $\tan^2 (45 + \phi_{\mu}'/2)$, which relates stress ratio to the rate of dilation. Of course this relation is not directly helpful in the sense, both stress ratio and rate of dilatancy are responses, not independently variable but rather are dependent on the current state. Hence a relation among themselves is not helpful where as relating either of them with the state parameter would be more appropriate.

Rowe (1962) had ofcourse further given a relation between dilatancy factor and relative density of the soil. But as already seen, relative density is not a sufficient parameter and $\psi$is a better parameter since it reflects the state of stress also.

With all this, it appears that the parameter ψ is not the final word. First of all, the data for which the generlisations are shown to be possible are very limited and are for the same type of sand, and show quite large scatter. The generalizations pertain only to the behaviour at peak or at phase transformation stage. The complete stress-strain response or the change of the state with loading itself is not defined. Also there is no logical basis as to how the behaviours of different sands can be generalized with just the ψ parameter even though their steady state lines have significantly varying slopes (or rate of change of ψ with stress). Perhaps extensive experimental verification is required before such generalizations can be used in practice.

8.8 Liquefaction

Liquefaction is the phenomenon where the soil drastically loses its strength to very low value, as a result of increase in pore water pressure usually under repeated or cyclic loading. It is generally observed in saturated loose, very fine or silty sands. The behaviour of very loose sands under static loading has already been discussed. Under cyclic loading, the soil shows similar behaviour with cumulative effects. i.e., in an undrained test, there would be a gradual build up of positive pore pressure to an extent that the effective stress may even reduce to zero. OR, under strain controlled cyclic tests, with the increase in pore pressure after each cycle, the strength of the soil would decrease for the same strain amplitude but the stress ratio q/p would gradually increase till the failure stress ratio is reached. Typical stress paths and stress strain curves leading to liquefaction are shown in Figs. 8.17 and 8.18.

It can be seen that pore pressure continuously increases with loading cycles, though the amount or magnitude for each cycle, decreases. Also the behaviour reverses to a dilating one once the stress path crosses the phase transformation angle, while it is the opposite i.e., compressive during unloading. The strain amplitudes are very small till this stress ratio is reached, and become appreciably large beyond this, and within a few more cycles the soil fails.

8.8.1 Factors affecting liquefaction

Sands with fine and uniform grain size are found to be more susceptible to liquefaction since the permeability of such soil can be low enough to prevent dissipation of the pore pressures developed under each cycle. This depends on the duration of the dynamic load also, i.e., a dynamic load acting over a longer period would increase the possibility of pore pressure build up. Infact many other factors, such as extent of the deposit leading to a longer drainage path, presence of air voids, etc., which contribute to a lower rate of dissipation of pore pressure, increase the chances of liquefaction.

Relative density is the most important factor. Soil at a high relative density is less compressible or may even be dilative and hence would never mobilize positive pore pressures.

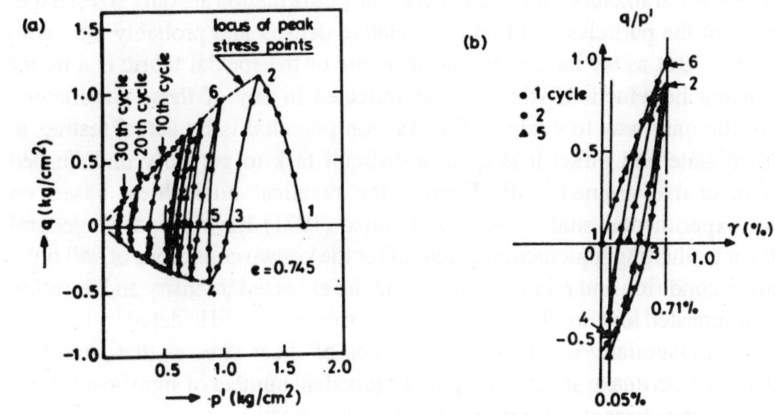

Fig. 8.17 Strain controlled undrained cyclic stress paths leading to liquification (After Ishihara et al., 1975).

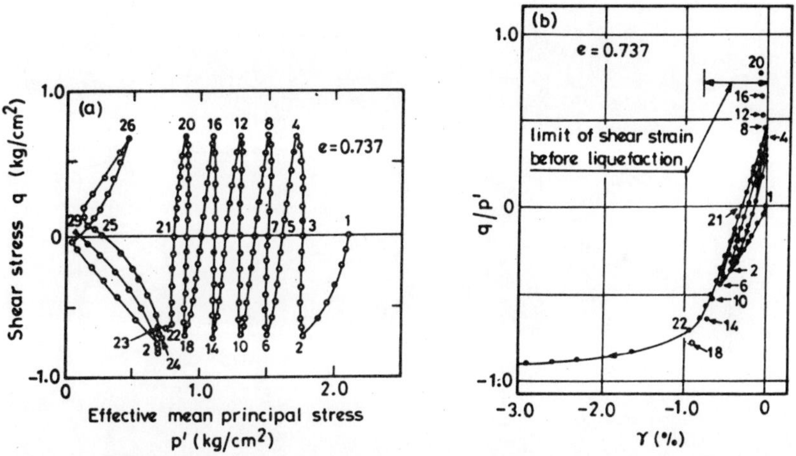

Fig. 8.18 Stress controlled undrained cyclic stress paths leading to liquification (After Ishihara et al., 1975).

Fabric or the structure imparting an additional stability to a loose state is already shown to be a factor responsible for liquefaction. What makes a soil attain such a fabric is an aspect not very well understood. It is found that method of sample preparation (or soil depositional environment) plays an important role, moist tamping method generally yielding liquefiable fabric while pluviation methods result in dense samples which seldom liquefy.

It has been observed that a soil liquefies for a less number of cycles when the cyclic stress ratio or strain amplitude is high.

Thus the task in practice is to identify whether the given soil is liquefiable, and if so under what stress ratio and what number of cycles are required for liquefaction.

There are several factors which influence the liquefaction potential of a soil. The measurable soil parameters which can provide any information are only size, shape, distribution of the particles, void ratio or relative density and probably the insitu state of stress. But as we have seen, the structure or the special fabric is a major factor causing liquefaction, which is not reflected in any of these parameters. Therefore the only way to evaluate liquefaction potential is by actual testing in undisturbed state and infact it is quite a difficult task to get truly undisturbed samples in coarse grained soils. Despite the practical difficulties, based on extensive experimental studies Seed and Idriss (1971) have proposed. general methods for evaluating liquefaction potential for the known conditions of soil type, initial stress condition and relative density and the expected intensity and duration of cyclic or repeated loading. The procedure has been explained in detail by Prakash (1981) which essentially involves a comparison of shear stresses that would be induced by an earth quake and the computed equivalent number of significant stress cycles against the shear stress required to cause liquefaction.

BS1377, 1975. Methods of test for soils for civil engineering purpose. British Standards Institution, London, pp.46.

Burland, J.B., 1990. On the compressibility and shear strength of natural clays. 13th Rankine Lecture, *Geotechnique,* **40**(3):329-378.

Carman, P.E., 1956. *Flow of Gases Through Porous Media.* Academic Press, New York.

Carpenter, G.W. and Stephenson, R.W., 1985. Permeability testing in the triaxial cell. *Geotech. Test. J.,* **9**(1):3-10.

Carroll, D., 1970. *Rock Weathering.* Plenum Press, New York, pp.200.

Casagrande, A., 1932. Research on the Atterberg limits of soils. Public Roads, **12**(8):121-126.

Casagrande, A., 1936. The determination of the preconsolidation load and its practical significance. *Proc. 1st Int. Conf. Soil Mech.,* Harvard, 3.

Casagrande, A., 1938. Notes on soil mechanics — First Semester. Harvard Univ (Unpublished) pp.129.

Casagrande, A., 1958. Notes on the design on liquid limit device. *Geotechnique,* **8**:84-91.

Casagrande,A. and Wilson,S.D., 1953. Effects of stress history on the strength of clays. *Harvard Soil Mechanics,* No 43, Cambridge, M.A.

Castro, G., 1969. Liquifaction of sands. *Harvard Univ. Soil Mechanics Series No.81,* Pierce Hall, Cambridge, USA.

Cedergren, H.R., 1967. *Seepage, Drainage and Flow Nets.* John Wiley and Sons Inc., New York, p.23.

Chandler, R.J., 1988. The in-situ measurement of the undrained shear strength of clays using the field vane. In A.F. Richards (editor) Vane Shear Strength Testing in Soils: Field and Laboratory Studies. *Am. Soc. Test. Mater. Spec.* Tech. Publ., **1014**:13-44.

Chang, M.F., 1988. Some experiences with the dilatometer test in Singapore. In J. de Ruiter (Editor) *Penetration Testing,* 1988, ISOPT-1 Balkema, Rotterdam, 1:489-496.

Chapman, D.L., 1913. A contribution to the theory of electro- capillarity. *Phylosophical Magazine,* **25**(6):475-481.

Charlie, W.A., Von Gunten, M.W. and Doehring, D.O., 1982. Temperature controlled microwave drying of soils, Geotech. Test. *J. Am. Soc. Test. Mater.,* **5**(3/4):68-75.

Chen, W.F. and Mizuno. E., 1990. *Nonlinear Analysis in Geotechnical Engineering.* McGraw Hill, New York p.413.

Clarke, B.G., Carter, J.P. and Wroth, C.P., 1979. In-situ determination of the consolidation characteristics of saturated clays. *Proc. 7th European Conf. on Soil Mechanics and Foundation Engineering,* Brighton, England, **2**:207-211.

Coleman, J.D., 1962. Stress – strain relations for partly saturated soils. Correspondance, *Geotechnique,* **12**(4):348-350.

Cox, J.B., 1970. Shear strength characteristics of the recent marine clays in South East Asia. *J.S.E. Asian Soc. Soil Eng.,* **1**:1-28.

Coyle, H.M. and Shiffert, J.B., 1968. Manufactured soil samples for laboratory research. *J. Mater. Am. Soc. Test. Mater.,* **3**(2):277-293.

Croney, D., Coleman, G.D. and Black, W.P.M., 1958. The movement and distribution of water in soil in relation to highway design and performance. Road Res. Lab. Note No. RN/3209:1-14.

Cundall, P.A. and Strack, O.D.L., 1982. Modelling of microscopic mechanisms in granular materials. In *Mechanics of Granular Materials - New Models and Consitutive Relations :* Jenkis J T and MaSatake (Editors), Elsevier, Amsterdam 1-8.

Dafalias, Y.F, 1986. On elasto plastic – visco plastic constitutive modelling of cohesive soils. *Geomechanical Modelling in Engineering Practice.* Dungar, R. and Studer, J.A.,(editors) Balkema, Rotterdam, 313-350.

Dafalias, Y.F., Hermann, L.R. and Scott, J.D., 1980. Description of natural clay behaviour by a simple bounding surface plasticity formulation. *Proc. Workshop on Limit Equilibrium, Plasticity and Generalised Stress-strain in Geotech. Eng., ASCE* : 711-744.

Daniel, D.E., 1989. In-situ hydraulic conductivity tests for compacted clay. *Proc. Am. Soc. Civ. Eng.*, Geotechnical Engineering Div. **115**(9):1205-1226.

Davidson, J.L. and Boghart, A., 1983. Flat dilatometer testing in Florida USA. *Proc. Int. Conf. on Soil and Rock Investigation by In-situ Testing,* Paris, France, **2**:251-255.

Da-Cruz, P.T., 1963. Shear strength characteristics of some residual compacted clays. *Proc. 2nd Panamerican Conf. Soil Mech. Found. Eng.*, **1**:73-102.

Darcy, H.,1856. *Historie Des Foundatanies Publique de Dijon,* Dalmount. 590-594.

Delege, P. and Lefebvre, G., 1984. Study of the structure of a sensitive Champlain clay and its evolution during consolidation. *Can. Geotech. J.*, **21**:21-25.

Demartinecourt, J.P. and Bauer, G.E., 1983. The modified borehole shear device, *Geotech. Test. J. Am. Soc. Test. Mater.*, **6**(1):24-29.

Dick, R.C., 1982. Discussion on paper The Plummet balance — A potential tool for subsieve particle size analysis by Nagaraj and Sivapullaiah. *Geotech. Test. J.*, **5**(2):56.

Dickson, J.W. and Smart, P., 1978. Some interactions between stress and microstructure of Kaolin. *Modification of Soil Structure* (Editors) Emerson, Bond and Texter.

Donald, I.B., Jordan, D.O., Parkere, R.J. and Toh, C.T., 1977. The vane test – a critical appraisal. *Proc. 9th Int. Conf. on Soil Mechanics and Foundation Engg.* Tokyo **1**:81-88.

Durgunoglu, H.T. and Mitchell, J.K., 1975. Static penetration resistance of soils. *Proc. Am. Soc. Civ. Eng. Speciality Conf. on In-situ Measurement of Soil Properties,* Raleigh, N.C., **1**:151-188.

Edlefsen, N.E. and Cole, R.C., 1936. A comparison of the specific gravity balance and the pipette methods of determining density of soil suspension. *Soil Sci,* **42**(2):131-137.

Ekman, V.E., 1959. Determining the lower plastic limit of soil by pressing, *Soils Found.* (Transl) **2**:74-77.

Feda, J., 1982. *Mechanics of Particulate Materials — The Principles.* Elsevier, Amsterdam pp.477.

Federico, A., 1983. Relationships ($C_u - w$) and ($C_u - s$) for remoulded clayey soils at high water content. *Riv. Ital. Geotech.* **XVII**(1):38-41.

Fernuik, N. and Haug, M., 1990. Evaluation of in-situ permeability testing methods. *Proc. Am. Soc. Civ. Eng. Geotech. Eng.* Div., **116**(2):297-311.

Folks, d.J. and Crooks, J.H.A., 1985. Effective stress paths and yielding in soft clays below embankments, *Can. Geotech. J.*, **22**:357-374.

Fredlund, D.G., Hasan, J.U. and Filson, H.L., 1980. The prediction of total heave. *Proc. 4th Intl. Conf. on Expansive Soils,* Denever, Colorado, USA.

Fredlund, D.G., Morgenstern, N.R. and Widger, A., 1978. Shear strength of unsaturated soils. *Can. Geotech. J.*, **15**(3):313-321.

Fredlund, D.G. and Morgenstern, N.R., 1977. Stress-strain variables for unsaturated soils. *J. Geotech. Eng.*, Am. Soc. Civ. Eng., **103**(5):447-466.

Gibbs, J.H. and Holtz, W.G., 1956. Engineering properties of expansive clays. *Transactions Am. Soc. Civ. Engrs.* **121**:641-643.

Gens, A., 1982. Stress-strain and strength characteristics of a low plasticity clay. Ph.D. thesis, University of London.

Gouy, G., 1910. Surla constitution de la charge electrique a la surface d'un electrolyte. *Anniue Physique (Paris) Serie 4* **9**:457- 468.

Grace, H. and Henry, J.K.M., 1957. Discussion on the planning and design of the new Hong Kong Airport. *Proc. Inst. Civ. Eng.* (London), **7**:305.

Griffiths, F.J., Joshi, R.C. and Nagaraj, T.S., 1988. Removal of cementation bonds in stressed overconsolidated clays. Geotechnical Testing, *J., Am. Soc. Test. Mater.*, **11**(4):227-232.

Griffiths, F.J. and Joshi, R.C., 1989. Change in pore size distribution due to consolidation of clays. *Geotechnique*, **39**(1):159-167.

Grim, R.E., 1969. *Clay Mineralogy*, Mc Graw Hill, New York, 2nd Edition.

Habib, P. and Luong, M.P., 1978. Soils pulerulents sous changement cyclique. Materiaux et. Structures sous Chargement Cyclique. Assoc. Amicale des Ingenieurs Anciens Eleves de l'Ecole Nationale des ponts et. chaussees (Palaiseau 28-29 Sept) (In French), 49-79.

Handy, R.L., 1975. Measurement of in-situ shear strength. *Proc.* Am. Soc. Civ. Eng. Conf. on In-situ Measurement on Soil Properties, Raliegh, NC, **2**:143-149.

Handy, R.L. and Fox, N.S., 1967. A soil bore hole direct shear test device. *Highway Res. News No.* **27**:42-57.

Hansbo, S., 1957. A new approach to the determination of the shear strength of clay by fall cone test. *Swedish Geotech Institute Proc.* **14**:1-48.

Hanzawa, H., Itoh, Y. and Suzuki, K., 1979. Shear characteristics of a quick sand in the Arabian Gulf. *Soils and Foundations*, **19**, 4: 1-15.

Head, K.H., 1986. *Mannual of Soil Laboratory Testing Vol.3, Effective Stress Tests.* Pentech Press, London, 743-1238.

Henkel, D.J., 1959. The relationships between the strength, pore water pressure and volume change characteristics of saturated clays, *Geotechnique*, **9**(1):119-135.

Henkel, D.J., 1960. The shear strength of saturated remoulded clays. *Proc. Res. Conf. on Shear Strength of Cohesive Soils*, Am. Soc. Civ. Eng., Colorado 533-544.

Hettler, A. And Vardoulakis, I., 1984. Behaivour of dry sand tested in a large triaxial apparatus, *Geotechnique*, **34**(1):183-198.

Hilf, J.W., 1956. An investigation of pore water pressure in compacted cohesive soils, Technical Memorandum 654, US Dept. of Interior Bureau of Reclamation, Denver, Colorado.

Hodek, R.J. and Lovell, C.W., 1979. Soil aggregates and their influence on compaction and swelling, Transportation Research Record No. 733 :94-99.

Hogentogler, C.A., 1936. Essentials of soil compaction, *Proc. Highway Research Board*, NRC, Washington DC :309-316.

Holtz, R.D., Jamiolokowski, M.B. and Lancellotta, R., 1986. Lessons from oedometer tests on high quality sample. Proc. Am. Soc. Civ. Engrs. *Geotech Engg. J.*, **112**(8):768-776.

Holtz, W.G. and Gibbs, H.J., 1956. Shear strength of pervious gravelly soils. *Proc. Am. Soc. Civ. Engrs.* Paper No. 867.

Hough, B.K., 1957, Basic Soil Engineering, Ronald, New York, pp.513.-

Houston, W.N. and Mitchell, J.K., 1969. Property interrelationships in sensitive clays. *J. Soil Mech. Found. Div.* Am. Soc. Civ. Eng., **95**(4):1032-1062.

Hovayni, P., 1958. A new grooving tool. *Geotechnique*, **8**:79.

Hvorslev, M.J., 1937. Uber die Festigkeitseigenschaften Gestorter Bindiger Boden. Denmarks Naturvidenskabelige Samfund. *Ingeniorvidenskabelige Skrifter*, A, No. 45.

Ishihara, K., Tatsuoka, F. and Yasudar, S., 1975. Undrained deformation and liquefaction of sand under cyclic stresses. *Soils and Foundations*, Japan, **15**(1):29-44.

Ishihara, K. and Okoda, S., 1978. Yielding of overconsolidated sand and liquefaction model under cyclic stresses. *Soils and Foundations*, **18**(1):57-72.

Jain, S.K., 1981. *Fundamental Aspects of the Normality Rule*, Engineering Publications, Blacksburg, Virginia.

Jamiolkowski, M., Ladd, C.C., Germaine, J.T. and Lancellolta, R., 1985. New developments in field and laboratory testing of soils. Theme Lecture. *Proc. 11th Int. Conf. on Soil Mech. and Found. Eng.* San Francisco, Balkema **1**:57-114.

Jardine, R.J., 1985. Investigation of pile-soil behaviour with special reference to the foundations of offshore structures. Ph.D. thesis, University of London.

Jennings, J.E., 1961. A revised effective stress law for use in the prediction of the behaviour of unsaturated soils. *Proc. Conf. Pore Pressure and Suction in Soils,* London, Butterworths :26-30.

Jones, D.E. and Holtz., W.G., 1973. Expansive soils — The hidden disaster. *Civil Eng.* Am. Soc. Civ. Eng., **43**(8):49-57.

Josa, A., Alonso, E.E. and Lloret, A., 1987. Stress-strain behaviour of partly saturated soils. *Proc. 9th European Conf. Soil Mech. Found. Eng.*, Dublin, **2**:561-564.

Jose, B.T., Sridharan, A. and Abraham, B.M., 1989. Log – Log method for determination of preconsolidation pressure, *Geotechnical Testing Journal,* Am. Soc. Test. Mater., **12**(3).

Judson, S., Deffeyes, K.S. and Hargraves, R.B., 1976. *Physical Geology.* Prentice Hall, Englewood Cliffs, N.J., pp.560.

Karube, D., Shibuya, S., Baba, T. and Kotera, Y., 1988. Analysis of a vane test based on effective stress. In A.F. Richards (Editor) Vane Shear Strength Testing in Soils : Field and Laboratory Studies. Am. Soc. Test. Mater., Spl. Tech. Publ., 1014:131-149.

Kenney, T.C., 1959. Discussion on Geotechnical properties of Glacial Lake clays. *Proc. Am. Soc. Civ. Eng.*, Soil Mech. Found. Div., **85**(3):67-79.

Kenney, T.C., 1960. Discussion on A preliminary study of geotechnical properties of varved clays as reported in Canadian Engineering Case Records. *Proc. 14th Can. Soil Mech. Conf.*, Ottawa, 242-251.

Kenney, T.C., 1963. Atterberg limits, Correspondance Geotechnique, 13:159-162.

Kenney, T.C., 1968. A review of recent research on strength and consolidation of soft and sensitive clays. *Canadian Geotechnical Journal,* **5**:97-119.

Kenney, T.C., Moum, J. and Berre, T., 1968. An experimental study of bonds in a natural clay. *Proceedings of 1st Geotechnical Conference,* Oslo, 65-69.

Kenney, T.C. and Landva, A., 1965. Vane - triaxial apparatus. *Proc. 6th Int. Conf. on Soil Mech. and Found. Eng.*, Montreal, Canada, **1**:269-272.

Klausner, Y., 1991. *Fundamentals of Continuum Mechanics of Soils,* Springer-Verlag, London Ltd. :8-9.

Klausner, Y. and Shainberg, I., 1967. Consoldiation properties of arid-region soils. *Proc. 3rd Asian Reg. Conf. Soil Mech. Found. Eng.*, Haifa **1**:17-19.

Klausner, Y. and Shainberg, I., 1971. Consolidation properties of Na^+ and Ca^{++} adsorbed montmorillonites. *Proc. 4th Asian Reg. Conf. Soil Mech. Found. Eng.*, Bangkok, **1**:373-378.

Komornik, A. and David, D., 1969. Prediction of swelling pressure of clays, *Proc. Am. Soc. Civ. Engrs.* **95**(1):209-215.

Koppula, S.D., 1981. Statistical estimation of compression index. *Geotech. Test. J.*, **4**(2):68-73.

Kozeny, J., 1927. Uber kapillare teitung des wassers in boden. Wein, Akad Wiss., 136 Part 2a:271.

Krishnamurthy, M., Nagaraj, T.S. and Sridharan, A., 1979. Laboratory method for prepartion of layered soil system. *Geotech Test. J.*, Am. Soc. Test. Mater. **2**(3):163-167.

Lacasse, S., 1985. In-situ determination of deformation parameters, Panel Discussion Section 2A. *Proc. 11th Int. Conf. on Soil Mech. and Found. Eng.* San Francisco, Vol.**5**.

Lacasse, S. and Lunne, T., 1988. Calibration of dilatometer correlations. In J. de Ruiter (Editor) *Penetration Testing 1988,* ISOPT-1 Balkema, Rotterdam, **1**:539-548.

Ladd, C.C. and Lambe, T.W., 1963. The shear strength of undisturbed clay determined from undrained tests. Am. Soc. Test. Mater., Spec. Tech. Publ., **361**:342-370.

Lambe, T.W., 1951. *Soil Testing for Engineers,* Wiley, New York.

Lambe, T.W., 1960a. A mechanistic picture of shear strength in clay. *Proc. Am. Soc. Civ. Engr. Res. Conf. on Shear Strength of Cohesive Soil,* Boulder, Colorado :555-580.

Lambe, T.W., 1960. Structure of compacted clay, Transactions Am. Soc. Civ. Eng., **125**:682-705.

Lambe, T.W., 1973. General report on "Upto date methods of investigating the strength and deformability of soils. *Proc. 8th Int. Conf. Soil Mech. Found. Eng.,* Moscow, **3**:3-43.

Lambe, T.W. and Whitman, R.V., 1969. *Soil Mechanics,* John wiley and Sons Inc., NY, p.320.

Lapierre, C., Leroueil, S. and Locat,J., 1989. Mercury intrusion and permeability of Louiseville clay. Canadian Geotechnical Conference, 23-35.

Larsson, R. and Sallfors, G., 1986. Automatic continuous consolidation testing in Sweden. Consolidation of Soils, Teting and Evaluation. *Am. Soc. Test. Mater.,* Spl. Tech. Publ. **892**:299-328.

Leonards, G.A., 1962. *Foundation Engineering.* McGraw Hill, New York, pp.1136.

Leonards, G.A. and Altschaeffe, A.G., 1964. Compressibility of clay. J. of Soil Mech. and Foundation Division, Am. Soc. of Civil Engineers, 90(5):133-155.

Little, A.L., 1969. The engineering classification of residual tropical soils. *Proc. Spl. Sess. on Engg. Properties of Lateritic Soils.* Spl. 7th Int. Conf. Soil Mech. Found. Eng., Mexico City **1**:1-10.

Liu, T.K. and Thornburn, T.M., 1964. Study of the reproducibility of Atterberg limits. *Highway Res. Rec.,* **63**:2-28.

Lloret, A. and Alonso, E.E., 1985. State surfaces for partly saturated soils. *Proc. 11th Int. Conf. on Soil Mech. Found. Eng.,* San Francisco, **2**:557-562.

Lo, K.Y., 1972. The influence of mechanical disturbance on consolidation of clays. *Proceedings of South East Asian Regional Conference on Soil Mechanics and Foundation Engineering,* Hong Kong, **1**:223-232.

Lowe III, J., Jonas, E. and Obrician, V., 1969. Controlled gradient consolidation test. *J. of Soil Mech. and Foundation Dn.,* Am. Soc. Civ. Eng., **95**(1):77-97.

Lumb, P. and Holt, J.K., 1968. The undrained strength of a soft marine clay from Hong Kong. *Geotechnique,* **18**:25-36.

Luong, M.P., 1980. Stress-strain aspects of cohesionless soils under cyclic and transient loading. In: *Proc. Int. Sym. Soils under Cyclic and Transient Loading,* Sivansea. Pande, G.N. and Zienkiewicz (Editors), Balkema, Rotterdam. **1**:77-87.

Lutenegger, A., Remmes, B.D. and Handy, R.L., 1978. Borehole shear test for stiff soil. *Proc. Am. Soc. Civ. Eng.,* Geotech. Eng. Div., **104**(11):1403-1407.

Lutenegger, A.J., and Saber, R.T., 1988. Determination of collapse potential of soils. *Geotechnical Testing Journal,* Am. Soc. Test. Mater, : **11**(3):173-178.

Lutenegger, A.J. and Kabir, M.G., 1988. Dilatometer C-reading to help determine straitigraphy, In: J. De Ruiter (Editor), *Penetration Testing 1988, ISOPT-1,* Balkema, Rotterdam, **1**:549-554.

Madhav, M.R. and Ramakrishna, K.S., 1977. Undrained modulus from vane shear test. Proc. Am. Soc. Civ. Eng., *Geotech. Eng.,* **103**(11):1337-1340.

Mahmoud, M., 1988. Vane testing of soft-clays, *Ground Eng.,* **21**(7):36-40.

Marchetti, S., 1980. In-situ tests by flat dilatometer. *Proc. Am. Soc. Civ. Eng.,* Geotech Eng. Div., **106**(3):299-321.

Marsal, R.J., 1969. A statistical model for granular materials. Spl. Sess. 13, 7th Int. Conf. Soil Mech. Found. Eng., Mexico, 115-122.

Marshall, T.J., 1956. A plummet balance for measuring the size distribution of soil particles, *CSIRO Abstr.* :142-147.

Marsland, A. and Quaterman, R.S., 1982. Factors affecting the measurements and interpretation of quasi-static penetration tests in clays. *Proc. 2nd European Symp. on Penetration Testing ESOPT-II,* Amsterdam, 697-702.

Matsuoka, H., 1982. A stress strain model for granular materials considering the mechanism of fabric change. In *Mechanics of Granular Materials – New models and constitutive relations.* Jenkins, J.T. and Satake,M., (Editors) Elsevier, Amsterdam, 99-116.

Matyas, E.L. and Radhakrishna, H.S., 1968. Volume change characteristics of partly saturated soils. *Geotechnique,* **18**(4):432-448.

McNeal, B.L., Norwell, W.A. and Coleman, N.T., 1966. Effect of solution composition on the swelling of extracted soil clays. *Soil Sci. Soc. Am.;* **30**:313-317.

Menzies, B.K., 1988. A computer controlled hydraulic triaxial testing system. In:Advanced Triaxial Testing of Soil and Rock, *Am. Soc. Test. Mater.,* Spec. Tech. Publ., **977**:82-94.

Menzies, B.K. and Merrifield, C.M., 1980. Measurement of shear stress distribution on the edges of a shear vane blade. *Geotechnique,* **30**:314-318.

Mesri, G. and Godlewski, P.M., 1977. Time- and stress- compressibility interrelationship. *Proc. Am. Soc. Civ. Eng.,* Geotech Eng. Div., **103**(5):417-430.

Mesri, G. and Olsen, R.E., 1971. Mechanisms controlling the permeability of clays. *Clays and Clay Minerals,* **19**:151-157.

Metcalf, J B. and Townsend, J.B., 1960. A preliminary study of the geotechnical properties of varved clays as reported in Canadian Engineering Case Records. *Proc. 14th Canadian Soil Mechanics Conf.,* Ottawa, :203-225.

Milovic, D.M., 1970. Effect of sampling on some soil characteristics. Sampling of Soil and Rock, *Am. Soc. Test. Mater.,* Spec. Tech. Publ., **483**:164-179.

Mitchell, J.K., 1976. *Fundamentals of Soil Behaviour.* John Wiley, New York, pp.422.

Mitchell, J.K., Campanella, R.G. and Singh, A., 1968. Soil creep as a rate process. *J. Soil Mech. Found. Eng. Div.,* Am. Soc. Civ. Eng., **94**(1):231-253.

Mitchell, J.K., Hooper, D.R. and Campanella, R.G., 1965. Permeability of compacted clay. *J. Soil Mech. Found. Eng. Div.,* Am. Soc. Civ. Eng., **91**(4):41-65.

Moelwyn-Hughues, E.A., 1957. *Physical Chemistry.* Pergamon Press, London.

Mogami, T., 1965. A statistical approach to the mechanics of granular materials. *Soils and Foundations.* **5**(2): 26-36.

Mohan, D. and Goel, R.K., 1959. Correspondance, The one point method of determining the value of liquid limit of soil. *Geotechnique,* **9**(3):143-144.

Moum, J. and Rosenquist, I.Th., 1967. On the weathering of young marine clay. *Proc. 4th Int. Conf. on Soil Mech. Found. Eng.,* London. **1**:77-79.

Mroz, Z., Norris, V.A. and Zienkiewicz, O.C., 1979. Application of an anisotropic hardening model in the analysis of elasto-plastic deformation of soils. *Geotechnique,* **19**(1):1-34.

Nagaraj, T.S., 1963. A remoulding device for soil samples. *J. Cent. Board Irr. Power,* **20**(2):139-143.

Nagaraj, T.S., 1981. Effective stress considerations in non- interacting and interacting systems. *Geotech. Eng. J.* Asian Institute of Technology, Bangkok.

Nagaraj, T.S., 1992. Analysis and prediction of soil and concrete behaviour - Particualte and non particulate considerations, Vol.1. Doctor of Science Thesis, Indian Institute of Science, Banglaore, Section 3.1-3.5.

Nagaraj, T.S., 1993. *Principles of Testing of Soils, Rocks and Concrete,* Elsevier Science Publishers, Amsterdam, pp.708.

Nagaraj, T.S., Pandian, N.S. and Narasimha Raju, P.S.R., 1993. Stress-state-permeability relationships for fine grained soils, *Geotechnique,* **43**(2):333-336.

Nagaraj, T.S., Sreenivasulu, V. and Narasimha Rao, S., 1976. Stress deformation and Failure mechanisms in remoulded and compacted kaolinite clays. *Proc. 5th Int. Conf. on Soil Mech. Found. Eng.*, Budapest, **1**:557-569.

Nagaraj, T.S., Srinivasa Murthy, B.R. and Bindumadhava, 1987. Liquid limit determination further simplified. Geotech. *Test. J.*, Am. Soc. Test. Mater., **10**(3):142-145.

Nagaraj, T.S., Srinivasa Murthy, B.R. and Vatsala, A., 1990. Discussion on change in pore size distribution due to consolidation of clays by Griffiths and Joshi, *Geotechnique*, **40**(2).

Nagaraj, T.S., Srinivasa Murthy, B.R. and Vatsala, A., 1991. Prediction of soil behaviour Part III - Cemented saturated soils. *Indian Geotechnical J.*, **21**(2):169-186.

Nagaraj, T.S. and Bindumadhava, 1992. Analysis of compcation characteristics of fine grained soils. Indian Geotech. Conf. 1992, Calcutta, **1**:441-444.

Nagaraj, T.S. and Jayadeva, M.S., 1981. Reexamination of one point methods of liquid limit determination. *Geotechnique*, **31**(3):413-425.

Nagaraj, T.S. and Jayadeva, M.S., 1983. Critical reappraisal of plasticity index of soils. *Proc. Am. Soc. Civ. Eng. Geotech. Eng. Div.*, **109**(7):994-1000.

Nagaraj, T.S. and Sivapullaiah, P.V., 1981. Plummet Balance - A potential tool for subsieve particle size analysis, *Geotech. Test. J.*, **4**(1):36-40.

Nagaraj, T.S. and Srinivasa Murthy, B.R., 1983. Rationalization of Skempton's compressibility equation. *Geotechnique*, **33**(4):433- 443.

Nagaraj, T.S. and Srinivasa Murthy, B.R., 1985. Compressibility of partly saturated soils, *J. of Geotech. Engg.*, ASCE, **111**(7):937-942.

Nagaraj, T.S. and Srinivasa Murthy, B.R., 1985. Prediction of preconsolidation pressure and recompression index of soils. *Geotech. Testing J. Am. Soc. Test. Mater.*, **8**(4):199-203.

Nagaraj, T.S. and Srinivasa Murthy, B.R., 1985. Rational approach to predict swelling soil behaviour. Transportation Research Record, TRB No. 1032:1-7.

Nagaraj, T.S. and Srinivasa Murthy, B.R., 1986. A critical reappraisal of compression index equations. *Geotechnique*, **36**(1):27-32.

Nagaraj, T.S. and Srinivasa Muthy, B.R., 1986a. Prediction of compressibility of overconsolidated uncemented soils, *J. Geotech. Engg.*, ASCE, **112**(4):484-488.

Nagaraj, T.S. and Srinivasa Murthy, B.R., 1988. Consistency Limits - Principles and Potentials. Ch 2, Vol.3. Geotechnical/ Ocean Eng. Handbook of Civil Engineering Practice Eds P N Chermisinoff and N P Cheemisinoff Technomic Pub. USA :23-51.

Nagaraj, T.S. and Tiwari, S.S., 1985. Prediction of strength of partly saturated soils. Tropical S'85. First Int. Conf. on Geomechanics in Tropical Laterite and Saprolitic Soils, Brazil, 359-370.

Nagaraj, T.S., Srinivasa Murthy, B.R., Vatsala, A. and Joshi, R.C., 1990. Analysis of compressibility of sensitive soils. *J. Geotec. Eng.*, Am. Soc. Civ. Eng. **116**(1):105-119.

Narasimha Rao, S. and Kodandaramaswamy, K., 1982. Method to predict ultimate compression in clays. *Proc. Am. Soc. Civ. Eng.*, Geotech Eng. Div., **108**(2):310-314.

Nemat Nasser, S. and Mehrabadi, M., 1982. Stress and fabric in granular masses. *In Mechanics of Granular Materials — New models and constitutive relations*, Jenkins, J.T. and Satake, M., (Editors) Elsevier, Amsterdam, 1-8 .

Nelson, J.D., Brand, E.W., Moh, Z.C. and Mason, I.D., 1971. The use of residual stress to define sample quality. Proc. 4th Asian Conf. Int. Group on Soil Sampling on Quality in Soil Sampling, 82-87.

Newland, P.L. and Allely, B.H., 1956. Results of some investigations of two sensitive clays. Proc. 2nd Aust-NZ Conf. Soil Mech., 39-45.

Nishida, Y., 1956. A brief note on compression index of soil. *Proc. Am. Soc. Civ. Engrs.* Paper 1027, **82**(3):1-14.

Norman, L.E.J., 1958. A comparison of values of liquid limit determined with apparatus having bases of different hardness. *Geotechnique*, **8**:79-83.

Nuyens, J.G.E. and Kockaerts, R.F., 1967. Reliable technique for determining plastic limit. *Mater. Res. Stand.*, **7**:295-299.

Oda,M., 1972. Initial fabrics and their relations to mechanical properties of granular materials. Soils and Foundations. **12**(1):17-36.

Okumura, T., 1971. The variation of mechanical properties of clya samples depending on the degree of disturbance. Proc. Spl. Session, Quality in Soil Sampling. Fourth Asian Conf. Int. Soc. Soil Mech. and Found. Engg., Bangkok, 73-81.

Olson, R.E., 1963. Effective stress theory of soil compaction, *J. Soil Mech. and Found. Engg. Div.*, Am. Soc. Civ. Eng., **89**(2):27-45.

Olson, R.E. and Mesri, G., 1970. Mechanisms controling the compressibiility of clay. Am. Soc. Civ. Eng., Soil Mech. Found. Eng. Div., **96**(6):1863-1878.

Olsson, J., 1921. Method of investigating the strength properties of clays as used by Swedish State Railways. *Tek. Ukebl.*, **41**(37):310-311,340.

Osterman, J., 1960. Notes on shearing resistance of soft clays. *Acta Polytech*, Scand.

Oswald, R.H., 1980. Universal compression index equation. Geotech. Eng. Div., *Proc. Am. Soc. Civ. Eng.*, **106**(11):1179-1200.

Palmer, A.C. and Pearce, J.A., 1973. Plasticity theory without yield surfaces. Symp. on Plasticity and Soil Mech., Cambridge, 188-200.

Pandian, N.S., Nagaraj, T.S. and Siva Kumar Babu, G.L., 1992. Generalized state parameter for partly saturated soils. *J. of Geotechnical Engineering*, Am. Soc. Civ. Eng., **118**(4):622-627.

Pandian, N.S., Nagaraj, T.S. and Sivakumar Babu, G.L., 1993a. Behaviour of tropical clays. Part 1 - Index properties and microstructural considerations. *J. Geotech. Eng.*, Am. Soc. Civ. Eng., **119**(5):826-839.

Pandian, N.S., Nagaraj, T.S. and Sivakumar Babu, G.L., 1993b. Behaviour of tropical clays. Part 2 - Engineering behaviour. *J. Geotech. Eng.*, Am. Soc. Civ. Eng., **119**(5):840-861.

Pandian, N.S., Sridharan, A. and Kumar, K.S., 1992. A new method for the determination of coefficient of consolidation. *Geotech. Testing. J.*, Am. Soc. Test. Mater., **15**(1):74-79.

Pandian, N.S. and Nagaraj, T.S., 1990. Critical reappraisal of colloidal activity of clays. *Proc. Am. Soc. Civ. Eng. Geotech Eng.* Div., **116**(2):285-296.

Parry, R.H.G., 1960. Triaxial compression and extension tests on remoulded saturated clay. *Geotechnique*, **10**:166-180.

Parry, R.H.G. and Amerasinghe, S.F., 1973. Components of deformation in clays. Symp. on Plasticity and Soil Mechanics, Cambridge, 108-126.

Paul Tamas, 1984. Discussion on Critical reappraisal of plasticity index of soils. *Am. Soc. Civ. Eng.*, **10**(9):1367.

Pellettier, J.H., Olson, R.E. and Rixner, J.J., 1979. Estimation of consolidation properties of clay from field observations, *Geotech Testing J.*, Am. Soc. Test. Mater., **2**(1):34-44.

Pietruszczak, S. and Poorooshasb, H.B., 1985. On modelling of cyclic behaviour of soils: Developments in Soil Mechanics and Foundation Engg. - 2. *Stress-strain Modelling of Soils.* Edited by P.K. Banerjee, Elsevier Applid Science Publishers, London - New York :139-184.

Poorooshasb, H.B. and Pietruszczak, S., 1985. On yielding and flow of sands: A generlised two surface model, *Computers and Geotechnique*, **1**:33-58.

Prakash, S., 1981. *Soil Dynamics*, McGraw Hill Book Co., Chapter on Liquifaction of soils, 274-336.

Prevost, J.H., 1980. Constitutive theory for soils. Proc. Conf. on Limit Equilibrium, Plasticity and Stress-strain in Geotech. Eng., Am. Soc. Civ. Eng. :745-814.

Proctor, R.R., 1933. Fundamental principles of soil compaction, Engineering News Record, 111:9-10,12-13.

Quigley, R.M. and Thomson, C.D., 1966. The fabric of anisotropically consolidated sensitive marine clay. *Canadian Geo J.* **3**(1):61-73.

Ramiah, B.K., 1959. Time effects of consolidation properties of clays. Ph.D. thesis, Purdue University.

Raymond, C.P. and Wahls, H.E., 1976. Estimating one dimensional consolidation including secondary compression of clay loaded from overconsolidated to normally consolidated state. *Transportation Res. Board Spl.* Report **163**:17-23.

Raymond, G.P., Gaskin, P.N. and Addo-Abedi, F.Y., 1979. Repeated compressive loading on Leda Clay, *Can. Geotech. J.*, **16**(1):1-10.

Robertson, P.K., Campanella, R.G. and Gillespie, D., 1988. Excess pore pressures and the flat dilatometer test. In:J. De Ruiter (Editor), *Penetration Testing 1988, ISOPT-1,* Balkema, Rotterdam, **1**:567-576.

Robertson, P.K. and Campanella, R.G., 1983. Interpretation of cone penetration tests, Part 1: Sand. *Can. Geotech. J.*, **20**:718-733.

Roscoe, K.H., Schofield, A.N. and Wroth, C.P., 1958. On the yielding of soils. *Geotechnique*, **8**(1):22-53.

Roscoe, K.H. and Burland, J.B., 1968. On the generalised stress- strain behaviour of wet clay. In: *Engineering Plasticity Heyman and Leekie* (Editors).

Rosenquist. I. Th., 1955. Investigations in the clay – electrolyte – water system. Pub. 9. Norwegian Geotechnical Institute :1-125.

Rothenberg, L. and Bathrurst, R.J., 1990. Analytical study of induced anisotropy in idealised granular materials, *Geotechnique,* **39**(4):601-614.

Rowe, P.W., 1962. The stress-dilatancy relation for static equilibrium of an assembly of particles in contact. *Proc. Roy. Soc. A.*, **269**, 500-527.

Rowe, P.W., 1972. The relevance of soil fabric to site investigation practice. *Geotechnique,* **22**:193-300.

Rowe, P.W. and Barden, L., 1966. Importance of free ends in triaxial testing. *Proc. Am. Soc. Civ. Eng.*, Soil Mech. Div., **90**(1):1-27.

Russel, E.R. and Mickle, J.L., 1970. Liquid limit values of soil moisture tension. Proc. Am. Soc. Civ. Eng., *J. Soil Mech. Found. Eng. Div.*, **96**():967-987.

Saito, T. and Yanai, K., 1969. Some swelling characteristics of compacted clays. Transactions Japan Soc. Civ. Engr., 172:11-18.

Sandbaekken, G., Berre, T. and Lacasse, S., 1986. Oedometer testing at Norwegian Geotechnical Institute. Consolidation of Soil, Testing and Evaluation, *Am. Soc. Test. Mater.*, Spec. Tech. Publ., **892**:329-353.

Satake, M., 1982. Fundamental Quantities in the graphical approach to granular materials. In *Mechanics of Granular Materials - New models and constitutive relations,* Jenkins, J.T. and Satake, M., (Editors) Elsevier, Amsterdam, 9-20.

Schmertmann, J., 1975. The borehole shear test. Proc. Am. Soc. Civ. Eng. Conf. on In-situ Measurements of Soil Properties, New York, **2**:124-127.

Schofield, A.N. and Wroth, C.P., 1968. *Critical State Soil Mechanics.* Mc Graw Hill, London.

Scott, R.F., 1963. *Principles of Soil Mechanics.* Addison-Wesley Pub. Co., Mass.

Scott, R.F., 1989. Constitutive relations for soil: Present and future. In: *Constitutive Equations for Granular Non-cohesive Soils.* Saada, A. and Bianchini, G. (Editors). Balkema, Rotterdam 723-725.

Seed, H.B., Woodward, R.J. and Lundgren, R., 1964. Fundamental aspects of Atterberg limits. Proc. Am. Soc. Civ. Eng., *J. Soil Mech. Found. Eng.*, **90**(6):75-105.

Seed, H.B., Woodward, R.J. and Lundrgen, R., 1962. Prediction of swelling potential for compacted clays. *Proc. Am. Soc. Civ. Engrs,* **88**(3):53-87.

Seed, N.B. and Idriss, I.M., 1974. Simplified procedure for evaluating soil liquefacton potential. *J. of Soil Mech. Found. Eng. Div.*, Am. Soc. Civ. Eng., **97**(9):1249-1273.

Senneset, K., 1989. A new odedometer with splitted ring for the measurement of lateral stress. *Proc. 12th Int. Conf. on Soil Mech. Found. Eng.*, Rio de Janerio, Balkema, Rotterdam, **1**:115-118.

Shahinpoor, M., 1982. Frequency distribution of voids in randomly packed monogrganular layers. In *Mechanics of Granular Materials - New models and constitutive relations.* Jenkins, J.T. and Satake, M., (Editors) Elsevier, Amsterdam, 173-186.

Shainberg, I., Bresler, E. and Klausner, Y. (1971). Studies of Na-Ca Montomorillonite Systems - I. The Swelling Pressure. *Soil Science J.*, **111**:214-219.

Shainberg, I. and Klausner, Y., 1971. Consolidation properties of Na^+ and Ca^+ adsorbed montmorillonites. *Proc. 4th Asian Reg. Conf. on SM&FE,* Bangkok, **1**:p.373.

Sheeran, D.E. and Krizek, R.J., 1971. Preparation of homogenous soil samples by slurry consolidation. *J. Mater.*, **6**(2):356-373.

Sherwood, P.T. and Ryley, M.D., 1970. An investigation of a cone penetrometer method for determination of liquid limit. *Geotechnique,* **20**:203-208.

Silvestri, V. and Aubertin, M., 1988. Anisotropy and in-situ vane tests. In:A.F. Richards (Editor), Vane Shear Strength Testing in soils: Field and Laboratory Studies, *Am. Soc. Test. Mater.*, Spec. Tech. Publ., **1014**:88-103.

Sinha, N.K., 1981. Deformation behaviour of ice like materials in engineering applications. Int. Sym. on Mech. of Structural Media. Octawa, 1:419-430.

Sivakumar Babu, G.L., Pandian, N.S. and Nagaraj, T.S., 1993. A reexamination of permeability index of clays. *Canadian Geotechnical Journal,* 30(1):187-191.

Skempton, A.W., 1944. Notes on compressibility of clays. *Qly. J. Geo. Soc.* London :119-135.

Skempton, A.W., 1953. Colloidal activity of clays. *Proc. 3rd Int. Conf. Soil Mech. Found. Eng.* **1**:57-61.

Skempton, A.W., 1954. Discussion on the structure of inorganic soil. *Proc. Am. Soc. Civ. Eng., Soil Mech. Found.* Div., **80**(1):19-22.

Skempton, A.W. and Henkel, D.J., 1953. The post glacial clays of the Thames Estury at Tilbury and Shellhaven. *Proc. 3rd Int. Conf. on Soil Mech. Found. Eng.*, Zurich, Switzerland, **1**:302-308.

Skempton, A.W. and Northey, R.D., 1953. The sensitivity of clays, *Geotechnique,* **3**:30-53.

Sloane, R.L. and Kell, T.R., 1966. The fabric of mechanically compacted Kaolin. *Clays and Clay Minerals,* **14**:289-295.

Smart, P., 1967. Particle arrangements in kaolin. 15th Natl. Conf. on Clays and Clay Minerals, Pergamon Press, 241-249.

Smith, R.E. and Wahls, H.E., 1969. Consolidation under constant rate of strain. *Proc. Am. Soc. Civ. Eng., Soil Mech. Found.* Div., **95**(2):519-539.

Smith, W.O., Foote, P.D. and Busang, P.F., 1929. Packing of homogenous spheres, *Phys. Rev.*, **34**:1271-1274.

Snethan, D.R., Johnson, L.D. and Patrick, D.M., 1977. An investigation of the natural microscale mechanisms that cause volume change in expansive clays. Interim Report FHWA - RD - 77 - 75, US Department of Transport.

Som, N.N., 1968. The effects of stress path on the deformation and consolidation of London Clay. Ph.D. thesis, University of London.

Sowers, G.F., 1981. Rock permeability or hydraulic conductivity - An overview. Permeability and Groundwater Contamination Transport, *Am. Soc. Test. Mater.*, Spec. Tech. Publ., **476**:65-83.

Spangler, M.G., 1953. Soil Engineering, Int. Text Book, Scranton, Penn.

Sridharan, A., 1968. Some studies on the strength of partly saturated soils. Ph D Thesis. Purdue Univ. Lafayette, Indiana, USA.

Sridharan, A., Murthy, N.S. and Prakash, K., 1987. Rectangular hyperbola method of consoliation analysis, *Geotechnique,* **37**(3):335-368.

Sridharan, A., Sreepada Rao, A. and Sivapullaiah, P.V., 1986. Swelling pressure of clays, *Geotechnical Testing Journal,* Am. Soc. Test. Mater., **9**(1):24-33.

Sridharan, A. and Jayadeva, M.S., 1982. Double layer theory and compressibility of clays. *Geotechnique,* **32**(2):133-144.

Sridharan, A. and Madhav, M.R., 1964. Time effects on vane shear strength and sensitivity of clay. *Am. Soc. Test. Mater., Proc.* **64**:958-967.

Sridharan, A. and Narasimha Rao, S., 1973. The relationship between undrained strength and plasticity index. *Geotechnical Engineering,* **4**:141-153.

Sridharan, A. and Rao, G.V., 1971. Effective stress theory of shrinkage phenomena. *Can. Geotech. J.,* **8**(4):503-513.

Sridharan, A. and Rao, G.V., 1973. Mechanisms controlling volume change of saturated clays and the role of the effective stress concept. *Geotechnique,* **23**(3):359-382.

Sridharan, A. and Rao, G.V., 1979. Shear strength behaviour of saturated clays and the role of effective stress concept, *Geotechnique,* **29**(2):177-193.

Sridharan, A. and Sreepada Rao, 1982. Rectangular hyperbola fitting method for one dimensional consolidation. *Geotech. Test. J.,* **4**(4):161-168.

Srinivasa Murthy, B.R., Nagaraj, T.S., Balakrishna, C.K. and Bindumadhava, 1985. Control of mechanical stabilization, Indian Geotechnical Conference, Roorkee, 1:171-176.

Srinivasa Murthy, B.R., Nagaraj, T.S. and Bindumadhava, 1987. Influence of coarse particles on compressibility of soils. *Proc. Int. Symp. on Prediction and Performance in Geotech. Eng.,* Calgary, Alberta, Canada, Balkema, Rotterdam, 195-200.

Srinivasa Murthy, B.R., Nagaraj, T.S. and Vatsala, A., 1986. Critical reappraisal of empirical shear strength equations. *Proc. Asian Regional Symp. on Geotechnical Problems and Practices in Foundation Engineering,* Colombo, Sri Lanka, **1**:284-292.

Srinivasa Murthy, B.R., Vatsala, A. and Nagaraj, T.S., 1988. Can Cam-clay model be generalised? *J. Geotech. Eng.,* Am. Soc. Civ. Eng., **114**(5):601-613.

Srinivasa Murthy, B.R., Vatsala, A. and Nagaraj, T.S., 1991. Revised Cam-clay model. *J. Geotech. Eng.,* Am. Soc. Civ. Eng., **117**(6):851-871.

Srinivasa Murthy, B.R. and Nagaraj, T.S., 1987. Prediction of heave in swelling soils. *Proc. 6th Int. Conf. Expansive Soils.* New Delhi, India, Dec. 1987 :201-206.

Stepkowska, T., 1982. Study in physics of clay behaviour. *Geotechnique,* Archiwum Hydrotechniki, **29**(4):469-489.

Tavenas, F., Leblond, J., Jean, P. and Leroueil, S., 1983. The permeability of natural soft clays. Part II. Permeability Characteristics, *Canadian Geotechnical J.,* **20**(4):600-645.

Taylor, D.W., 1948, reprinted in 1959. *Soil Mechanics.* Asia Publishing House, New Delhi, 1st Edition, pp.700.

Terzaghi, K., 1936. The shearing resistance of saturated soil and angle between planes of shear. *Proc. I Int. Conf. Soil Mech. and Found. Engg.,* Harvard Mass, **1**:54-56.

Terzaghi, K., 1925. Erdban Mechanik auf bodenphysikalischer Grundlage. Deuticke, Leipzig/ Vienna, pp.399.

Terzaghi, K., 1926. Simplified soil tests for subgrade and their physical significance. *Public Roads,* **7**(8):153-162.

Terzaghi, K. and Peck, R.B., 1967. *Soil Mechanics in Engineering Practice,* Wiley, New York, 2nd Ed, pp.729.

Thorley, A., Braise, Y., Calhoon, M.L., Zeman, Z.P. and Watt, W.G., 1970. Borehole instruments for economical strength and deformation in-situ testing. *Proc. Br. Geotech. Soc. Conf. on In-situ Investigations in Soils and Rocks,* London, 155-165.

Tjelta, T.L., Tieges, A.W.W. and Geise, J.M., 1987. In-situ density measurements by nuclear backscatter for an offshore soil investigation. Norwegian Geotechnical Institute Publication No.169, pp.5.

Tolls, D.G., 1990. A framework for unsaturated soil behaviour, *Geotechnique,* **40**(1):31-44.

Torstensson, B.A., 1975. Pore pressure sounding instrument. Proc. Am. Soc. Civ. Eng. Speciality Conf. on In-situ Measurement of Soil Properties, Raleigh, N.C., **2**:48-54.

Torstensson, B.A., 1977. Time dependent effects in the field vane test. *Proc. Int. Symp. on Soft Clays,* Bangkok, 387-397.

Trollope, D.H., 1960. On the structural arrangement of fine grained sediments. Conf. on Interparticle Forces in Clay-Water Electrolyte Systems, Melbourne.

Uppot, J.O. and Stephenson, R.W., 1989. Permeability of clays under organic permeants. *Proc. Am. Soc. Civ. Eng.,* Geotech. Eng. Div., **115**(1):115-131.

van Olphen, H., 1963. *An Introduction of Clay Colloid Chemistry.* Inter Science Publ., New York.

Vaughan, P.R., Maccarini, M. and Mokhtar, S.M., 1988. Indexing the engineering properties of residual soils. *Q.J.Eng. Geol.,* London, **21**:69-84.

Verwey, E.J.W. and Overbeek, J.Th. G., 1948. *Theory of the Stability of Lyophobic Colloids,* Elsiever Publ. Co., New York.

Vesic, A.S. and Clough, G.W., 1968. Behaviour of granular materials under high stresses. *Proc. Am. Soc. Civ. Engrs.* **94**(3):661-688.

Vijayavergiya, V.N. and Ghazzaly, O.I., 1973. Prediction of swelling potential for natural clays. *Proc. 3rd Int. Conf. on Expansive Soils,* Haifa, Isreal **1**:227-236.

Wadia, D.N., 1957. *Geology of India.* Macmillan, London.

Warkentin, B.P., Bolt, G.H. and Miller, R.D., 1957. Swelling Pressure of Montmorillonite. *Proc. Soil Sci. Soc. Am.,* **21**:495-497.

Wernick, E., 1977. Stresses and strains on the surface of anchors. Proc. 9th Int. *Conf. on Soil Mech. Found. Eng.,* Tokyo, Speciality Session 4.

Whyte, I.L., 1982. Soil plasticity and strength — A new approach using extrusion. *Ground Eng.,* **15**(1):16-24.

Wiesel, C.E., 1973. Some factors influencing in-situ test results. *Proc. 8th Int. Conf. on Soil Mech. Found. Eng.,* Moscow. **1**:475-479.

Wineland, J.D., 1975. Borehole shear device. Proc. Am. Soc. Civ. Eng. Conf. on In-situ Measurements of Soil Properties, New York. 1:511-522.

Wissa, A.E.Z., Christian, J.T., Davis, E.H. and Sigurd Heiberg, 1971. Consolidation at constant rate of strain. *Proc. Am. Soc. Civ. Eng.,* Soil Mech. Found. Div., **97**(10):1393-1413.

Wood, D.M., 1982. Laboratory investigations of the behaviour of soils under cyclic loading - A review. In *Soil Mechanics, Transient and Cyclic Loads.* Pande, G.N. and Zienkiewicz (Editors), John Wiley and Sons Limited.

Wroth, C.P., 1979. Correlations of some engineering properties of soils. 2nd International Conference on BOSS, Imperial College, London, 121-132.

Wroth, C.P. and Hughes, J.M.O., 1973. An instrument for the in- situ measurement of the properties of soft clays. *Proc. 8th Int. Conf. on Soil Mech. Found. Eng.,* Moscow. **1**:487-494.

Wroth, C.P. and Wood, D.M., 1978. The correlation of index properties with some basic engineering properties of soils. *Canadian Geotechnical J.,* **15**(2):137-145.

Wu, T.H., 1958. Geotechnical properties of Glacial Lake clays, *Proc. Am. Soc. Civ. Eng.,* Soil Mech. Found. Div., **84**(3):1-29.

Yong, R.N. and Nagaraj, T.S., 1977. Investigation of fabric and compressibility of a sensitive clay. *Proc. Int. Sym. Soft Clay*, Asian Institute of Tech. Bangkok, 327-333.

Yoshida, R.T., Fredlund, D.G. and Hamilton, J.J., 1983. The prediction of total heave of a slab on grade floor on Regina Clay. *Can. Geotech. J.*, **20**(1):69-81.

Young, A.G., Mc Clelland, B. and Quiros, G.W., 1988. In-situ shear testing at Sea. A.F.Richards (Editor), Vane Shear Strength Testing in Soils: Field and Laboratory Studies, Am. Soc. Test. Mater., Spec. Tech. Publ., **1014**:46-47.

Youssef Abdel-Fattah, A., 1989. The vane cone, A new device for soil shear strength measurement. *Geotech. Test. J.*, **12**(1):60-68.

Zeevart, L., 1973. Foundation Engineering for Difficult Subsoil Conditions, Van Nostrand, New York.

AUTHOR INDEX

SUBJECT INDEX

EPILOGUE

In retrospect, it is thus possible to develop a systematic approach in analysis, and prediction of soil behaviour. The consequent acquired ability to assess engineering properties viz., strength, compressibility and permeability of soils forms an important link in the overall chain of events like study of the field situations, analysis, design and construction in the practice of geotechnical engineering. How and to what extent the prediction of the engineering properties can be made, within the limits of accuracy acceptable at engineering level, has been the focal theme of presentation.

Generalized Approach

Recognizing that effective stress $(\sigma - u)$ in fine grained soils to be equal to the mobilized internal stress $(R - A)$, we have explored a means for generalization and prediction of their behaviour within the framework of a scientific base. The macroparameter that can account for the physico-chemical activity due to interparticle forces (R and A) is the liquid limit water content of fine grained soils. The water holding capacity at liquid limit is dominantly influenced by the specific surface area.

At this stage, the following interpretation of the state of soil at its liquid limit state has been possible.

At liquid limit water content of fine grained soils:

(i) The soil moisture potential, a reflection of energy level is of the same order.
(ii) This state can be linked to the consolidation/ suction pressure of the same order (5 – 6 kPa).
(iii) The shear strength is in the range of 1.7 – 2.8 kPa with the value being 2.2 – 2.4 for most of the soils.

As a consequence of the above as well as the evidence provided by pore size distribution and the observation of the same order of permeability at liquid limit water contents for fine grained soils, the complex microstructure of clays can be reasoned out to be of the same pattern. This is yet another step towards linking the effects of the microstructure of clays to the behaviour at macrolevel, beyond classification, quantification and scaling at microlevel hitherto attempted. From the above observations, it is clear that liquid limit water content forms a reference state in relation to which other states as realized due to the effects of stress can be characterized and generalized. For saturated, uncemented normally consolidated soils, an idealized system, the generalized state-effective stress relation is of the form

$$e/e_L = a - b \log (\sigma - u)$$

where (e/e_L) is the generalized soil state parameter and $(\sigma - u)$ is the conventional effective stress.

The above mode of generalization is not adequate in all cases since the inherent nature as well as the diversity of the processes dictated by stress, time and environment are responsible for the formation of a wide spectrum of soil states. Although the above generalization for the ideal state still forms the basis, it is necessary to develop appropriate predictive models to cope up with the complexity of soil systems. One possible way is to categorize the in-situ soil with increasing degrees of variation from the ideal system based on the number of independent parameters required to predict their behaviour, viz.,

(i) normally consolidated uncemented saturated soils (reference state),
(ii) overconsolidated uncemented saturated soils,
(iii) partly saturated soils,
(iv) naturally cemented saturated soils,
    a. soft and sensitive soils,
    b. stiff cemented soils,
(v) cemented, partly saturated residual soils and
(vi) layered/ fissured/ varved in-situ soils.

The independent parameters required to identify and predict the behaviour of each of these types would be progressively more as the degree of variability increases. For example, preconsolidation pressure or overconsolidation ratio would be necessary in addition to in-situ density and liquid limit of soils to model the behaviour of overconsolidated soils. Cementation bond strength or yield stress would be the additional parameter to predict soft, sensitive soils and stiff clays. Degree of saturation will be another additional parameter needed for the development of predictive physical models to encompass partly saturated residual and compacted soils. The behaviour of layered/ varved and fissured soils cannot be predicted from micromechanistic considerations alone. The minimum input parameters for assessment of engineering properties are summarized in the tabular form.

Table : Analysis and Prediction of soil Behaviour with Minimum Input Parameters

	Soil type		Parameters
(1)	Normally consolidated uncemented saturated soils. (Reference state)	—	w_L or e and σ'_v
(2)	Over consolidated uncemented saturated soils.	—	w_L, e and σ'_v
(3)	Partly saturated uncemented soils	—	w_L, e, and S_r and σ'_v
(4)	Naturally cemented saturated soils		
	a. Soft and sensitive	—	w_L, e, and S_u
	b. Stiff cemented	—	w_L, e s_u and σ'_v
(5)	Partly saturated cemented soils	—	w_L, e, S_r and s_u and σ'_v
(6)	Layered or fissured in combination with any one or more of the above complexities	—	Not possible from micro-mechanistic considerations

Prediction of Soil Behaviour

In the predictive methods developed so far, the computable parameter is the generalized state parameter (e/e_L). If the soil is partly saturated, it has been observed that $(e/e_L)\sqrt{S_r}$, as generalized state parameter, would reflect the combined effects of total stress and matrix potentials. The above parameters would suffice if particulate mechanics govern engineering behaviour. Often environmental factors cause chemical bondings between soil particles or their aggregated units imparting non-particulate characteristics to the in-situ soils. To separate the zones of dominance of nonparticulate and particulate responses it would be necessary to determine the bond strength or yield stress upto which the responses are independent of the physical state of the soil. In-situ vane shear strength, cone penetration resistance or unconfined compressive strength of undisturbed samples would be required to assess cementation bond strength or yield stress for separation of zones of nonparticulate and particulate responses.

The basis for development of predictive models is to develop correlations between the generalized state parameters and the measured engineering behaviour such as consolidation pressure, unconfined compressive strength, deviatoric stress in triaxial compression test and permeability. Once the functional relationships between the generalized state parameters and the measured responses are established the resulting phenomenological models can form the basis to predict the soil behaviour for other different states encountered in practice. The use of appropriate predictive models can be done after identification of the nature of the soil system. It has also been possible to critically examine the predictive models within the framework of critical state concepts.

Limitations

It has not yet been possible to develop similar predictive capability in the case of granular particulate media due to the fact that their behaviour is significantly affected by micro-fabric features which cannot be reflected with measurable macroparameters such as relative density, shape size and distribution of particles. Another area where extensive work has to be done to extend the generalized approaches is in in-situ soil testing. The recent developments in the in-situ testing of soils have been phenomenal. The assessment of compressibility and deformation modulus can be done by screw plate tests, pressuremeters and flat plate dilatometers. In-situ strength measurements can be made, apart from cone penetration and vane shear tests by borehole shear tests. The predictive models based on input parameters such as water content, overburden pressure and liquid limit water contents, at this stage cannot be extended to in-situ tests due to the lack of appropriate functional relationships. By extensive, planned field testing programmes at different sites of known soil conditions it would be possible to embark on this task. The functional relationships between generalized state parameters or their incremental changes and measured parameters in in-situ tests have to be developed. For example, relationship between cone resistance, incremental pressures in pressuremeter tests, material index and horizontal stress index and generalized state parameters. Another difficulty that would arise in developing appropriate correlations is in assessing the zone of influence for purposes of determining changes in water content or void ratios for computing the incremental changes in generalized state parameters.

Concluding Remarks

The geotechnical engineering practice may undergo a major change with time. Ground improvement techniques in general and soil reinforcement technique in particular using geosynthetics or geogrids will be the answer to most of the problems. It is desirable to have simple and quick methods to assess engineering properties by appropriate phenomenological models with minimum input parameters of routine tests. The approaches so far presented in this treatise are intended to achieve substantial savings in time and cost which are very much in tune with this trend. It would be possible to assess the properties of soils as reliably and as confidently as needed by the engineering profession.

APPENDIX 1

Formulation of Double Layer Equations

The basic interparticle interaction forces in a clay electrolyte suspension vary over a wide range depending on the following factors:

(a) Properties of the pore fluid;
   (1) ion concentration,
   (2) valency of adsorbed cation,
   (3) temperature and
   (4) dielectric constant of the pore fluid.
(b) Properties of the clay mineral;
   (1) specific surface of the clay particles,
   (2) charge density of the clay and
   (3) dielectric constant of the solids (for attractive forces only).

Since, the effects of each of the factors involved is difficult to isolate experimentally, a theory relating these factors and the resultant force fields, is of immense value. Such a theory was developed independently by Gouy (1910) and Chapman (1913) and is generally known as Gouy-Chapman theory. Many investigators (Verwey and Overbeek, 1948; Bolt, 1956; Babcock, 1963; Van Olphen, 1963; Sridharan and Jayadeva, 1982 etc) have given solutions in different forms for the Poisson-Boltzman equation. This will lend the necessary theoretical basis for modelling the behaviour of clays.

It is attempted in the next section to give an account of the diffuse double layer theory and derive double layer equations.

DOUBLE LAYER EQUATIONS

Assumptions

The mathematical description of the diffuse double layer has been developed for the cases of both planar and spherical surfaces. Only the planar case is treated here. The following idealizing assumptions are made (Mitchell, 1976).

(1) Ions in the double layer can be considered as point charges with no interactions between them.
(2) Charge on the particle surface is uniformly distributed.
(3) The particle surface is a plate that is large relative to the thickness of the double layer (one dimensional condition).
(4) The static dielectric constant of the medium is independent of position.

Poisson-Boltzmann Equation

Figure A.1 indicates the potential versus distance plot of a diffuse double layer.

The Poisson equation relates potential, charge and distance in the form:

$$d^2\phi/dx^2 = -(4\pi\rho/\varepsilon) \qquad \text{A.1}$$

where x is the distance from the surface

ρ is the charge density

ε is the dielectric constant of the medium

ϕ is the electrical potential at the point

The derivation of Boltzmann equation describing the variation of ion concentration with potential is given in detail by Moelwyn-Huges (1957). The equation is given as

$$n = n_{io} \exp(-v_i e \phi/kT) \qquad \text{A.2}$$

where n_i is true number of ions of i type per unit volume near the surface

n_{io} is true number of ions of i type per unit volume far away from the surface i.e., the concentration of bulk solution

v_i is true valency of the cation of i type

e is true unit electro static charge (4.8×10^{-10} esu)

k is true Boltzmann constant (1.38×10^{-16} erg/°K)

T is true temperature in degree kelvin (°K).

The charge density in the double layer is contributed by the ions so that

$$\rho = e.\Sigma\, v_i\, n_i \qquad \text{A.3}$$

From equations A.2 and A.3

$$\rho = e\,\Sigma\, v_i\, n_{io} \exp(-v_i e \phi/kT) \qquad \text{A.4}$$

Substituting equation A.4 in equation A.1 yields

$$d^2\phi/dx^2 = -(4\pi e/\varepsilon)\,\Sigma\, v_i\, n_{io} \exp(-v_i e\phi/kT) \qquad \text{A.5}$$

Equation A.5 is the general differential equation for the electric double layer adjacent to a planar surface. It is known as general Poisson-Boltzmann equation.

SOLUTION TO DIFFUSE DOUBLE LAYER EQUATION

For the case of a single cation and anion species of equal valance, that is, $i = 2$

$$|v_+| = |v_-| \equiv v;\; n_0^+ = n_0^- \equiv n$$

Equation A.5 simplifies to

$$d^2\phi/dx^2 = (8\pi n v e/\varepsilon)\sinh(v e\phi/kT) \qquad \text{A.6}$$

for which solutions are available (Verwey and Overbeek, 1948; Bolt and Peech, 1953; Bolt, 1955; 1956; Babcock, 1963; Van Olphen, 1963).

Generally, depending on the boundary conditions two types of solutions are possible:

(1) Constant surface potential conditions and

(2) Constant surface charge condition

In the case of clay surfaces, which provide by far the major part of the surface area, (when compared to the edges) a constant surface charge is exhibited. Thus only solutions for a constant surface charge condition are further dealt with.

Solutions to the equation A.6 are commonly given in terms of the dimensionless quantities.

$$y = \nu e \phi / k T ; \; z = \nu e \phi_0 / k T \qquad \text{A.7}$$

and $$\xi = K x ; \; K^2 = (8 \pi n e^2 \nu^2 / \varepsilon k T) \; (\text{cm}^{-2}) \qquad \text{A.8}$$

In terms of these variables equation A.6 becomes

$$\frac{d^2 y}{d \xi^2} = \sinh y \qquad \text{A.9}$$

The boundary conditions for the integration are that for $\xi = \infty$, $y = 0$ and $dy/d\xi = 0$. Thus,

$$dy/d\xi = -2 \sinh (y/2) \qquad \text{A.10}$$

The boundary condition for the second integration is that for $\xi = 0$, $y = z$ (i.e, $\phi = \phi_0$), which leads to

$$\tanh (y/4) = (\tanh (z/4)) \exp (-K x) \qquad \text{A.11}$$

This is the equation derived by Babcock (1963). Van Olphen (1963) derived the solution in another form which can be transformed into the equation.A.11. Equation A.11 describes roughly an exponential decay of potential with distance from a surface at given potential (y_0 or ϕ_0) (Fig. A1). The total double layer charge per unit surface area is given by

$$\sigma = -\int_0^\infty \rho \; dx \qquad \text{A.12}$$

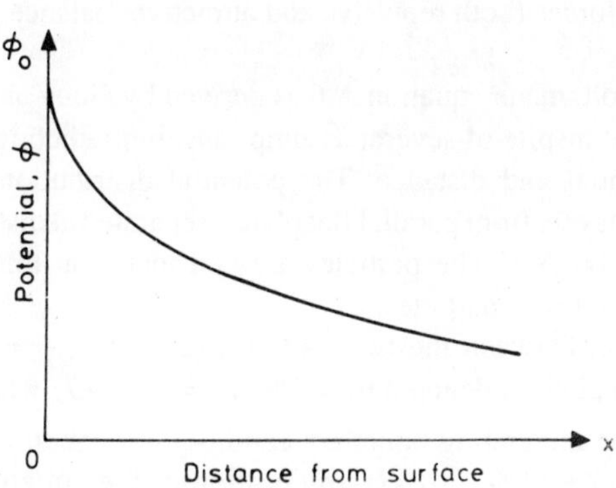

Fig. A.1. Potential distribution with distance from a changed surface

and, using the Poisson equation (Eqn. A.1),

$$\sigma = (\varepsilon/4\,\pi)\int_0^\infty (d^2\phi/dx^2)\;dx = -\left(\frac{\varepsilon}{4\,\pi}\right)(d\,\phi/dx)_{x=0} \qquad \text{A.13}$$

The slope of the potential function at the surface $(d\phi/dx)_{x=0}$, can be found from equation A.10, leading to

$$\sigma = \left(\frac{2\,\varepsilon\,n\,k\,T}{\pi}\right)^{1/2} \sinh(z/2) \qquad \text{A.14}$$

With the relationship between concentration and potential defined by equation A.2, that between potential and distance by equation A.11, and that between charge and surface potential by equation A.14, the diffuse double layer is completely described, provided the solution concentration n, cation valance v, and either the surface charge σ or surface potential ϕ_0 are known.

Sridharan and Jayadeva (1981) have developed a simplified method similar to that of Bolt (1955), for computing the potential-distance relationship, by choosing the ordinate at an imaginary plane, a distance x_0 behind the true surface, where the potential reaches infinity. If x' is the distance measured from this imaginary plane, then $x' = x + x_0$ and the equation A.11 can be reduced to the form

$$\tanh\,(y/4) = \exp\,(-K\,x') \qquad \text{A.15}$$

To evaluate the value of x_0, put $x' = x_0$, and $y = z$ at the surface in equation A.15

$$\tanh\,(z/4) = \exp\,(-K\,x_0) \qquad \text{A.16}$$

Equation A.16 gives the exact value of x_0 unlike the ones suggested by Bolt (1955) and Mitchell (1976), which are only approximate.

INTERACTING DOUBLE LAYER EQUATIONS

As discussed in earlier sections, the clay particles carrying net negative charges develop electrical diffuse double layers, on their surfaces. These double layers of two adjacent particles will usually overlap and interact with each other, under the influence of external force system and result in an equilibrium state such that the mobilized surface forces (both repulsive and attractive) balance with the external force system.

The Poisson-Boltzmann equation A.6 as derived by Gouy and Chapmann has proved to be useful inspite of several assumptions, in predicting the relationship between the potential and distance. The potential distribution for the case of interacting double layers from parallel flat plates, separated at distance 2d is shown schematically in Fig. A.2. The particles are considered at this stage as single particles and not groups of particles.

In the midplane, between the two particles $(d\,\phi/dx)_{x=d} = 0$. The potential function at the mid plane is denoted by u. Then $u = ve\phi_m/kT$; where ϕ_m is the mid plane potential. The integrating boundary condition for equation A.9 are that for $\xi = Kd$, $y = u$ and $(dy/d\,\xi)_{x=d} = 0$. Integration of the equation A.9 gives

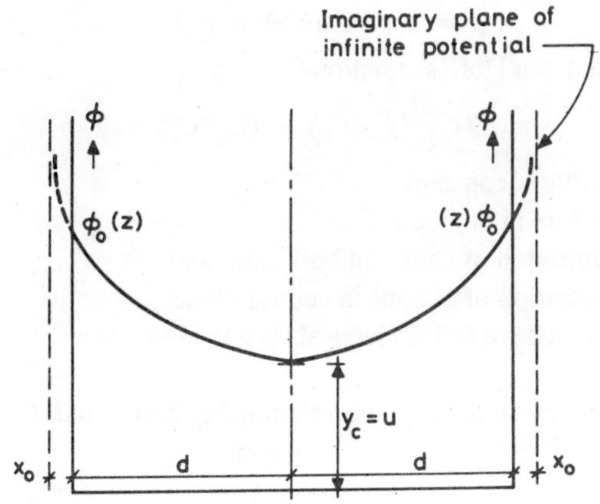

Fig. A.2. Potential vs. distance relation for an interacting double layer between two particles

$$dy/d\xi = -(2\cosh y - 2\cosh u)^{1/2} \qquad \text{A.17}$$

Integrating for the second time between the limits 0 and *d*, with the boundary conditions for $x = 0$, $y = z$, and for $x = d$, $y = u$,

$$\int_z^u (2\cosh y - 2\cosh u)^{-1/2}\, dy = \int_0^d d\xi = -Kd \qquad \text{A.18}$$

The surface potential z is related to the surface charge density σ by equation A.13. Now $y = ve\phi/kT$ and $\xi = Kx$,

$$(dy/d\xi)_{x=0} = (ve/KkT)\,(d\phi/dx)_{x=0}$$

or

$$(d\phi/dx)_{x=0} = (KkT/ve)\,(dy/d\xi)_{x=0} \qquad \text{A.19}$$

Substituting equation A.19 in equation A.13

$$\sigma = -(KkT\varepsilon/4\pi ve)\,(dy/d\xi)_{x=0} \qquad \text{A.20}$$

From equation A.17, at $x = 0$, $y = z$

$$(dy/d\xi)_{x=0} = -(2\cosh z - 2\cosh u)^{1/2} \qquad \text{A.21}$$

From equation A.8, $K = (8\pi ne^2v^2/\varepsilon kT)^{1/2}$. Substituting these values in equation A.20, and on simplification leads to,

$$\sigma = ((n\varepsilon kT)/2\pi)^{1/2}\left\{(2\cosh z - 2\cosh u)^{1/2}\right\}$$

or

$$(dy/d\xi)_{x=0} = \sigma\left\{\frac{2\pi}{(n\varepsilon kT)}\right\}^{1/2} \qquad \text{A.22}$$

Based on equations A.18 and A.22 van Olphen (1963) prepared tables giving sets of values of u and z for various values of *Kd*, for specific values of $(dy/d\xi)_{x=0}$. This enables to compute the mid plane potential *u* for known values of the soil and pore fluid parameters. For computing the osmotic repulsive pressure, p between the two particles, Langumir's equation (eqn. A.23) has been suggested.

$$p = 2\,n\,k\,T\,(\cosh u - 1) \qquad \text{A.23}$$

Bolt (1955) has used van Hoff's equation

$$p = RTC_0\left\{(C_c/C_0) + (C_0/C_c) - 2\right\} \qquad \text{A.24}$$

Where R is universal gas constant,
T is the temperature in degree kelvin,
C_0 is the concentration of cation in bulk fluid and
C_c is the concentration of cations in central plane.

Sridharan and Jayadeva (1982) have shown that equations A.23 and A.24 are identical.

Bolt (1955) used the following interrelationship between d, C_c and C_0

$$v\,(\beta\,C_0)^{1/2}\,(x_0 + d) = 2\,(C_0/C_c)^{1/2}\int_0^{\pi/2}\frac{d\psi}{(1-(C_0/C_c)^2\sin^2\psi)^{1/2}} \qquad \text{A.25}$$

where $\beta = (8\,\pi\,F^2)/1000\,R\,T\,\varepsilon = 1\times10^{15}$ cm/m.mole @ 20° C
F = Faraday's constant
= variable related to the central concentration

Bolt has suggested the value for x_0 as

$$x_0 = 4/v\,\beta\,\Gamma \approx (1/v)\ \text{Å for illite}$$
$$\approx (2/v)\ \text{Å for kaolinite}$$
$$\approx (4/v)\ \text{Å for montmorillonite}$$

Γ = surface charge density in meq/cm^2

For saturated clays the distance between the parallel clay platelets can be expressed in terms of void ratio, e_1 of the system as

$$e_1 = G\,\gamma_w\,Sd\times10^{-4} \qquad \text{A.26}$$

where G is the specific gravity of soil particles,
γ_w is the unit weight of water,
S is the specific surface of the solid particles in m^2/g and
d is the half space distance between the particles in Å.

Now using the interrelationships of equations A.24 to A.26, e vs p relationship can be computed.

It can be shown that for a given pressure, p both the types of analysis yield identical results. Sridharan and Jayadeva (1982) have also shown that by direct numerical integration of equation A.18 together with equations A.22 and A.26, the e vs p relationship can be computed in a more exact and continuous form. They have presented a set of graphs and tables which avoids the interpolation for intermediate values in van Olphen's approach..

Table A.1 presents the values of Kd for various values of u and $(dy/d\xi)_{x=0}$, as given by Sridharan and Jayadeva (1982). For known values of Kd and $(dy/d\xi)_{x=0}$, u can be estimated from Table A.1 and from equation, A.23, p can be computed. In this presentation the following simplified procedure is followed for computing the d vs p relationship:

1. Known data:

   Base exchange capacity in meq/100g

   Specific surface, S in m^2/g

   Specific gravity of soil solids, G

   Let $P = BEC/S$ in meq/100 m^2, then

$$\sigma = 289104\, P \text{ esu/cm}^2 \tag{A.27}$$

2. Since, $-(dy/d\xi)_{x=0} = \sigma(2\pi/\varepsilon n k T)^{1/2}$

$$-(dy/d\xi)_{x=0} = 2513.61\, P/(\varepsilon n T)^{1/2} \tag{A.28}$$

3. $K = (8\pi n e^2 V^2/\varepsilon k T)^{1/2} = 50.272\, \nu\, (n/\varepsilon T)^{1/2}$ (per Å) A.29

   $d = e_1/G\gamma_w S$ and for $\gamma_w = 1\, g/cc$, $d = e_1/GS$

$$Kd = K e_1/GS = 50.272 \left\{e_1\, \nu/(GS)\right\} \left\{(n/\varepsilon T)^{1/2}\right\} \tag{A.30}$$

4. Since, $p = 2nkT\,(\cosh u - 1)$,

$$p/nT = 0.1694544\,(\cosh u - 1) \tag{A.31}$$

Now knowing the solid and fluid properties viz. $P/(n\varepsilon T)^{1/2}$ and Kd, the nondimensional parameter u, being the function of the midplane potential, can be computed by numerical integration of equation A.18. Using equation A.31, p can be computed.

It is shown by Sridharan and Jayadeva (1982) that the influence of attractive force is very small in the range of void ratios occurring in practice. As such only repulsive forces are considered in further investigation.

The above critical examination of the state-of-the-art of the diffuse double layer theory, brings out that the double layer equations are easily amenable for computations.

VALUES OF PHYSICAL PROPERTIES AT FIELD CAPACITY

For computing the d vs p or e vs p relationship for a given soil from Gouy-Chapman diffuse double layer theory, the values of the following physical properties of the pore fluid are to be known:

(1) Valancy of the cation.
(2) Cation concentration.
(3) Temperature.
(4) Static dielectric constant.

It has been shown by several investigators that the d vs p or e vs p relationship is very sensitive to the above physical properties (Bolt, 1956; Olson and Mesri, 1970; Shainberg and Klaussner, 1971 etc.). As such, it is desirable to use an appropriate value of these parameters in the double layer computations.

As there is no ambiguity in the field values of the temperature and the static dielectric constant of water for a saturated system, a temperature of 298° K and a dielectric constant of 78.54 have been assumed for computations. Values of valancy and ion concentration to be chosen require further consideration.

Bolt (Bolt and Bruggenwert, 1976) has stated that to specify the soil solution with any precision, in normal natural soil systems of temperate regions, the soil solution at field capacity is around 0.01 molar in total electrolyte concentration. He has also specified that, at field capacity, the solution roughly consists of equal amounts of monovalent and divalent cations.

While studying the swelling pressure of Na-Ca bi-ionic montmorillonite system Shainberg et. al., (1971) have established that only for Na percentage less than twenty percent, the influence of Ca is reflected. For Na percentage greater than thirty-five percent the behaviour is identical to that of Na- montmorillonite. This is also indicated by McNeal and Coleman (1966) and Bar-On et al., (1970). As such, the presence of fifty percent monovalent ions at field capacity, tends the soil to behave as a totally monovalent system.

An assumption of concentration value of 0.01 molar and valancy value of 1.0 would reasonably represent the field conditions for a natural soil.

ANALYTICAL MODEL

The classical Gouy-Chapman diffuse double layer theory for planar surfaces has enabled various investigators to analyse the compressibility behaviour of pure clays. (Bolt and Miller, 1955; Bolt, 1956; Warkentin et al., 1957; Klausner and Shainberg, 1971; Shainberg et al., 1971; Sridharan and Jayadeva, 1982 and others).

In all the above investigations *e* vs log *p* relationships have been predicted from the known physical properties of the soil and are compared with experimental observations. These approaches indicate that the soil systems are distinctly different. The re- examination of the void ratio vs pressure relationship from micro-mechanistic considerations within the frame work of the Gouy-Chapman diffuse double layer theory, has revealed that *d* vs *p* relationship is unique for a given physico-chemical environment (Sridharan and Jayadeva, 1982). The type of clay is subdued by the specific surface of the clay.

Analytical model at micro level

To generate the possible unique *d* vs *p* relationship corresponding to the physico-chemcial environment of the natural soils, the three clays, having the physical properties listed in Table A.2, are selected for further computations.

Since, the selected clay minerals represent the extreme range, it is believed that the subsequent formulations are valid for any other type of clay minerals and natural soils. Using the physical properties of the pore fluid as discussed earlier, the *d* - *p* relationship is obtained by numerical integration of equation A.18, together with the derived inter-relationships of equations A.28, A.30 and A.31. Computations show that the *d* vs *p* relation practically follows the same path for all the three clay minerals, for a given physico chemical environment, which has been presented in Chapter 4 (Fig. 4).

Table A.1 Values of *Kd* for various values of *u* and $(dy/d\xi)_{x=0}$

u	Values of $(dy/d\xi)_{x=0}$							
	1	5	10	50	100	250	500	1000
0.1	2.938	3.989	4.181	4.339	4.359	4.371	4.375	4.377
0.5	1.351	2.346	2.536	2.695	2.715	2.727	2.731	2.733
1.0	0.735	1.590	1.777	1.935	1.955	1.967	1.971	1.973
1.5	0.435	1.125	1.306	1.463	1.483	1.495	1.499	1.501
2.0	0.264	0.794	0.966	1.121	1.141	1.153	1.157	1.159
2.5	0.161	0.553	0.709	0.862	0.882	0.894	0.898	0.900
3.0	0.098	0.375	0.513	0.662	0.681	0.693	0.697	0.700
3.5	0.060	0.249	0.365	0.506	0.526	0.538	0.542	0.544
4.0	0.036	0.161	0.253	0.385	0.405	0.417	0.421	0.423
4.5	0.022	0.102	0.171	0.292	0.311	0.323	0.327	0.329
5.0	0.013	0.064	0.113	0.219	0.238	0.250	0.254	0.256
5.5	0.008	0.040	0.073	0.162	0.181	0.193	0.197	0.199
6.0	0.005	0.024	0.046	0.118	0.137	0.148	0.152	0.154
6.5	0.003	0.015	0.029	0.085	0.102	0.114	0.118	0.120
7.0	0.002	0.009	0.018	0.060	0.076	0.087	0.091	0.093
7.5	0.001	0.006	0.011	0.041	0.055	0.066	0.070	0.072
8.0	0.001	0.003	0.007	0.027	0.039	0.050	0.054	0.056
8.5		0.002	0.004	0.018	0.027	0.037	0.041	0.043
9.0		0.001	0.002	0.011	0.019	0.027	0.031	0.033
9.5			0.001	0.007	0.012	0.020	0.023	0.025
10.0			0.001	0.004	0.008	0.014	0.017	0.019
10.5			0.001	0.003	0.005	0.010	0.013	0.015
11.0				0.002	0.003	0.007	0.009	0.011
11.5				0.001	0.002	0.004	0.006	0.008
12.0				0.001	0.001	0.003	0.004	0.006

Table A.2 Some properties of the Three Clay Minerals

Type of clay mineral	Montmorillonite	Illite	Kaolinite
BEC, in meq/100g	100	40	3
Specific surface, in m/g	800	100	15
BEC/S.A = P, in meq/100m	0.125	0.40	0.20
Surface charge in esu/cm	36138	115641.60	57820.8
Surface potential function z	8.349	10.672	9.287
Surface potential, ϕ_0 in mv	214.60	274.30	238.70

APPENDIX 2

Suggested Flow Chart for Prediction of Soil Behaviour

A. Classify the In-situ as:

(a) Cases wherein analysis and prediction are not tenable (e.g. Cohesionless soils, layered soils)
(b) normally consolidated : uncemented
(c) overconsolidated : uncemented
(d) cemented soft and sensitive soils
(e) stiff cemented soils
(f) partly saturated soils

Input data

(1) Homogeneity
(2) Liquid limit, w_L
(3) Soil fraction coarser than 425 μm . F
(4) Depth of sampling, D in m
(5) Water content, w
(6) Bulk density, γ_b
(7) Field vane strength, S_u or undrained shear strength of undisturbed soil reconsolidated to field state of stress
(8) Undrained strength of remoulded state of soil, S_{ur}

Compute

(i) w_{LM}
(ii) Degree of saturation, S_r
(iii) In-situ void ratio, e
(iv) Effective overburden pressure, σ_o in kPa
(v) (ω_{LM}) field

Classification

(i) If nonhomogeneous — (a)
(ii) If w_L is not determinable — (a)
(iii) If $w_{LM} < 25\%$ — (a)
(iv) If $S_r \approx 100\%$ — (b), (c) or (d)
(v) If $S_r < 95\%$ — (f)

(vi) If $(w/w_{LM}) \approx 1.122 - 0.234 \log \sigma_o$ — (b) or (d)

(vii) If $S_{ur} = S_u$ — (b)

(viii) If $S_{ur} < S_u$ — (d)

(ix) If $(w/w_{LM})_f > 1.122 - 0.234 \log \sigma_o$ — (d)

(x) If $(w/w_{LM})_f < 1.122 - 0.234 \log \sigma_o$ — (c) or (e)

(xi) If $S_u \approx S_{ur}$ — (c)

(xii) If $S_u > S_{ur}$ — (e)

B. Engineering Properties:

(b) Normally consolidated

(i) $C_c = 0.234\, e_L$

(ii) $\phi' = 21°,\ c' = 0$

(iii) $q_u = 10^{\left[\frac{0.932 - (w/w_{LM})}{0.21}\right]}$

(iv) $k = 10^{-\left[\frac{2.16 - (w/w_{LM})}{0.195}\right]}$

(c) Overconsolidated

(i) $\sigma_c = 10^{\left[\frac{1.122 - 0.046 \log \sigma_o - (w/w_{LM})}{0.188}\right]}$

(ii) $C_r = 0.046\, e_L$ upto σ_c

(iii) $C_c = 0.234\, e_L$ beyond σ_c

(iv) q_u at large strain $= 10^{\left[\frac{0.932 - (w/w_{LM})}{0.21}\right]}$

(v) $\phi' = 21°$ (Min) at large strain

(vi) $k = 10^{-\left[\frac{2.16 - (w/w_{LM})}{0.195}\right]}$

(d) Sensitive soil

(i) $\sigma_{yf} = 3.78\, S_u + 7$

(ii) $C_r = 0$ upto σ_{yf}

(iii) To obtain C_c beyond σ_{yf} plot the $e - \log \sigma$ curve by adding σ_{yf} to σ at respective void ratio of a remoulded $e - \log \sigma$ path (Eqn: w/w_L 1.122 – 0.234 log σ)

(iv) q_u = constant upto σ_{yf}. Beyond this predict as per text

(v) $k = 10^{-\left[\frac{2.16 - (w/w_{LM})}{0.195}\right]}$

(f) Partly saturated soils

(i) compute (e/e_L) at the in-situ stress state

(ii) If $(e/e_L) > (e_{sat}/e_L) = 1.122 - 0.234 \log \sigma_o$ collapse takes place upon inundation

(iii) $(e - e_{sat})$ is magnitude of collapse

(iv) If $(e/e_L) < (e_{sat}/e_L)$ upon inundation heave takes place. For the possible range of heave see text

(v) From the given water content obtain $e_{sat} = wG$. Minimum available shear strength is that computed at this void ratio as per the equation.

$$q_u = 10^{\left[\frac{0.932 - (w/w_{LM})}{0.21}\right]}$$

At $S_r < 70\%$ $q_u \approx \sigma_c \tan \phi$

where σ_c is the stress corresponding to $e/e_L \sqrt{S_r}$ and $\phi = 21°$

Note: In all the equations units of stress are in *kPa* and permeability is expressed in cm/sec.

User Friendly Software Program

To classify the insitu soil into any one of the categories

(a) cases wherein analysis and prediction of soil behaviour are not tenable (e.g. cohesionless soils, layered system).
(b) normally consolidated: cemented or uncemented.
(c) soft and sensitive soils.
(d) overconsolidated: uncemented or cemented.
(e) partly saturated uncemented soils, and

Then to assess by appropriate physical models the engineering properties of soils, viz

(i) Prediction of $e - \log \sigma'$ relationship for the entire pressure range, as well as prediction of compression and recompression indices.
(ii) Prediction of undrained strength as water content changes with effective overburden pressures.
(iii) Stress-state-permeability relationships of the soil.

With the following input data

1. Depth of sampling.
2. Bulk density and water content of the soil.
3. Soil fraction coarser than 425m and liquid limit water content of the soil.
4. Vane strength of undisturbed and insitu remoulded soil or undrained strength of soil reconsolidated to 0.7 times overburden pressure, as the case might be.

INPUT

w_L (%)	S_r (%)
F (%)	σ_o (kPa)
w_{LM} (%)	S_u (kPa)
D in m	S_{ur} (kPa)
w (%)	$(w/w_{LM})_f$
γ_d (g/cc)	

CLASSIFICATION

→ (a) Cohesionless and layered
→ (b) Normally consolidated
→ (c) Overconsolidated
→ (d) Soft sensitive
→ (e) Stiff cemented
→ (f) Partly saturated

ENGG. PROPERTIES

(b)

$$C_c = 0.234\ e_L$$

$$\phi' = 21^\circ\quad c' = 0$$

$$q_u = 10^{\left\{\frac{0.932 - w/w_{LM}}{0.21}\right\}}$$

$$k = 10^{-\left\{\frac{2.16 - w/w_{LM}}{0.195}\right\}}$$

(c)

$$\sigma_c = 10^{\left\{\frac{1.22 - 0.046 \log \sigma_0 - w/w_{LM}}{0.188}\right\}}$$

$$C_r = 0.046\ e_L$$

$$q_u = 10^{\left\{\frac{0.932 - w/w_{LM}}{0.21}\right\}}$$

$\phi' = 21^\circ$ (min) at large strains

$$k = 10^{-\left\{\frac{2.16 - w/w_{LM}}{0.195}\right\}}$$

(f)

$\left\{\frac{e}{e_L}\right\} > \left\{\frac{e_{sat}}{e_L}\right\}$ Collapse

$\left\{\frac{e}{e_L}\right\} < \left\{\frac{e_{sat}}{e_L}\right\}$ heave

to compute heave follow text

collapse = $(e - e_{sat})$

$$q_u = 10^{\left\{\frac{0.932 - w_{sat}/w_{LM}}{0.21}\right\}}$$

(d)

$$\sigma_{yf} = 3.57\ S_u + 7$$

$C_r = 0$ upto σ_{yf}

to compute follow text

C_c beyond σ_{yf}

q_u is const upto σ_{yf}

$$k = 10^{-\left\{\frac{2.16 - w/w_{LM}}{0.195}\right\}}$$

REFERENCES

Aas, G., 1965. A study of the effect of vane shape and rate of strain on the measured values of in-situ shear strength of clays. Proc. 6th Int. *Conf. on Soil Mech. Found. Engg.*, Montreal, Canada, **1**:141-145.

Aas, G., 1967. Vane tests for investigation of anisotropy of undrained shear strength of clays. *Proc. Geotechnical Conf.* Oslo, Norway, **1**:3-8.

Abdun-Nur, E.A., 1959. Plastic limit comparison of cube and standard thread test methods. *Am. Soc. Test. Mater.*, Spec. Tech. Publ. **245**:212-216.

Alonso, E.E., Gens, A., and Josa, A., 1990. A constitutive model for partially saturated soils. *Geotechnique*, **40**(3):405-430.

Al-tabbaa and Wood, D.M., 1987. Some measurements of the permeability of Kaolin. *Geotechnique*, **37**(4):499-503.

Andersen, A., 1981. Exploitation sampling and in-situ testing of soft clay. In· E W. Brand and R.P. Brenner (Editors) *Soft Clay Engineering*, Elsievere, Amsterdam :241-304.

Andersen, A. and Sollie, S., 1966. An inspection vane. Vane Shear and Cone Resistance Testing of In-situ Soils. *Am. Soc. Test. Mater.*, Spec. Tech. Publ., **399**:3-7.

Anderson, D.C., Crawby, W. and Zabick, J.D., 1985. Effects of various liquids on clay soil : Bentonite slurry mixtures. Hydraulic Barriers in Soil and Rock, *Am. Soc. Test. Mater.*, Spec. Tech. Publ. **874**:93-101.

Andresen A. AA. and Kolstad, P., 1979. The NGI 54mm sampler for undisturbed sampling of clays and representative sampling of coarser materials. *Proc. Int. Symp. on Soil Sampling*. State-of- the-art on Current Practice of Soil Sampling, Singapore, 13-21.

Armstrong, J.C. and Petry, T.M., 1986. Significance of specimen preparation upon soil plasticity. Geotech. Test. *J., Am. Soc. Test. Mater.*, **V. 9**(3):147-153.

ASTM Subcommittee - D18.02, 1987. Suggested method for performing the borehole shear test. Geotech. Test. *J., Am. Soc. Test. Mater.*, **V. 10**(1):19-25.

Atchinson, G.D. and Donald, I.B., 1956. Effective stresses in unsaturated soils, Proc. 2nd Aust. New Zealand Conf. on *Soil Mech. Found. Eng.*, 192-199.

Atkinson, J.H. and Bransby, P.L., 1978. *The Mechanics of Soils - An Introduction to Critical State Soil Mechanics.* McGraw Hill Book Co (UK) Ltd., pp.375.

Atterberg, A., 1911. The behaviour of clays with water, their limits of plasticity and their degrees of plasticity, *Kungliga Lantberuks Handl Tidskr*, **50**(2):132-158.

Aylmore, L.A.G. and Quirk, J.P., 1960. Domain or Turbo-stratic structure of clays. *Nature*, **187**:1046.

Aylmore, L.A.G. and Quirk, J.P., 1962. The structural status of clay systems. *9th Natl. Conf. on Clays and Clay Minerals*, Pergamon Press, 104-130.

Babcock, K.L., 1963. Theory of the chemical properties of soil colloidal systems at equilibrium, *Hilgardia*, **34**(11).

Baguelin, F., 1973. Measurement of coefficient of consolidation by pressuremeter test (In French). Discussion *Proc. 8th Int. Conf. on Soil Mechanics and Foundation Engg.*, Moscow. **4.2**:22-23.

Baguelin, F., Jazequel, J.F. and Shields, D.H., 1978. *The Pressuremeter and Foundation Engineering.* Trans. Tech. Publications, Germany, pp.617.

Baladi, G.Y. and Rohani, B., 1977. Liquefaction potential of dams and foundations - Development of an elastic plastic constitutive relationship for saturated sand. Research Report 5-76-2, Report 3 1977, US Army Engg., CE, Vicksburg, Miss.

Baldi, G., Hight, D.W. and Thomas, G.E., 1988. State-of-the-art paper, A revaluation of conventional triaxial test methods. Advanced Triaxial Testing of Soil and Rock. *Am. Soc. Test. Mater.*, Spl. Tech. Publ. **977**:219-263.

Ballard, G.E.H. and Weeks, W.F., 1963. Human error in determining plastic limit of cohesive soils. Mater. Res. Stand. *Am. Soc. Test. Mater.*, **V. 3**(9):726-729.

Barden, L. and Sides, G.R., 1970. Engineering behaviour and structure of compacted clay, *J. Soil Mech. and Found. Engg. Div.*, Am. Soc. Civ. Eng., **94**(4):1171.

Barros, P.L.A. and Barros, C.L.A., 1989. Vane shear test application on cohesionless soils. *Proc. 12th Int. Conf. on Soil Mechanics and Foundation Engineering,* Rio de Janerio, Balkema, Rotterdam, **1**:171-174.

Bar-On, P., et. al., 1970. The electrophoretic mobility of Na/Ca montmorillonite particles. *J. Colloid Interface Science,* **33**:471- 472.

Bauer, G.E., Selvadurai, A.P.S. and Demartinecourt, J.P., 1980. Measurement of in-situ strength properties of Champlain Sea clay with modified shear device. Proc. 33rd Canadian Geotechnical Conf., Calgary, Alberta, Canada.

Been, K. and Jefferies, M.G., 1985. A state parameter for sands. *Geotechnique,* **35**(2):99-112.

Belvisco, R., Clampoli, S., Cotecchia, V. and Federico, A., 1985. Use of cone penetrometer to determine consistency limits. *Ground Eng.,* **18**(5):21-22.

Bernal, J.D. and Mason, J., 1960. Coordinates of randomly packed spheres. *Nature,* **188**:910-911.

Biarez, J. and Wiendieck, K., 1963. La comparison entre l'anisotropie mechanique et l'anisotropie de structure des milieux pulverulents, *C.R. Acad. Sci.,* **256**:1217-1220.

Bindumadhava, 1985. Prediction of compressibility behaviour of natural soils. M.Sc thesis, Indian Institute of Science.

Bishop, A.W., 1961. The measurement of pore pressure in triaxial test. Conf. on Pore Pressure and Suction in Soils, Butterworths, 38-46.

Bishop, A.W. and Henkel, D.J., 1962. *The Measurement of Soil Properties in the Triaxial Test.* Edward Arnold London 2nd ed, pp.228.

Bjerrum, L., 1954. Geotechnical properties of Norwegian marine clays, *Geotechnique,* **4**:49-69.

Bloch, P., 1978. Discussion on 'Stress state variables for unsaturated soils' by Fredlund, *Proc. Am. Soc. Civil Eng.* Geotech. Eng. Div., **104**(GT2):303-304.

Boghrat, A., 1987. Dilatometer testing in highly overconsolidated soils. *Proc. Am. Soc. Civil Eng.* Geotech. Eng. Div., **113**:516-519.

Bolt, G.H., 1956. Physico-chemical analysis of the compressibility of pure clays. *Geotechnique,* **6**(2):86-93.

Bolt, G.H. and Bruggenwert, M.G.M., 1976. *Soil Chemistry - A Basic Elements.* Elsevier Sci. Publ. Co., New York, 11.

Bolt, G.H. and Miller, R.D., 1955. Compression studies of illite suspensions. *Proc. Soil Sci. Soc. of Am.,* **19**:285-288.

Bolt, G.H. and Peech, M. 1953. The application of Gouy-Chapman theory to soil water systems. *Proc. Soil Sci. Soc. Am.,* **17**:210- 213.

Borg, L., Friedman, M., Handlin, J. and Higgs, D.V., 1960. Experimental deformation of St.Peter Sand — a study of cataclastic flow, *Mem. Geol. Soc. Am.,* **79**:133-191.

Boussinesq, J., 1885. *Application des potentials a L'etude de L'equibibre et duo mouvment des Solids Elastiques,* Gauthier Villars, Paris.

Bowles, J.W., 1979. *Physical and Geotechnical Properties of Soils.* McGraw Hill, New York.

Bozozuk, M., 1984. Study on Gloucester test fill. Report Div. Building Res., Nat. Res. Council, Ottawa, Canada.